P9-APG-332

WITHDRAWN FROM LIBRARY

Columbia Univ. Library

MONTGOMERY COLLEGE
ROCKVILLE CAMPUS LIBRARY
ROCKVILLE, MARYLAND

Applied Illumination Engineering

APPLIED ILLUMINATION

ENGINEERING

By Jack L. Lindsey, FIES

Published by
THE FAIRMONT PRESS, INC.
700 Indian Trail
Lilburn, GA 30247

MAY 1 2 1992

A AY 8 30 4

Library of Congress Cataloging-in-Publication Data

Lindsey, Jack, 1940-
    Applied illumination engineering / by Jack Lindsey.
       p. cm.
    Includes index.
    ISBN 0-88173-060-2
    1. Electric lighting. I. Title.

TK4161.L54   1991   621.32-dc20   88-45787
                                                        CIP

Applied Illumination Engineering / By Jack Lindsey.
©1991 by The Fairmont Press, Inc. All rights reserved. No part of this publication may be reproduced or transmitted in any form or by any means, electronic or mechanical, including photocopy, recording, or any information storage and retrieval system, without permission in writing from the publisher.

Published by The Fairmont Press, Inc.
700 Indian Trail
Lilburn, GA 30247

Printed in the United States of America

10 9 8 7 6 5 4 3 2 1

ISBN 0-88173-060-2   FP

ISBN 0-13-040726-7   PH

While every effort is made to provide dependable information, the publisher, authors, and editors cannot be held responsible for any errors or omissions.

Distributed by Prentice-Hall, Inc.
A division of Simon & Schuster
Englewood Cliffs, NJ 07632

Prentice-Hall International (UK) Limited, London
Prentice-Hall of Australia Pty. Limited, Sydney
Prentice-Hall Canada Inc., Toronto
Prentice-Hall Hispanoamericana, S.A., Mexico
Prentice-Hall of India Private Limited, New Delhi
Prentice-Hall of Japan, Inc., Tokyo
Simon & Schuster Asia Pte. Ltd., Singapore
Editora Prentice-Hall do Brasil, Ltda., Rio de Janeiro

DEDICATION

---

*This book is dedicated to Marge, Shawna and Jackson.
Thanks for putting up with the hundreds of hours I spent
at the computer, and not with you.*

ERRATA SHEET
For
APPLIED ILLUMINATION ENGINEERING
By Jack L. Lindsey, FIES

CONTENTS

Chapter 1

The Nature of Light

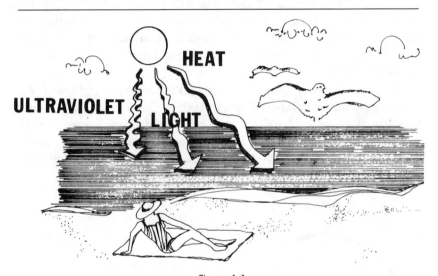

Figure 1-1
Radiant energy from the sun takes many forms. Heat, light, and ultra-violet energy are among the more common.

Imagine for a moment that you are basking on the beach on a balmy spring day. Although the air is cool you are not cold. The feeling of warmth comes from radiant energy–heat–from the sun. Later you will develop a tan, the result of another form of the sun's energy, ultraviolet radiation. You can also see a variety of sights–seagulls circling lazily overhead, the foaming surf, and a ship arriving from a distant port. The ability to see these objects is dependent upon yet another form of the sun's radiant energy–light.

Light, then, is a form of radiant energy. It is similar to radiant heat and ultraviolet, yet slightly different. Light is defined as radiant energy which is capable of exciting the retina and producing a visual sensation.

Over the centuries scientists have developed many theories about the nature of light. An early belief was that light was shot out of the eye, and made objects visible when it struck them. This theory was discarded when Aristotle asked why, if light is shot out of the eye, it was impossible to see in the dark.

Research and theoretical analysis provided no real insight into the nature of light until 1678, when Christian Huygens, a Dutch physicist, demonstrated that light is wavelike in nature (Figure 1-2). This, and the work of other researchers, formed the basis for Maxwell's electromagnetic wave theory in 1873. Maxwell postulated, in part, that electromagnetic energy travels in the form of waves, and that light is an electromagnetic wave.

Maxwell's theory appeared to explain the nature of light, and was accepted by most scientists. Then, in 1905, a German physicist named Max Planck presented the Quantum Theory, which was to earn him the 1915 Nobel prize in physics. Planck's theory postulated that radiant energy travels in the form of discrete packages which he called"quanta,"

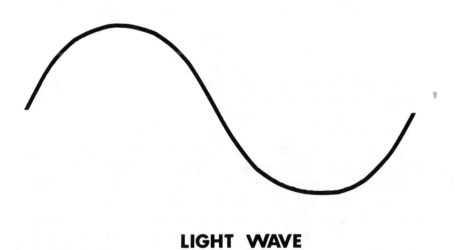

LIGHT WAVE

Figure 1-2.
Light travels in the form of waves, similar to the waves which are formed when a pebble is dropped into a pond of water.

that the energy contained in each quanta was related to the frequency of its wave, and that high frequency waves contain more energy than low frequency waves. "Frequency" is the number of complete cycles a wave makes in one second. If the wave shown in Figure 1-2 occurred over a period of one second, the wave would have a frequency of one cycle per second, or one hertz. Note that "hertz" means "cycles per second."

Planck also provided a means of calculating the amount of energy in each quanta:

$Q = h\nu$

Where

Q is energy, in Joules

h is Planck's Constant, 6.626×10^{-34} J·s

ν is the frequency, in Hz

It can readily be seen from Planck's equation that waves of high frequency contain more energy than waves of low frequency.

Later work by such distinguished scientists as Einstein, Rutherford, and Bohr has lead to a general acceptance of the "Unified Theory," which states that light is dualistic in nature; that it behaves in accordance with both the Electromagnetic Wave Theory and the Quantum Theory.

A popular current theory is that light behaves as quanta at the points of emission and absorption, but travels in the form of waves. All radiant energy can then be explained as waves, and the length of the wave determines the form which the energy takes.

The length of an individual wave may vary from 10^{-16} meters to about 10^7 meters. This length is used to classify it as a cosmic ray, X-ray, light, heat, radio or television wave, or any of the other subdivisions in the electromagnetic spectrum, as shown in Figure 1-3. For example, a wave with a length of about 3100 miles is 60 Hz electric power. If the length of the wave is shortened to 160 meters, the radiant energy takes the form of a radio wave which might be broadcast by a commercial radio station, while a wave 18 inches in length might be transmitted by a television station. Figure 1-4 compares light waves and electric power.

The unit of measure of the length of electromagnetic waves is the meter (about 39.37"). Since waves of light are very short it is more convenient to express them in smaller sub-units of the meter such as the nanometer (one billionth of a meter), the Ångstrom (one one-hundred

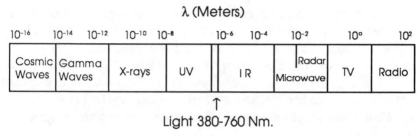

λ (Meters)

Cosmic Waves	Gamma Waves	X-rays	UV	I R	Radar / Microwave	TV	Radio

↑
Light 380-760 Nm.

Figure 1-3

The electromagnetic spectrum. Note that only a very small portion of the spectrum is defined as light. Other important areas of the spectrum are infrared heat, television and radio waves, and X-rays.

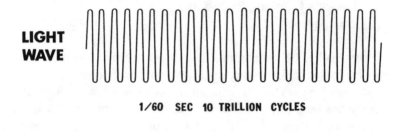

LIGHT WAVE

1/60 SEC 10 TRILLION CYCLES

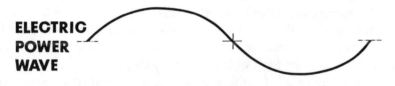

ELECTRIC POWER WAVE

1/60 SEC ONE CYCLE

Figure 1-4

The difference between high frequency waves and low frequency waves is the number of oscillations in a second of time. For comparision, electric power waves oscillate 60 times per second while light waves oscillate about 600 trillion times per secone.

millionth of a meter), or the micron (one millionth of a meter). The nanometer is the currently commonly used unit for light. Note, however, that the Ångstrom and micron have been used extensively in the past, and many references to these units will be found. In fact, some equations use these units, and care must be taken to assure that the appropriate units are used in calculations. For comparison, one inch equals about 25.4 million nanometers, 2.54 million Ångstroms, or 254 thousand microns.

Light, the visible portion of the electromagnetic spectrum, ranges from 380 nanometers to 760 nanometers in wavelength. Note that these limits are approximate since the sensitivity of the human eye varies slightly between individuals. From the Quantum Theory it can now be seen that a wave of blue light contains more energy than a wave of green light, and a wave of green contains more energy than a wave of red.

LIGHT AND COLOR

The color of a light wave is determined by its length. A wave 380 nanometers long is violet light while a wave of green might be 500 nanometers long. Yellow light is about 580 nanometers and red ranges from about 620 to 760 nanometers. Figure 1-5 illustrates the relationship between wavelength and color.

Light which contains a relatively balanced combination of wavelengths is considered to be "white" light. If a ray of this light is passed through an optical prism the light will be refracted into its component colors and will appear as illustrated in Figure 1-6.

This shows that white light is composed of a combination of different colors of light. In fact, white light can be created by blending only red, green, and blue light, which are called the "primary colors of light." This is called the additive property of light: colors are added together to produce white light. By contrast, colored pigments follow the subtractive principle: if the primary colors, yellow, cyan, and magenta are added, the result is black. Other colors are created by subtract-colors from black, and white is the absence of color. See Figure 1-7.

When light strikes an object the light can be reflected or absorbed. Colored surfaces will absorb all colors of light except their own, which will be reflected, as illustrated in Figure 1-8. A pure white surface will reflect all colors of light, while a black surface will absorb all colors of light.

The perceived color of an object is thus dependent upon the light source under which the object is viewed. In order to see a color, that color must be present in both the object and the light source. If a color

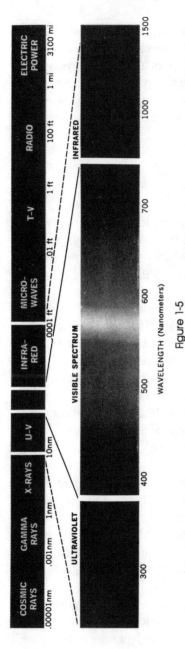

Figure 1-5

The length of a wave of light determines the color. Waves 380 nanometer waves are yellow, and 760 nanometers is the upper end of the red region for most eyes. *(Courtesy GE Lighting)*

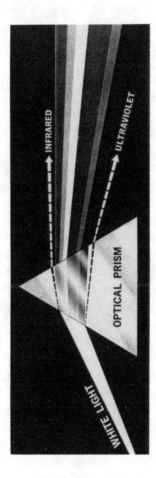

Figure 1-6

White light contains all colors. If a beam of white light is diffracted by a prism, a rainbow of colors will be seen. If these colors are combined through a second prism, the white light beam will be reconstructed and the individual colors will no longer be visible. *(Courtesy GE Lighting)*

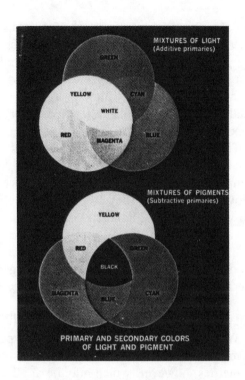

Figure 1-7
Colors of light are called additive since the combining of the primary colors blue, green, and red produces white. Colors of pigments are the opposite of light: the primary colors magenta, cyan, and yellow combine to produce black. *(Courtesy GE Lighting)*

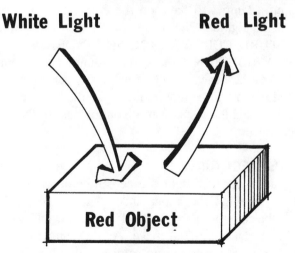

Figure 1-8
Objects absorb all colors of light except their own. If white light strikes a red object, only red light is reflected. All other colors are absorbed.

is missing from either the object of the light source, that color will not be seen in the object. For example, low pressure sodium lamps emit only yellow light at wavelengths of 589 and 589.6 nanometers. The color bands in a color chart viewed under low pressure sodium lamps will appear to be either yellow or a shade of gray regardless of their actual color. The same chart, viewed under sunlight, would exhibit a rainbow of color.

The principles of selective reflection and absorption have important effects which must be considered in lighting design and application. For example, high pressure sodium lamps produce large quantities of green, yellow, and red light. They are somewhat deficient in blue, thus a blue wall will reflect little light, and blue objects will appear dull and will have distorted color.

It should be noted that light waves themselves are not visible. We do not see the waves as they travel through space. We see only the reflection or emission of light from a surface.

COLOR AND THE LIGHTING DESIGNER

When specifying a particular lamp type and color to achieve a desired visual effect, there is simply no substitute for experience. There are metrics, however, which may be applied to predict the probable effect which a lamp will have on the color and visual warmth or coolness of the lighted environment. The metrics are the COLOR TEMPERATURE, COLOR RENDERING INDEX, and SPECTRAL POWER DISTRIBUTION of the lamp. It is important to remember that an evaluation of these metrics will provide only general guidelines, and a visual evaluation will be required to make the final determination if color rendering and the visual warmth or coolness of the environment are a major concern.

COLOR TEMPERATURE

The visual color of a light source can be described by its color temperature. Imagine the appearance of a bar of steel. At room temperature the color of the bar is a dark gray. Now imagine that the bar is placed in a blacksmith's furnace. As the temperature of the bar increases, a gradual change in its color can be observed. As the bar absorbs heat it begins to take on a dull red color. As heat is applied and the bar's temperature increases the color begins to change to a bright cherry red. As more heat is applied the color shifts to orange, then yellow, and finally to a bluish white. If the temperature of the bar were measured at various points and its color noted, we could then describe the color of the bar in terms of its temperature. This is the principle used to describe the color of a lamp.

In practice, the steel bar is replaced by a theoretical object called a "blackbody radiator." The blackbody is capably of absorbing all of the energy which strikes it and then perfectly re-radiating that energy. An ordinary incandescent lamp exhibits characteristics which are nearly identical to the blackbody in the visible portion of the spectrum, and is called a "gray-body radiator."

The discussion of color temperature thus far has simply referred to the temperature of the light source, with no reference to the scale that is used. An American, if asked the temperature at which water freezes, would quickly respond "32 degrees." A European, if asked the same question, would answer "0 degrees." A scientist from any country might reply "273 Kelvins." Who is right? The answer is that they are all correct, depending upon the frame of reference of the respondent. The American uses the Fahrenheit scale, the European uses the Celsius scale, and the scientist uses a variety of scales, including the Kelvin scale as cited. The Kelvin scale was named after its founder, Baron Kelvin of Largs, a British physicist. It is also known as the Absolute scale, and has its zero point at –273 degrees Celsius. It uses the same gradients as Celsius, thus water freezes at 273 Kelvins and boils at 373 Kelvins. See Figure 1-9 for a comparison of these temperature scales.

The Kelvin scale has wide application in the scientific and engineering worlds and is used by the lighting community to describe the color temperature of lamps.

To interpret the meaning of color temperatures assigned to various lamps, simply remember that the lower temperatures refer to lamps with a large red component which will generally create a feeling of warmth in a space. As the temperature increases the lamp will generally contain more blue and the space will become progressively cooler in appearance. For example, a blackbody at normal room temperature will appear black, at 800 K it will be a dull red, at 3,000 K it is yellow, white at 5,000 K, pale blue at 8,000 K, and a bright blue at 50,000 K.

Theoretically, the term "color temperature" applies only to incandescent sources. When discussing discharge lamps (fluorescent and HID)

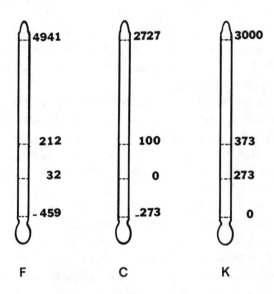

Figure 1-9
Comparison of three temperature scales. The Fahrenheit scale is used primarily in the United States. The Celsius scale is used by most other countries. Scientists use a variety of scales, including Kelvin and Celsius. The Kelvin scale is also called the Absolute scale.

the term "apparent color temperature" or "correlated color temperature" is used. This distinction is necessary since discharge lamps do not exhibit radiation characteristics which coincide with a blackbody radiator.

The color temperatures associated with specific lamps will be discussed in detail in later chapters of this text as the light sources are covered. While no specific cutoff points exist for the classification of color temperatures as warm, neutral, cool, or cold, the following ranges should be generally acceptable to describe the effects of the source on the visual environment.

Less than 3500 K   Warm
3500 K to 4000 K   Neutral
4000 K to 6000 K   Cool
Over 6000 K   Cold

It is important to understand that color temperature alone does not describe the ability of a lamp to render color. Color temperature only describes the apparent color of the source and the degree of warmth or coolness which it will impart to the space. Another metric, color rendering index, is required to describe the way in which the source might be expected to render color.

COLOR RENDERING INDEX

A lamp's color rendering index (CRI) provides a means of comparing the shift in color of a test object when viewed under the lamp, as compared to the color when viewed under a "reference" lamp of the same correlated color temperature as the test lamp. It is significant to note that both lamps must have the same color temperature. If different temperature sources are used the comparison is meaningless. In general, the test lamp is compared to the reference lamp at eight standardized colors, and a correlation drawn between the two lamps. A test lamp with a perfect correlation would have a CRI of 100. The CRI will drop as the differences widen between the two lamps at the eight specified colors. In general, lamps with high CRI's will provide better color rendering than lamps with low CRI's. Unfortunately, the CRI system does not indicate the magnitude of the differences at specific colors, nor does it indicate the direction of the color shift.

SPECTRAL POWER DISTRIBUTION

Most lamp manufacturers publish pictorial representations of the relative quantities of light which their lamps produce as functions of wavelength. These pictures are called "Spectral Power Distribution Curves." Typical curves for warm white and warm white deluxe fluorescent lamps are shown in Figure 1-10. The warm white lamp has very strong green and yellow components, but drops off rapidly in the red region. The warm white deluxe has substantially lower green and yellow components, and is much stronger throughout the red region. We can therefore conclude that the warm white deluxe lamp will render most red colors better than the warm white lamp. Curves for specific lamps may be obtained from lamp manufacturers and examined to determine the relative quantities of light produced at various colors. Lamps which have strong components of desired colors can generally be expected to render those colors in an acceptable manner. Lamps

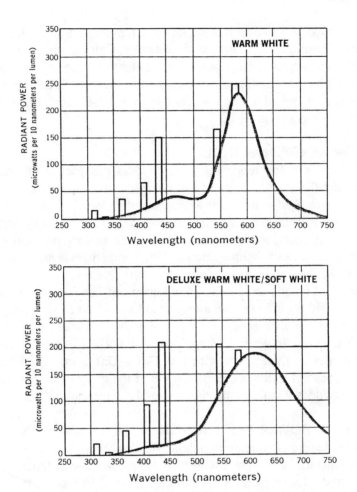

Figure 1-10
Spectral power distributions for Warm White and Warm White Deluxe
fluorescent lamps. See test for explanation. *(Courtesy GE Lighting)*

which are deficient in the desired colors will seldom be acceptable. A
notable exception to this rule, however, are the new "multi-phosphor"
lamps, which appear to have somewhat distorted spectral distributions
according to spectral power distribution curves, yet produce quite
acceptable color rendering when viewed by people with normal color
vision. Color blind individuals, particularly those with red-green
deficiencies, will usually prefer conventional lamps. This phenomenon

can be explained by briefly discussing the TRI-STIMULUS EYE SEN-SITIVITY curves.

The human eye contains three separate types of color receptors, called "cones," which respond to the three primary colors of light: red, green, and blue. The relative sensitivity of each type of cone is shown in Figure 1-11. It is believed that, in individuals with normal color vision, the visual sensations of color from the cones is blended, and colors are perceived as being well rendered. Individuals with deficiencies in one or more types of cones will therefore perceive colors as unbalanced under multi-phosphor lamps, and seldom consider these lamps acceptable for tasks requiring a high degree of color rendering.

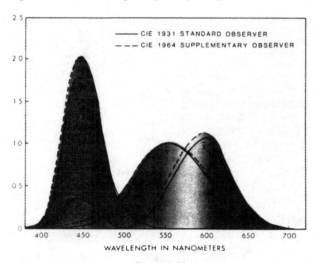

Figure 1-11
Tri-stimulus eye sensitivity curves. See text for explanation. *(Courtesy IES Lighting Handbook, IESNA)*

In summary, color temperature, color rendering index, and spectral power distribution data can provide an indication of the visual appearance and color rendering capabilities of a light source, but are not absolute metrics. While they can be of use in selecting a light source, they should be applied with caution when dealing with an unknown lamp. There is no substitute for experience when selecting a source for a specific color rendering requirement. When in doubt, obtain samples of various lamps and compare them to obtain a first-hand knowledge of their color rendering capabilities.

FUNDAMENTAL LIGHTING UNITS
AND RELATIONSHIPS

The basic units of measure used in lighting are the candela, the lumen, the footcandle, and the footlambert. Figure 1-12 shows the relationships between the first three of these units. The footlambert is gradually being phased out, however it is a highly useful metric and will be discussed along with its replacement, candelas per unit area. These are the fundamental units and form the basis for virtually all of the lighting calculations which are commonly performed.

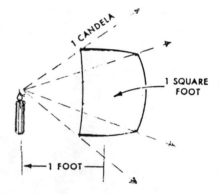

Figure 1-12
The relationships between the fundamental units used in lighting. If a point source of 1 candela is placed in the center of a sphere having a radius of 1 foot, one lumen of light will fall on each square foot. This will produce an illuminance of 1 footcandle since, by definition, 1 footcandle is 1 lumen per square foot.

The CANDELA is the unit of luminous intensity. Imagine a flashlight pointed at a wall. The beam of light has a magnitude (or intensity) of some number of candelas directed at the wall. An ordinary wax candle has an intensity of about 1 candela in all directions as measured from the flame. Candela is abbreviated cd and is represented in formulae by the letter I.

The term "CANDLEPOWER" is frequently used instead of candela. From a theoretical viewpoint this is incorrect, since candlepower is defined as luminous intensity, and the units of candlepower are candelas. For example, the wax candle noted in the previous paragraph has a candlepower (or luminous intensity) of one candela.

The LUMEN is the unit of luminous flux. One lumen is the amount of light striking a one-square-foot area, all points of which are 1 foot away from a point source of 1 candela intensity. Lumen is abbreviated Lm and is represented by the Greek letter Φ (phi).

The FOOTCANDLE is the unit of measure of the density of light striking a surface, and is measured in lumens of light per square foot of area. One footcandle is, by definition, equal to 1 lumen striking 1 square foot. The footcandle is the commonly used unit of illuminance. The SI unit of illuminance is the LUX, which is defined as lumens per square meter, and is gradually coming into use in the United States. One footcandle is approximately 10.76 lux, however for purposes of simplicity it is common practice to use a 10-to-1 conversion. For example, 10 footcandles would be expressed as 100 lux. Footcandle is abbreviated fc and is represented by the letter E.

The FOOTLAMBERT is the unit of measure of light exiting (leaving) a surface. See Figure 1-13. The light may be reflected by the surface, as in the reflection of light by a wall or piece of paper, transmitted through a material and emitted by the surface of the material, such as a diffuser in a light fixture, or generated by a surface such as a fluorescent lamp. One footlambert is equal to one lumen of light emitted by one square foot of surface area. For example, if 2 lumens per square foot strike a diffuser which transmits 50% of the light, then 1 lumen per square foot will be emitted by the diffuser and the diffuser will have a luminance of one lumen per square foot, or 1 footlambert. For reflecting surfaces the footlamberts will equal the illuminance times the reflectance of the surface. Footlambert is abbreviated fL.

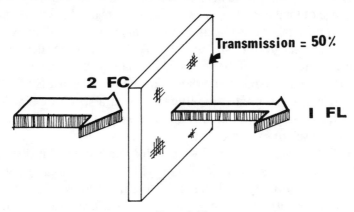

Figure 1-13

The footlambert is a unit of luminance expressing the exitance of light from a surface in lumens per square foot. One footlambert is one lumen per square foot leaving a surface. If two footcandles strike one side of a diffuser which transmits 50% of the light, the other side of the diffuser will have a luminance of 1 footcandle.

The footlambert is currently being replaced by candelas per unit area, with the unit of area generally being the square inch or square meter. Candelas per square inch may be converted to footlamberts by the relationship: 1 cd/sq in = 452 fL.

FUNDAMENTAL
LIGHTING CALCULATIONS

There are two basic types of lighting calculations which are of particular use. The first is the average illuminance calculation, which predicts the average footcandles in a space, and comes from the definition of the footcandle.

Footcandles are, by definition, lumens of light per square foot of area. Assume that 5,000 lumens of light are evenly distributed over an area of 100 square feet. The illuminance will, by definition, be equal to 5000 lm/100 sq. ft., or 50 lumens per square foot. Since footcandles are lumens/square foot, the illuminance is 50 footcandles. Note that this example deals only with the lumens actually reaching the work surface, not the total number of lumens produced by the lighting system. The percentage of total lamp lumens which actually reaches the work surface will be considerably less than the total number of installed lumens, and will typically range from 40% to 70%. The procedure for determining the actual percentage will be discussed in detail in later chapters.

The second calculation method is called the INVERSE SQUARE LAW. It is used to predict the illuminance at a specific point in the space. It is the most commonly used of several methods, and also forms the basis for photometric testing of luminaires.

As light travels away from a source it spreads out and is distributed over a wider area, as shown in Figure 1-14. While the total quantity of light remains virtually unchanged, the distribution over a wider area reduces the density of light when it strikes a surface. This reduction is equal to $1/distance^2$, thus the name "inverse square law."

The illuminance at a point which is at right angles to the fixture, as shown in Figure 1-15, may be found from:

$$FC = \frac{Candlepower}{Distance^2} \text{ or } E = \frac{I}{D^2}$$

For example, if the intensity is 2000 candelas and the distance is 10 feet, the illuminance will be:

$$FC = \frac{Candlepower}{Distance^2} = \frac{2000 \text{ Cd}}{10Ft^2} = 20 \text{ FC}$$

The inverse square law assumes that the surface upon which the illuminance is calculated is normal to the ray of light. If the surface is not at a right angle the equation must be modified to reflect the further spread of light due to the angle. This procedure will be discussed in detail in Chapter 8.

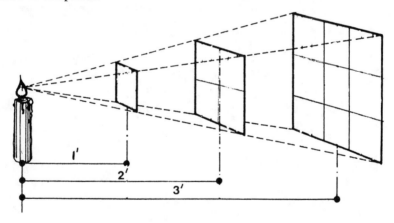

Figure 1-14

Light spreads out to cover a wider area as the distance from the source increases. The decrease in illuminance is inversely proportional to the square of the distance from the source. If the light which strikes a 1-square-foot area placed 1 foot from the source is allowed to travel 2 feet, it will cover an area of 4 square feet. At 3 feet the area covered is 9 square feet; at 4 feet the area expands to 16 square feet.

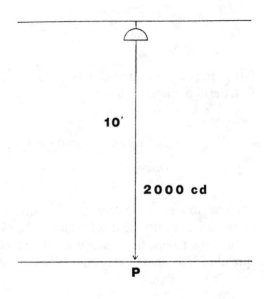

Figure 1-15
Diagram used for the ex-
ample calculation of the
illuminance at a point with
the inverse square law.

Chapter 2

LAMPS FOR GENERAL LIGHTING: INCANDESCENT

Woe be it to the lighting designer who recommends:
>   standard fluorescent lamps in a freezer...
>   standard sodium lamps for merchandise lighting...
>   low wattage metal halide lamps in open fixtures...
>   HID lamps as the sole source of light in a building...

Why?

>   Because:
>   standard fluorescent lamps won't start at low temperatures...
>   the color of standard sodium lamps isn't acceptable for merchandising...
>   low wattage metal halide lamps can shatter in operation and scatter very hot glass...
>   in the event of a power interruption HID lamps can take up to 20 minutes to re-strike...

These are only a few of the traps into which an inexperienced lighting designer or specifier can fall. Any of them will be embarrassing at the least and can be expensive to correct. Only by understanding the basics of lamp characteristics and applications can the designer avoid improper applications and assure that the lamp which has been specified is suitable for the job requirements.

The lamp industry produces over 14,000 different lamps, each with a purpose, and many with characteristics which make them unsuited for most applications. Fortunately, the significant characteristics apply to most or all of the lamps within a generic category, thus many decisions can usually be based on a knowledge of lamp fundamentals. The examples cited above serve to illustrate this statement.

Lamps for general lighting systems fall into three basic categories: incandescent, fluorescent, and high intensity discharge. Each category is distinctly different from the others in many respects. Each has specific characteristics which make it particularly well suited for some lighting tasks and marginal or unacceptable for others.

This chapter will examine incandescent lamps, while the following two chapters will explore fluorescent and high intensity discharge lamps, respectively.

INCANDESCENT LAMPS

In 1879 the "Wizard of Menlo Park," Thomas Alva Edison, produced the first commercially viable incandescent lamp. At last it was possible to bring the sun indoors, without regard to whether it was day or night, and light up the world. No longer was it necessary to contend with candles, gas flames, or kerosene lamps, which produced noxious odors, smoke, or irritating fumes, and gave off little light. Edison's latest invention was to create a new way of life.

Thomas Edison was not the first to attempt to develop an electric light source; he was the first to develop one which worked. In keeping with his axiom that "genius is 1/10th inspiration and 9/10ths perspiration," he tirelessly researched the efforts of earlier researchers who had failed to produce a workable lamp. Like them, his early efforts centered on the quest for a substance which could be heated to a very high temperature at which it would "incandesce," or produce usable quantities of visible light. At high temperatures, however, oxidation occurs rapidly and the material burns up. This necessitated the operation of the lamp in a vacuum to exclude oxygen and preclude oxidation. This was no simple task during the 19th century, and added substantially to the time required to complete an experiment. The early experiments centered primarily on the use of metals, which tended to melt before reaching a high enough temperature to produce usable light. Then, after numerous experiments with metals had failed, Edison spied a spool of thread in his wife's sewing basket. He carefully shaped a piece of the thread, Coat's No. 29, into a filament and baked it in an oven to carbonize it. The first eight attempts to bake and attach the fragile filament to the lead in wires failed, but the ninth attempt met with success and the historic test began. After 10 hours the mercury vacuum pump had removed sufficient air from the lamp envelope and power was applied. The filament glowed. It continued to glow for about 40 hours, much longer than any previous test, before the thin thread burned through. At this point Edison knew that he could produce a commercially viable product.

The search for a suitable filament now turned to carbon-based materials. The list of trials was almost unlimited: wood, rope, animal hooves, horns, and hides, lemon rind, and even macaroni, to mention a few. No organic material was sacred. Unfortunately, none was suitable. Then, early in 1880, Edison tried a sliver of bamboo from a fan used in the lab to cool experimental mixtures. The results held promise, and a massive worldwide search began. Over 6,000 varieties of bamboo were procured from China, Japan, the South Pacific Islands, and the jungles of South America. After exhaustive tests, 3 of the varieties were found acceptable. One of these, obtained from a small grove near a Shinto Temple in Kyoto, Japan, was far superior to the others. This grove was to supply filament for all of the lamps which Edison produced for the next 14 years.

Bamboo filaments worked; however, they had limitations. A major drawback was that the maximum length of the filament was governed by the distance between the joints on the bamboo cane. This was overcome by the development of an extruded cellulose fiber by an English inventor, Leigh S. Powell, in 1888. Powell's process consisted of dissolving cotton in hot zinc chloride, extruding the mixture through a die to form a long, thin thread, and washing away the zinc chloride. The remaining thread was continuous and without the joints found in bamboo, thus it could be cut into any desired length. By 1894 the use of cellulose was almost universal, and would continue to be so until the introduction of the tungsten filament in 1910. While the incandescent lamp has undergone a number of major improvements since that time, it is significant to note that the base filament material, tungsten, is unchanged.

NOMENCLATURE

While incandescent lamps vary widely in size, shape, type of base, and wattage, they consist of the same basic components. Figure 2-1 shows the common parts. Their functions are:

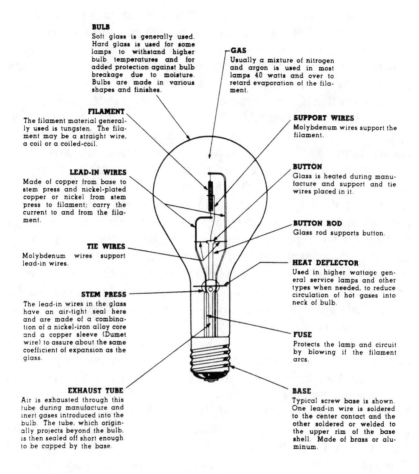

BULB
Soft glass is generally used. Hard glass is used for some lamps to withstand higher bulb temperatures and for added protection against bulb breakage due to moisture. Bulbs are made in various shapes and finishes.

GAS
Usually a mixture of nitrogen and argon is used in most lamps 40 watts and over to retard evaporation of the filament.

FILAMENT
The filament material generally used is tungsten. The filament may be a straight wire, a coil or a coiled-coil.

SUPPORT WIRES
Molybdenum wires support the filament.

LEAD-IN WIRES
Made of copper from base to stem press and nickel-plated copper or nickel from stem press to filament; carry the current to and from the filament.

BUTTON
Glass is heated during manufacture and support and tie wires placed in it.

BUTTON ROD
Glass rod supports button.

TIE WIRES
Molybdenum wires support lead-in wires.

HEAT DEFLECTOR
Used in higher wattage general service lamps and other types when needed, to reduce circulation of hot gases into neck of bulb.

STEM PRESS
The lead-in wires in the glass have an air-tight seal here and are made of a combination of a nickel-iron alloy core and a copper sleeve (Dumet wire) to assure about the same coefficient of expansion as the glass.

FUSE
Protects the lamp and circuit by blowing if the filament arcs.

EXHAUST TUBE
Air is exhausted through this tube during manufacture and inert gases introduced into the bulb. The tube, which originally projects beyond the bulb, is then sealed off short enough to be capped by the base.

BASE
Typical screw base is shown. One lead-in wire is soldered to the center contact and the other soldered or welded to the upper rim of the base shell. Made of brass or aluminum.

Figure 2-1
The major parts of a typical incandescent lamp. *(Courtesy GTE Sylvania)*

BASE – Holds the lamp in the socket and provides a means of making electrical connections. Bases come in a variety of types as shown in Figure 2-2. The medium base is the most common, and is used on general service lamps of 300 watts or less. It is typically made of aluminum or brass. Aluminum bases should be used indoors or in dry locations, while the brass base can be used in either indoor or outdoor environments.

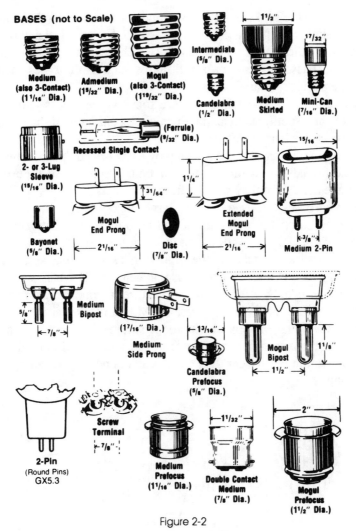

BASES (not to Scale)

Medium
(also 3-Contact)
(1 ¹/₁₆" Dia.)

Admedium
(1⁵/₃₂" Dia.)

Mogul
(also 3-Contact)
(1¹⁹/₃₂" Dia.)

Intermediate
(⁵/₈" Dia.)

Candelabra
(¹/₂" Dia.)

Medium
Skirted

Mini-Can
(⁷/₁₆" Dia.)

(Ferrule)
(⁹/₃₂" Dia.)

Recessed Single Contact

2- or 3-Lug
Sleeve
(¹⁵/₁₆" Dia.)

Bayonet
(⁵/₈" Dia.)

Mogul
End Prong

Disc
(⁷/₈" Dia.)

Extended
Mogul
End Prong

Medium 2-Pin

Medium
Bipost

Medium
Side Prong

Candelabra
Prefocus
(⁵/₈" Dia.)

Mogul
Bipost

2-Pin
(Round Pins)
GX5.3

Screw
Terminal

Medium
Prefocus
(1¹/₁₆" Dia.)

Double Contact
Medium
(⁷/₈" Dia.)

Mogul
Prefocus
(1¹/₂" Dia.)

Figure 2-2

Typical bases used on incandescent lamps. *(Courtesy GE Lighting)*

The mogul base is used on general service lamps of 300 watts or higher. It is also used on some special lower wattage lamps such as some low voltage types to carry the higher current associated with low voltage operation. The mogul base is capable of safely carrying 35 amperes, as opposed to the 25-ampere rating of medium bases. The larger physical size of the mogul base makes it more effective in dissipating heat to the socket, and helps support the added weight of large lamp envelopes. Mogul bases are brass and are frequently nickel plated, which gives them a silvery appearance. The candelabra and mini-candelabra bases are used on small lamps such as the types used in chandeliers and some appliances.

Specialized bases such as the pre-focus types are used in some projectors and spotlights to maintain precise positioning of the filament with respect to a lens or reflector.

Screw-in type bases normally employ a right-hand thread; however, left-hand threads are available on some lamp types. These are used to discourage theft since they will not fit most sockets.

Bi-post bases are used on very large, heavy lamps and for high wattages to carry the high current and help dissipate heat. They are commonly found on lamps designed for theatrical use.

Most bases are attached to the lamp with a special "basing cement." Standard general service lamps use a cement that can withstand temperatures up to 340°F. A high-temperature silicone cement rated for 500°F is used for special applications. Mogul bases that are used on some higher wattage lamps which produce more heat normally employ a mechanical connection instead of cement, as illustrated in Figure 2-3. This connection is stronger than cement and is not weakened by high temperatures.

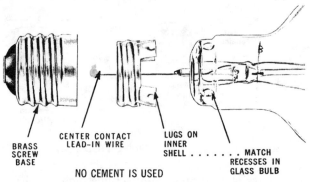

BRASS
SCREW
BASE

CENTER CONTACT
LEAD–IN WIRE

NO CEMENT IS USED

LUGS ON
INNER
SHELL . . . . MATCH
RECESSES IN
GLASS BULB

Figure 2-3

Typical mechanical connection used to attach some mogul bases to lamps. The lugs on the inner shell are positioned over the recesses in the bulb and locked into place by dimpling. The outer base is then screwed onto the inner shell. This method of base attachment provides a strong connection which is not affected by heat. *(Courtesy GE Lighting)*

BULB – The bulb, or envelope, is the glass enclosure which contains the working parts of the lamp. Many different glass formulations are used, however they may be classified into three broad categories: lime glass, heat resistant glass, and quartz glass. Lime glass is the most common and is used for most incandescent lamps. Heat resistant glass is used for lamps of high wattage or when the lamp is intended for outdoor use. Quartz glass is typically used for a special family of lamps called "halogen cycle." These will be discussed in detail later in this chapter.

FILAMENT – The filament is the light-producing element of the lamp, and is made of tungsten wire. The wire is thinner than a human hair and is usually wound into a coil. Some lamps use filaments which have been double coiled as shown in Figure 2-4. Coiling helps concentrate heat and causes the filament to become hotter, thus improving efficacy. Filaments are identified by the letter C (coiled) or CC (coiled coil), followed by a number. The number specifies the configuration and mounting arrangement. Filament shape, size, and configuration (Figure 2-5) are often dictated by the intended use of the lamp.

Specialized reflectors and optical systems require a very small source of light, so filaments for these lamps are compact. Lamps which will be subjected to vibration or sudden shock use C-9, C-17, or C-22 filaments which are supported at frequent intervals by wires that are designed to absorb the vibration or shock. These lamps should not be used for general service since the added support wires draw heat away from the filament and reduce efficacy.

FILL GAS – Most lamps rated at 40 watts or higher are filled with a mixture of inert gases. Nitrogen and argon are the most common. The fill gas helps retard the rate of filament evaporation. This principle can be explained with a simple analogy which is illustrated in Figure 2-6: if a container of water is heated the water will evaporate. As more heat is applied the rate of evaporation increases rapidly. The evaporation rate can be slowed if a tight-fitting lid is placed on the container since the lid will trap the expanding gases which causes an increase in the pressure on the surface of the water. This pressure on the water surface opposes the forces which cause evaporation. The fill gases act to increase the

Figure 2-4
Most filaments are coiled or
double coiled. This magnification
shows a coiled coil filament.
(Courtesy GTE Sylvania)

TYPICAL FILAMENTS (not to Scale)

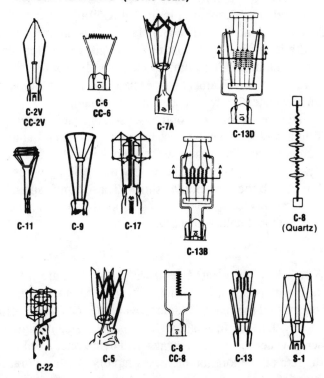

Figure 2-5
Many different filament configurations and mounting arrangements are used.
Each filament shown has a specific use. Some arrangements are designed to
position the filament at a specific location with respect to a reflector. Others are
designed to absorb physical shock or vibration. *(Courtesy GE Lighting)*

pressure on the filament in the same manner as the expanding gases in the container, and retard the rate of filament evaporation. This extends lamp life and improves efficiency. Lamps are normally filled to slightly less than 1 atmosphere pressure when cold, and designed to operate at about 1 atmosphere when hot.

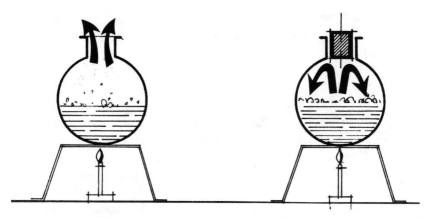

Figure 2-6

A close-fitting lid on a container of heated water traps expanding gases and causes an increase in the internal pressure on the surface of the water. This retards the rate of evaporation of the water and allows it to attain a higher temperature before boiling. The inert gases used in incandescent lamps act in a similar way to increase the pressure on the hot filament and retard the rate of filament evaporation.

Fill gases conduct heat away from the filament and to the bulb wall where it is dissipated as heat instead of light. In small lamps, those of less than 40 watts, the loss of heat through conduction offsets the benefits of reduced filament evaporation gained through the use of the gas. For this reason lamps of less than 40 watts generally utilize a vacuum instead of inert gases.

Krypton gas has lower heat conductivity than argon or nitrogen and is sometimes used to improve efficacy and life. The use of krypton permits improvements in efficacy of about 10% with no change in lamp life, or an improvement in life of 50% to 60% with no change in efficacy. Krypton is an expensive gas and is used only in some higher-priced specialty lamps where the improvement in efficacy is sufficient to justify the added cost, or in some very long life lamps used in special applications with high replacement labor costs. Note that krypton, by itself, can only extend lamp life by 50% to 60%. Lamps with very long life ratings produce substantially less light than "standard" life lamps, so more lamp watts will be required to provide equal light.

Gas-filled lamps are designated as Class C and vacuum lamps are Class B.

HEAT DEFLECTING DISK – Used in high wattage lamps to reduce the circulation of hot gases into the neck of the lamp and prevent over-heating of the base.

LEAD-IN WIRES – Carry current to the filament.

FUSE – Built into one of the lead-in wires, the fuse protects the lamp and prevents cracking or breakage of the bulb in the event of an arc.

BULB SHAPES

Lamps are produced in a wide variety of shapes. Some are utilitarian while others are decorative. The shape is designated by one or two letters followed by a number. The letters denote the shape, as shown in Figure 2-7.

In many cases the letters are simply abbreviations for the shape, such as "T" for tubular or "G" for globe. "A" and "B" designations are arbitrary and have no literal meaning. The numeric part of the lamp code denotes the maximum diameter of the lamp in 1/8ths of an inch. For example, a lamp designated as an A 19 is an arbitrary shape and is 19/8ths of an inch (2-3/8 inches) in diameter, as shown in Figure 2-8.

EFFICACY

Edison's first lamp had an efficacy of 1.4 lumens per watt. This was increased to about 4 lumens per watt in 1888 with the introduction of the extruded cellulose fiber and to 10 lumens per watt in 1910 when tungsten replaced cellulose. The introduction of fill gas in 1913 brought efficacy up to 13 lumens per watt, and additional improvements have provided further gains.

The theoretical maximum efficacy of tungsten is about 52 lumens per watt and occurs at 6120°F, the temperature at which tungsten melts. In practice, the highest efficacy standard lamp available is the 10,000

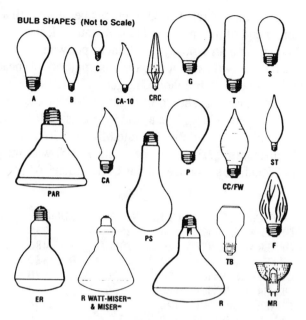

BULB SHAPES (Not to Scale)

Figure 2-7

Typical bulb shapes used for lamps. In most cases the letters have meaning, as discussed in text. The designations "A" and "B" are arbitrary and do not indicate a specific shape. *(Courtesy GE Lighting)*

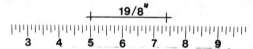

Figure 2-8

Lamps are identified by a letter and number. The letter designates the bulb shape, while the number denotes the maximum diameter of the lamp in 1/8ths of an inch. The "A-19" bulb shown is an arbitrary shape and is 19/8" in diameter.

watt studio lamp which is rated at 33.5 lumens per watt with a 75-hour life. At the other end of the spectrum is the 3-watt indicator lamp with a life of 3,000 hours and an efficacy of 4 lumens per watt.

In between these two extremes lie the vast majority of incandescent lamps used for general illumination. In general, efficacy increases as lamp wattage increases, as shown in Figure 2-9. Note, however, that efficacy is also related to lamp life; as life increases, efficacy decreases. For example, a 100-watt lamp with a rated life of 750 hours produces about 1750 lumens, or 17.5 lumens per watt. A 100-watt lamp with a rated life of 2500 hours produces about 1440 lumens, or 14.4 lumens

WATTS	DESCRIPTION	VOLTS	LIFE	LUMENS	LM/WATT
25	GENERAL SERVICE	120	2500	235	9.4
40	GENERAL SERVICE	120	1000	480	12.0
40	TRAFFIC SIGNAL	120	2000	380	9.5
60	GENERAL SERVICE	120	1000	870	14.5
75	GENERAL SERVICE	120	750	1190	15.9
100	GENERAL SERVICE	120	750	1750	17.5
100	GENERAL SERVICE	120	2500	1440	14.4
100	GENERAL SERVICE	120	4000	1290	12.9
100	TRAIN	34	1000	2160	21.6
100	GENERAL SERVICE	230	1000	1280	12.8
150	GENERAL SERVICE	120	750	2850	19.0
300	GENERAL SERVICE	120	750	6360	21.2
500	GENERAL SERVICE	120	1000	10850	21.7
1000	GENERAL SERVICE	120	1000	23740	23.7

Figure 2-9

Efficacy of typical incandescent lamps. Note that the efficacy increases as wattage increases. Note also that long life lamps have lower efficacy than standard life rated lamps, and that low-voltage lamps are more efficient than high-voltage lamps.

per watt, while an 8000-hour lamp of the same wattage produces only about 1080 lumens, of 10.8 lumens per watt.

Efficacy is also related to design voltage. Lamps designed to operate at high voltages have thinner filaments which are less efficient due to higher gas losses. As the design voltage decreases, filaments become progressively thicker and more compact. This reduces the gas loss and improves overall lamp efficacy.

OPERATION

In operation, an electric current is passed through the tungsten filament. Tungsten has a high electrical resistance, thus a voltage drop occurs which consumes power. This power produces heat which raises the temperature of the filament and causes it to incandesce. The hotter the filament, the more light is produced. The radiant energy produced by an incandescent lamp can be calculated from the Stefan-Boltzmann law which states that the radiated power is equal to a constant, 5.67×10^{-8}, times the temperature in Kelvins raised to the fourth power ($5.67 \times 10^{-8}T^4$). Thus it can be seen that small changes in temperature result in large changes in light output. In general, filament temperatures increase as lamp wattage increases, so high wattage lamps not only produce more total light, but their efficacy also increases. For example, a 10-watt lamp operates at a filament temperature of about 3900°F and produces 80 lumens, for an efficacy of 8 lumens per watt. A 100-watt lamp has a filament temperature of about 4900°F and produces 1750 lumens, for an efficacy of 17.5 lumens per watt.

Tungsten has a very low electrical resistance when cold, so the initial current can be very high at the instant the lamp is energized. This initial current, called "inrush," can range from 12 to 18 times normal operating current. Switches and contactors used to control large incandescent lighting loads must be designed to carry the inrush current. Devices meeting this requirement are "Tungsten" or "T" rated.

LAMP LIFE

The rated life of incandescent lamps is the point at which 50% of an infinitely large group of lamps will have failed. Rated life does not mean that every lamp will burn that long; in fact, failures can be expected to begin at about 40% of rated life and will increase in frequency so that half of the lamps have failed when rated life is reached. The remaining lamps will fail at some point beyond rated life. Figure 2-10 shows the mortality curve for incandescent lamps. Note the dramatic increase in failures at about 70% of rated life. If incandescent systems are to be group relamped this is normally the optimum economic time.

Standard incandescent lamps have life ratings ranging from 750 hours to 2,500 hours. This is an economic decision based on lamp cost versus power cost.

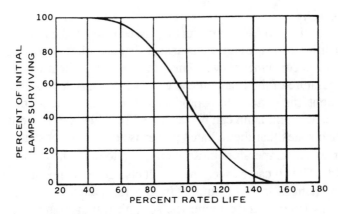

Figure 2-10
Mortality curve for typical incandescent lamps. Not all lamps will burn for full rated life. Failures begin at about 40% of life and increase as the accumulated burning hours approach rated life. At rated life, about one half of the lamps will have failed. *(Courtesy GTE Sylvania)*

As incandescent lamps operate, the tungsten filament evaporates or "boils off." The rate of evaporation is primarily a function of filament temperature. While the fill gas helps control the rate of evaporation, filament temperature is the most important factor in most lamps. Low-cost lamps such as the "A" lamp are designed for shorter lives than higher-cost lamps such as the "R" and "PAR" types. Some incandescent lamps are designed for very long lives. Traffic signal lamps, due to their conditions of use and high replacement labor costs, are typically designed to last 8,000 hours. Some lamps are rated for up to 20,000 hours; however, the filament temperature is greatly reduced with a commensurate reduction in light output and efficacy. Color rendering is also greatly reduced since the lamp's spectral distribution shifts toward the red region with a pronounced deficit in blue and green. The use of these lamps is seldom justified unless relamping labor costs are very high due to unusual conditions. For example, a 150-watt, 20,000-hour lamp produces less light than a 75-watt, 2,000-hour lamp.

Some specialized lamps used for photography and studio/stage applications are designed for very high filament temperatures which produce high efficacy and very good color. These lamps are typically rated at 50 to 300 hours life.

Lamp life can also be affected by vibration. Filaments are quite pliable until heated. After normal operation they become very brittle and, unless designed for vibration service, will break if subjected to repeated vibration. Lamps designed to operate at low voltages have thick filaments which can better withstand vibration. As the design voltage of the lamp increases, filament diameter decreases and vibration may become an important factor in lamp life.

Another major factor in lamp life is the voltage actually applied to the lamp, as discussed in the next section.

VOLTAGE

Incandescent lamps are designed to operate at a specified voltage. Common ratings range from 5.5 volts for specialized spot lights used for display lighting in retail stores, to 277 volts used for some industrial lamps. The most frequently used voltage is 120 since this is the most common power system voltage used for incandescent lighting. Lamps rated at 125, 130, and 125-130 volts are also produced for use on nominal 120-volt power systems. The 125-130 volt lamp is actually designed for 127.5 volts.

Other voltages are frequently used for specialized applications. Low voltage lamps provide better light control than normal line voltage lamps and are popular for display and accent lighting where the ability to precisely control the light and direct it to small areas is important. Lamps rated at 5.5, 6, and 12 volts are commonly used. Should locomotive lighting become your specialty, note that train lamps are normally rated at 32, 34, 60, and 75 volts.

It is important to note that lamps are designed to consume rated power, produce rated light, burn for the rated life, and attain rated color temperature only when operated at their rated voltage. Even minor variations will have pronounced effects on power consumption, light output, life, and color. These effects apply to both unintentional variations in voltage during normal operation, and to intentional variations when operating lamps on dimmers.

It is common practice in industry to use 130-volt lamps on 120-volt circuits as a means of increasing lamp life. The effects of this practice can be determined from Figure 2-11. The percent rated voltage is 120/130 = 92 percent. From the graph it can be determined that life will be about 2.9 times rated (290%), lumen output about 76% rated, and power consumption about 88% rated. The advisability of the practice is purely economic. Is it worth it to consume 88% of the power to get 76% light and 290% life? Each case requires a separate analysis based on the individual customer's cost for power, lamps, and relamping labor.

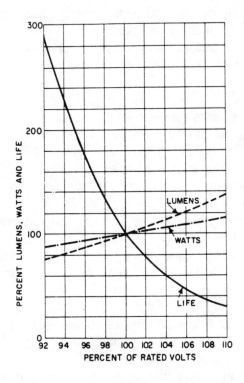

Figure 2-11
Effects of voltage variations on lamp life, lumen output, and watts. Small changes in voltage have pronounced effects. (Courtesy GTE Sylvania)

LAMP LUMEN DEPRECIATION

As lamps operate and tungsten evaporates from the filament, the lumen output will gradually decrease. This results from two factors. First, the tungsten that evaporates deposits on the inside of the bulb wall can be observed as a dark spot on the bulb (Figure 2-12). This absorbs light and reduces the number of lumens which pass through the glass. In gas-filled lamps convection currents in the gas carry the

Figure 2-12
As lamps operate, tungsten evaporates from the filament and deposits on the bulb wall. This reduces the amount of light produced by the lamp. Gas-filled lamps exhibit darkening on the bulb surface that was in the uppermost position while the lamp was operating. The lamp shown was operated in a vertical, base-down position. Vacuum lamps exhibit uniform darkening of all surfaces.

evaporated tungsten to the uppermost surface of the bulb. This can be observed by visual inspection of a burned-out lamp. If the lamp is operated in a vertical base-down position typically encountered in a table lamp, the darkening will be at the top of the bulb. Lamps operated horizontally will show a darkened area on one side of the envelope—the uppermost side. Vacuum lamps will exhibit a uniform darkening on the bulb since there are no convection currents in a vacuum and the tungsten radiates uniformly away from the filament.

The second factor which reduces light output is the reduction in power dissipated by the filament. This is caused by a reduction in the diameter of the filament as tungsten evaporates. As the filament becomes thinner, it has a higher resistance to current flow and the wattage consumed by the lamp decreases. This reduction in power further contributes to the reduction in light as the lamp ages.

Typical reductions in light output and power consumption are shown in Figure 2-13.

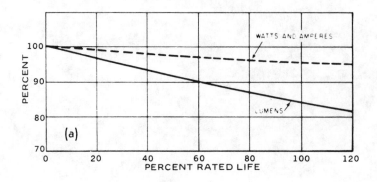

Figure 2-13
Reductions in light output and power consumption of a typical incandescent lamp over life. *(Courtesy GTE Sylvania)*

LAMP CLASSIFICATION

Lamps are classified by type and function into several broad categories. There is no firm dividing line between the categories and some lamps might be considered to fall into more than one category, depending upon the application.

General Service Lamps

These lamps are the most common type and are used to provide general illumination. They range in wattage from 10 to 1,500 watts, and are used in a wide variety of fixtures ranging from bare sockets and table lamps to sophisticated fixtures with esoteric reflector designs. The "A" and "PS" shapes are most common, with the "PS" favored for fixture designs requiring a long neck to achieve proper positioning of the filament with respect to the reflector.

Reflectorized Lamps

These lamps have a reflective coating on part of the inside of the bulb. This coating reflects light in a specific manner, depending on the design of the bulb. Reflectorized lamps are made in several different types:

Reflector Lamps. Reflector, or "R" lamps, have a distinctive shape as shown in Figure 2-14. They are designed to concentrate the light in front of the lamp. "R" lamps are available in two light distributions: flood and spot. Flood lights scatter the light to cover a large area in a somewhat uniform pattern. They are used to provide general illumination within an area.

Figure 2-14
Typical "R" lamps. *(Courtesy GE Lighting)*

Spot lights are designed to concentrate light into a narrow beam which can be directed to a specific location, and are used to highlight or accentuate the object being illuminated. Envelopes for "R" lamps are made in one piece by a "mold blowing" process where hot glass is blown into a mold of the desired shape. Lamps for indoor use are made of soft lime glass and are the most common. Specialized lamps made of heat-resistant glass are available for outdoor use or in applications where the lamp might come in contact with moisture.

After the envelope has been formed, it is coated with vaporized silver or aluminum, which acts as the reflector. The stem press is then inserted through the neck, the assembly is sealed, air is evacuated, and the fill gas inserted. The final steps are the addition of the base and printing of the "label" on the face of the lamp.

Reflector lamps use two common geometric shapes to reflect light. The most common is the parabola. If an infinitely small light source is placed at the focal point of the parabola, all rays of light which strike the parabola will be reflected as parallel rays as shown in Figure 2-15. In theory, this means that the rays will not diverge and a 6-inch reflector will project a 6-inch diameter beam of light over an infinite distance. In practice, this does not occur for two reasons: first, the filament is not an infinitely small source; in has dimension. Therefore, all of the light that strikes the parabola does not originate from the focal point so it cannot be reflected as parallel lines. Secondly, not all of the filament radiates light to the reflector. The balance is radiated towards the front of the lamp. This frontal light spreads as it leaves the lamp and diffuses rapidly (Figure 2-16). Spot light filaments are located near the focal point and direct most of the reflected light in nearly parallel lines. Flood light filaments are positioned in front of the focal point, which causes a rapid divergence of the reflected rays.

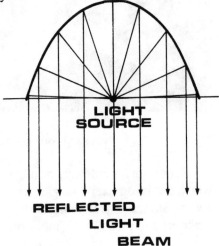

Figure 2-15
A parabola will reflect all rays of light on parallel lines if a point source is located at the parabola's focal point. This principle is used in reflector lamps to direct a beam of light.

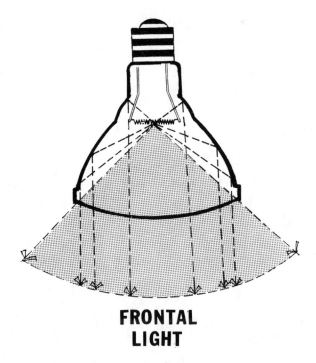

FRONTAL
LIGHT

Figure 2-16

In practice, filaments used in incandescent lamps are much larger than the theoretical point source used for reflector design. This causes some scattering of reflected light. Note also that a large portion of the light radiates directly out of the front of the lamp without striking the reflector. This light also scatters and widens the beam pattern.

The second common geometric shape used for reflector lamps is the ellipse. An ellipse has two focal points as shown in Figure 2-17. Lamps employing this shape are called "ellipsoidal reflector," or "ER" lamps. If a point source is located at one of the focal points, any ray of light that strikes the reflector will be reflected through the second focal point. If one-half of the ellipse is removed, the second focal point will remain as a point in space in front of the reflector. This principle allows light to be focused at a point in front of the lamp. As with "R" lamps, there is some scattering of light but the system directs the majority of light through the second focal point and the lamp works quite well in recessed deep-baffle type fixtures. Sufficient light is directed away from the baffle and through the focal point to permit a 50% reduction in wattage, yet allow about the same light out of the fixture as the higher wattage "R" lamp. The ellipsoidal reflector, or ER lamp, is considered to be a flood light.

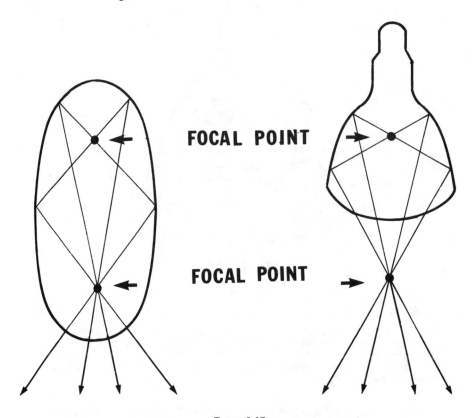

Figure 2-17
An ellipse has 2 focal points. A ray of light emanating from one focal point will be reflected through the second focal point as shown on the left. If the bottom half of the reflector is removed, the bottom focal point remains as a point in space, and reflected rays emanating from the top focal pass through the bottom point. This principle is used in ER lamps.

The front face of reflector lamps is frosted to even out the distribution of light from the lamp. Without this frosting the light pattern is a series of uneven concentric rings which imparts a non-uniform or mottled appearance to the object being illuminated. Flood lights have a heavy frosting to help diffuse the light while spot lights have a light frosting intended only to eliminate the uneven beam pattern.

Parabolic Aluminized Reflector Lamps (PAR Lamps). The PAR lamp, illustrated in Figure 2-18, differs from the "R" lamp in construction, performance, and diversity of applications. PAR lamps are made from heavy, heat-resistant glass shaped in a parabola. The envelope is made in two separate pieces through a molding process. Hot glass is placed in a mold and formed, under pressure, into precise shapes. The reflector portion is molded and then coated with vaporized aluminum. The lens is molded and embossed with a pattern, prisms for flood lights and a light stippling for spot lights. The filament is carefully positioned from the front of the lamp before the lens has been installed. This permits precise positioning of the filament to assure that it is at the focal point of the reflector. After this has been accomplished the lens is welded in place with a high-temperature flame. Air evacuation, gas fill, and base application complete the manufacturing process.

Figure 2-18
PAR lamps are molded from heavy, heat-resistant glass. The molding of the reflector shape and precise positioning of the filament improve performance over standard "R" lamps.

Since PAR lamps are made of heat-resistant glass, they may be used outdoors and in other areas where the lamp may contact moisture. PAR lamps provide much better light control than "R" lamps. This means that lower wattage PAR lamps can produce the same or more usable light than higher wattage "R" lamps. When using PAR lamps in lieu of "R" lamps it is normally possible to reduce wattage by one standard size: use a 100-watt PAR instead of a 150-watt "R" or a 75-watt PAR to replace a 100-watt "R."

Silver Bowl and White Bowl Lamps. These are typically PS or A shapes which have a silver or white coating on one-half of the envelope as shown in Figure 2-19. They are generally mounted in a vertical base-up position to direct the light upwards to the ceiling, where it is reflected back towards the floor. The silver or white finish serves two functions: to direct light towards the ceiling, and to prevent direct glare which could be present if the bright filament were exposed to view. Silver bowls reflect about 80% of the light upwards and allow the remaining 20% to pass through the coating. Silver and white bowl lamps were popular in the 1940's and 50's for classroom lighting.

Figure 2-19
Silver bowl lamps have declined in popularity due to low efficiency. *(Courtesy GE Lighting)*

Lamps of 500 to 1,500 watts were typically used. These generated large quantities of heat and were relatively inefficient producers of usable light due to absorption of light by the ceiling. The quality of light produced, however, was excellent and, to some extent, offset the lack of quantity.

Silver and white bowl systems are seldom used for general illumination in modern lighting systems due to their inefficiency and high operating cost.

Miscellaneous Reflectorized Lamps. Some other lamp types are occasionally reflectorized for specific applications. Tubular incandescent lamps used in jewelers' showcases frequently incorporate reflectors to direct light to the merchandise. Standard A or PS shapes with reflectors are produced by some specialty manufacturers and are useful where larger lamps will not physically fit, yet the benefits of a reflector are desired.

Halogen Cycle Lamps

Halogen cycle is the generic name applied to a family of lamps which have been in use for many years and known by a variety of names: quartz, quartz-iodine, and tungsten-halogen to name a few. Figure 2-20 shows some of the more popular shapes.

This unique group of incandescent lamps uses a special type of fill gas, normally iodine or bromine. These are members of a family of inert gases called "halogens," thus the use of the word halogen in the name. Halogen gases have a strong affinity for tungsten; they will chemically combine with hot tungsten molecules which have evaporated from the filament, and form tungsten-iodide or tungsten-bromide. This action takes place after the tungsten molecules have evaporated from the filament and have begun their journey to the bulb wall. The compounds are slippery; they will not adhere to the hot glass envelope of the lamp. As a result, convection currents within the lamp carry the tungsten-iodide (or bromide) back to the filament where the intense heat causes the elements to separate. The tungsten is deposited on the filament, and the halogen gas recirculates to capture another tungsten molecule. Thus, the lamp regenerates itself, and the process is called the "regeneration cycle." Unfortunately the tungsten does not necessarily re-deposit at the same spot from which it evaporated so the lamp does burn out.

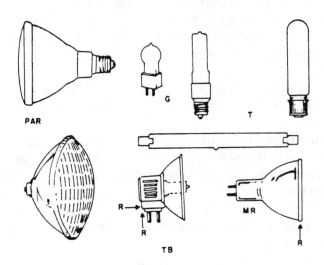

Figure 2-20
Halogen cycle lamps are available in a wide variety of sizes and shapes.
Large, line voltage lamps produce large quantities of light, and low initial cost
makes them popular for outdoor lighting when operating hours are very short.
Small low voltage lamps are more efficient than standard lamps and their
excellent color rendering characteristics make them good choices for display
and accent lighting. *(Courtesy GTE Sylvania)*

The significant result of the regeneration cycle is that the tungsten is deposited on the filament, not the bulb wall. For this reason halogen cycle lamps experience little lumen depreciation over life, and that which does occur is primarily the result of thinning of the filament which increases resistance and reduces current flow as previously discussed.

In order for the regeneration cycle to function, bulb wall temperature must be at least 500°C (932°F). This is accomplished by the use of small-diameter envelopes, generally 1/4 to 1/2 inch, and operation of the filament at temperatures of 3100 K (5120°F) or higher. Operation on dimming circuits may deactivate the regeneration cycle, resulting in bulb wall blackening. This may be reversed by operating the lamp at full power for a short time.

Until recently, halogen cycle lamps used quartz glass due to the requirement for high heat resistance and physical strength. New glasses with suitable characteristics are now in use for some lamp styles.

Quartz glass envelopes should not be touched with bare hands. Oil from the skin will cause the lamp to de-vitrify, or become porous, at normal operating temperatures. This admits air to the filament with a resultant early failure. Cotton gloves are recommended when handling quartz lamps. As an alternative, some manufacturers pack the lamp in a plastic or paper wrapper which may be used to avoid skin contact. If skin contact does occur, the lamp may be cleaned with a suitable solvent such as lighter fluid *before* the lamp is energized.

Halogen cycle lamps provide excellent color rendering due to their high operating temperature. They are popular for merchandise and display lighting in stores, and, in the large wattages, provide high lumen output from physically small sources. They are also well suited for outdoor lighting applications when operating hours are short, such as residential tennis courts, and for use with security systems which require instant full light output when the system is turned on.

Vibration and Rough Service Lamps

Vibration service lamps are designed for operation in applications where the lamp will be subjected to abnormal vibration. Examples are machine tools where supplemental light is desired; e.g., drill presses, lathes, and mills; and on traveling cranes used in industry. These lamps have specially designed filaments and support structures, shown in Figure 2-21a, which absorb vibration to prevent premature failure.

Rough service lamps use a different design for filaments and supports (Figure 2-21b), which will absorb a sudden shock. They are recommended for mechanics' trouble lights and similar applications where the lamp may be dropped or otherwise subjected to abnormal shock. As previously mentioned, they should not be used for general lighting since the added filament supports remove heat and reduce efficacy.

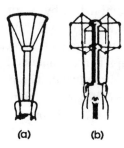

Figure 2-21
Filaments and support wires for vibration service lamps, left, are designed to absorb vibration. Rough service lamps use a different design, right, to absorb sudden shock. These lamps should not be used for general lighting since they produce less light than standard lamps of the same wattage. *(Courtesy GE Lighting)*

(a)   (b)

Miscellaneous and Specialty Lamps

There are many other classes of lamps whose uses are either so obvious as to not require explanation, or so specialized so as to not be considered for a general discussion of lamps. For lamps in the latter category it is suggested that lamp manufacturers' data be consulted for specific applications. For lamps in the former group, the fundamentals discussed in this chapter should suffice for most applications.

BULB FINISHES

Lamps are produced in three basic finishes: clear, inside frosted, and soft white.

Clear lamps have a clear glass bulb with the filament visible. The filament image is very bright and imparts a sparkle or glitter to shiny surfaces. Clear lamps are used to display jewelry and other sparkling items. In the past they were popular for automobile sales lots since they imparted a sparkle to glass, chrome trim, and polished surfaces. They have largely been replaced by more efficient sources on today's car lots. Clear lamps are also used with reflectors when it is desired to project a beam of light over long distances, since the small light source projected by the bare filament is relatively easy to control.

The inside frosted finish is the most common and is used to spread the brilliant filament image. Instead of the bright filament, the light source appears to be about the size of a golf ball, or slightly larger, in most lamps. This reduces the apparent brightness of the lamp to about 1/30th that of a clear lamp with a commensurate reduction in glare. Inside frosting does not reduce lumen output of the lamp even though it reduces brightness.

Soft white is the most recent development in lamp finishes. It is accomplished by coating the inside surface of an inside frosted bulb with a silica powder which further diffuses the filament image so that the entire envelope appears to be the light source. The use of soft white lamps is recommended if the lamp is exposed to view since glare is greatly reduced. The powder absorbs about 3% of the lumen output, but this slight reduction in efficacy is usually more than justified when it is desirable to improve visual comfort through glare reduction.

COLORED LAMPS

Colored lamps are produced by a variety of different methods. Each method produces lamps with different color characteristics and appearances. The common types of colored lamps utilize ceramic enamel, sprayed shellac or silicone, dip coatings covered with transparent plastic, and dichroic film. Ceramic enamels are fused to the bulb with heat to form a hard, glass-like surface which resists chipping and scratching. This type of finish is applied before the internal components of the lamp are installed. Sprayed shellacs or silicones are sprayed onto the exterior surface of the lamp after it has been assembled. These finishes can produce bright colors but, since they are not fired, are not as chip- or scratch-resistant as the ceramic enamels. Dip coatings have little scratch or scuff resistance so they are given a protective coating of clear plastic. These finishes produce brilliant colors and a visible filament image, and are widely used in outdoor signs. The plastic coating is somewhat tough but does not provide the same protection as ceramic enamel.

Dichroic coatings are generally applied to PAR lamps. They produce colors which are deeper and have more clarity than conventional colored lamps using the previously discussed coloring methods. These lamps have a series of very thin films applied to the inside surface of the lens, which selectively pass some wavelengths while reflecting all others, as shown in Figure 2-22.

Figure 2-22
Dichroic coatings can pass only selected colors of light through the front of the lamp. They can also pass infrared heat through the reflector, and reflect only visible light in the beam. This greatly reduces the heat projected by the lamp and reduces fading of merchandise.

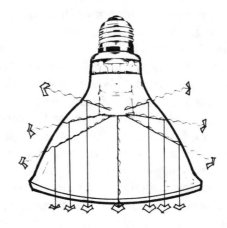

In fact, the coatings are only one wavelength thick. By varying the thickness of the layers the lamp can be made to selectively pass only specified colors, while reflecting all others. This means that only the desired color is transmitted by the lens while all other colors are reflected back into the lamp. As a result, the lamp produces more usable colored light than lamps using standard finishes which work by absorbing the unwanted colors, and the colors are more vibrant and alive.

Dichroic coatings which pass only infrared heat and reflect visible light are used on the reflector portions of some specialized PAR lamps for merchandising applications where it is desired to limit the amount of heat directed towards the items on display. These lamps are marketed under a variety of trade names and, for PAR lamps, usually include the word "cool" in the description of the lamp. The lamps reduce the heat content of the projected beam by about 2/3 while having no measurable effect on the quantity of "light" contained in the beam. They effectively reduce the projected heat by directing it back into the fixture.

Fixtures used with these lamps must be specifically designed for the lamp to avoid fire hazards.

LOW VOLTAGE LAMPS

Low voltage lamps are not actually a separate class of lamps, but recent advances in this technology have greatly increased their use and dictate a discussion of their characteristics and applications.

The major applications of low voltage lamps are display and accent lighting. Low voltage filaments are compact and permit precise light control if used with reflectors. Low voltage filaments are also more efficient than high voltage filaments and can achieve efficacies of up to 30 lumens per watt. When used in halogen cycle lamps the higher temperatures provide excellent color for merchandising.

Low voltage PAR lamps use filament caps, shown in Figure 2-23, to block frontal light. This reduces light scatter and permits placing more light at the desired location, an important feature to the retailer.

Low voltage lamps draw higher current than line voltage lamps of comparable wattage. In fact, a 12-volt lamp requires 10 times the current of a 120-volt lamp to consume the same watts. In most systems power is distributed to fixtures at 120 volts. Step-down transformers are contained in the fixture to reduce the voltage and increase current. A typical system is shown in Figure 2-24.

Low voltage systems for display and accent lighting typically use PAR or MR lamps.

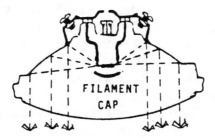

Figure 2-23
Filament caps are used on low voltage PAR lamps to cut off frontal light and reduce light scatter.

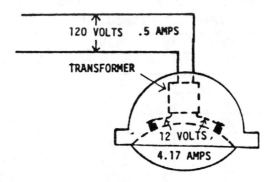

Figure 2-24
Typical low voltage system. Line voltage is supplied to the fixture. An integral transformer reduces voltage to meet the requirement of the lamp. High current is present only in the short leads which connect the lamp to the transformer.

LAMP TEMPERATURE

Incandescent lamps will operate at any temperature which is likely to be encountered under normal operating conditions. Under some abnormal conditions problems such as base loosening, cracking, or blistering can occur. These are almost always the result of improper application or poor equipment design. For example, lamps should not be installed in furnaces where temperatures may exceed the melting point of basing cements or glass envelopes.

Poorly designed fixtures for high wattage lamps can actually cause melting of the glass envelope. Figure 2-25 lists the actual bulb and base temperatures for several common A and PS lamps when burned in a vertical, base-up position in a 77°F ambient condition. As can be seen, these temperatures are well below the maximum ratings for glass and basing cements. These temperatures are for bare lamps operated in a 77°F ambient temperature. When operated in properly designed fixtures in normal environments, temperatures can be expected to increase slightly over those listed. This will not affect life or light output. An increase of several hundred degrees in bulb wall temperature will still be within the safe limits and will have virtually no effect on the filament, which typically operates at 4000°F to 5000°F.

MAXIMUM BARE-BULB TEMPERATURES OF STANDARD INCANDESCENT LAMPS*

Watts	Bulb	°Fahr.
25	A-19	110
40	A-19	260
60	A-19	255
100	A-19	300
150	A-23	280
200	A-23	345
300	PS-30	385
500	PS-35	415
1000	PS-52	480
1500	PS-52	510

*Bare lamp burning vertically, base up.

Figure 2-25
Typical maximum bulb and base temperatures for A and PS lamps operated bare in a 77°F ambient temperature. (Courtesy GTE Sylvania)

If temperatures rise to the point where they cause lamp failure, the lamp loss is usually the least of the problems. These temperatures will normally be high enough to cause insulation failure on the electrical conductors feeding the lamp. This can cause tripping of circuit breakers, blowing of fuses, personnel hazards and, in extreme cases, fire.

DIMMING

Incandescent lamps may be dimmed by a variety of means. The most common dimmers are variable transformers and solid state devices.

Transformers vary the voltage provided to the lamp and are usually used to dim an entire circuit. Transformers simply reduce the magnitude of the voltage supplied to the lamp. This results in a decrease in current, which reduces the power dissipated by the filament.

Solid-state devices typically employ electronic switches called triacs. Triacs are capable of turning the lamp on and off very rapidly–once every 1/2 cycle, as shown in Figure 2-26. This reduces the average power consumed by the lamp, which reduces light output. Solid-state dimmers which can replace a standard toggle switch are normally rated at 500 to 1,500 watts.

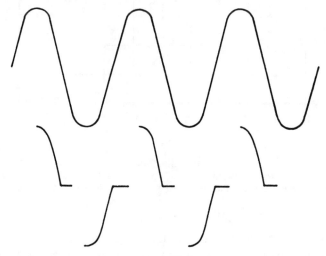

Figure 2-26
Waveform supplied to lamp by a typical solid state dimmer. Top diagram is for 100% setting; bottom is 50% power input.

Larger panel-mounted units with higher wattage ratings are available to control larger loads. Both transformer-type dimmers and solid-state dimmers reduce power consumption. The relationship between light output and power consumption is shown in Figure 2-27. Note that the reductions in power are much less than the reductions in light. A 25% reduction in power results in a reduction in light of about 50%. In the early stages of dimming the difference is even greater: light will be reduced by up to 3% for each 1% reduction in power.

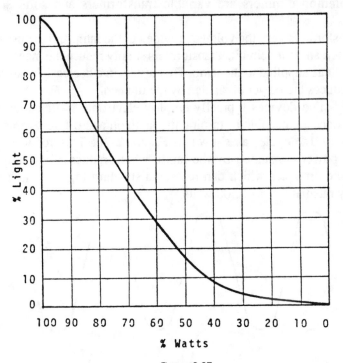

Figure 2-27
Relationship between power input and light output for incandescent lamps. This curve is for the lamp only and does not include losses within the dimmer.

Line-voltage, solid-state dimmers should not be used to dim low-voltage systems using transformers which are integral to the fixture. Transformers will overheat, causing premature failure. Transformer-type dimmers or special electronic dimming devices rated for operation with secondary low-voltage transformers may be used when dimming of low-voltage systems is desired.

Rheostats, also known as variable resistors, were used in the past to dim incandescent lighting systems. They are seldom used today since they do not reduce energy consumption, only light output of the lamps, and produce heat which must be dissipated.

Dimming of incandescent lamps, regardless of the means used to accomplish the dimming, results in the lowering of filament temperatures. This will cause a shift in the spectral distribution of the lamp into the red region. This means that red colors will be accentuated and blue will be weakened. This fact should be considered when dimming is specified.

As a final note, permanent dimming of incandescent lamps as a means of reducing power cost is seldom cost effective due to the decrease in efficacy. The use of a lower-wattage lamp will generally produce greater savings and be more cost effective. Dimming is typically done for aesthetic reasons, not economics.

SPECTRAL ENERGY DISTRIBUTION

Incandescent lamps radiate only a small portion of their total emissions in the visible part of the spectrum, as seen in Figure 2-28. The majority of the radiated energy is in the infrared region, with only a small radiation in the ultraviolet area. Most of the visible emissions fall into the red region; thus, incandescent lamps are strong in red and deficient in blue.

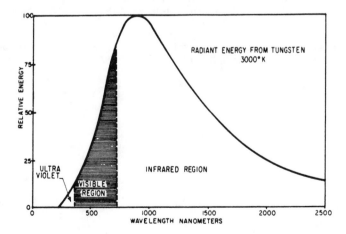

Figure 2-28
Spectral energy distribution for a typical incandescent lamp operating at a filament temperature of 3000 K. Note that most of the radiated power is in the form of infrared heat, not light. (Courtesy GTE Sylvania)

As discussed in the previous chapter, incandescent lamps are essentially gray bodies over the visible portion of the spectrum. For this reason their spectral energy distribution is a smooth curve which is a function of the filament temperature. As temperature increases the emissions in the shorter wavelengths increase which tends to improve color rendering of the cooler colors. Figure 2-29 shows the distribution curves for incandescent lamps at various operating temperatures.

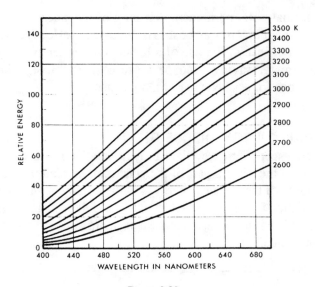

Figure 2-29
Relative energy radiated as light from incandescent lamps at various temperatures. *(Courtesy IES Lighting Handbook, IESNA)*

Chapter 3

LAMPS FOR GENERAL LIGHTING: FLUORESCENT

The fluorescent lamp is the most versatile package of light available in today's market. Ranging in light output from 115 lumens to 16,500 lumens in standard sizes, they are available in over 40 different wattages and numerous circuit types.

The forerunners of the fluorescent lamps were the Cooper-Hewitt and Steinmetz-Weintraub lamps–large, low-pressure mercury lamps which produced a bluish light with an efficiency far greater than incandescent lamps. These lamps, however, did not achieve popularity since the color of the light they produced was bluish, and little or no light of other colors was produced.

In 1896, Thomas Edison, the father of the incandescent lamp, applied for a patent for a fluorescent lamp, but did not develop it. Other researchers worked at producing a fluorescent lamp, but met with little success until 1935 when General Electric demonstrated a green fluorescent lamp at the annual Illuminating Engineering Society Conference. This was an indication of things to come and, in 1938 the first commercial fluorescent lamps were introduced. They were offered in 18-, 24-, and 36-inch lengths, and consumed 15, 20 and 30 watts, respectively. These lamps were available in six colors, plus white. The workhorse of fluorescent lamps, the 4-foot, 40-watt lamp, was introduced soon after and rapidly became the favored light source for office and industrial lighting. The first 4-foot lamp was produced in two basic colors: white, at a color temperature of 3500 K, and daylight, at 6500 K. They were rated at about 1400 lumens.

The benefits of fluorescent lamps were immediately obvious and, as the result of extensive research and development, they have expanded to become what many consider to be the most versatile light source available.

THEORY OF OPERATION

Fluorescent lamps are known as gaseous discharge lamps; they produce light by discharging an electric arc through a tube filled with low-pressure gas which contains mercury atoms. Some of the electrons in the arc collide with electrons in the mercury atoms. When collisions occur the mercury electrons are knocked out of orbit and, since they have absorbed energy as a result of the collision, jump to a higher energy level. They return to their normal orbit almost immediately and, in the process, give up the energy which was absorbed. The wavelength of this energy will vary according to the number of energy levels which have been transitioned by the electrons, and the number of steps which they take in returning to their original levels, but the primary emission is ultraviolet energy at a wavelength of 253.7 nanometers. The ultraviolet emissions are not visible, but they have the ability to stimulate the fluorescent powder which coats the inside surface of the lamp. The powders, called "phosphors," convert the high-frequency ultraviolet energy to visible light. Figure 3-1 illustrates this principle.

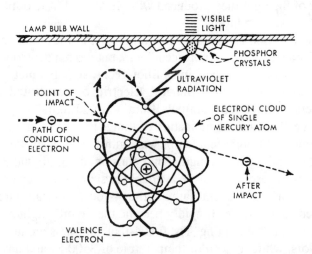

Figure 3-1.
Theory of fluorescent lamp operation. Free electron in arc collides with an electron in the mercury atom, giving up energy which causes the mercury electron to jump to a higher energy level. The electron returns to its original level almost immediately, giving up the absorbed energy in the form of ultraviolet radiation. The UV excites the phosphor coating on the lamp, and is converted to visible light. (*Courtesy Illuminating Engineering Society, IES Lighting Handbook, 1981 Reference Volume*)

NOMENCLATURE

The major components of fluorescent lamps are shown in Figure 3-2.

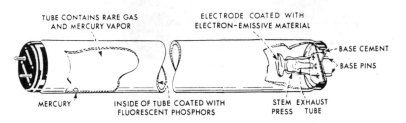

TUBE CONTAINS RARE GAS
AND MERCURY VAPOR

ELECTRODE COATED WITH
ELECTRON-EMISSIVE MATERIAL

BASE CEMENT

BASE PINS

MERCURY

INSIDE OF TUBE COATED WITH
FLUORESCENT PHOSPHORS

STEM EXHAUST
PRESS TUBE

Figure 3-2
Major parts of a fluorescent lamp. (*Courtesy Illuminating Engineering Society,
IES Lighting Handbook, 1981 Reference Volume*)

BASE – The base has two primary functions: to support the lamp in the fixture, and to provide a means of making electrical connections. There are several different types of bases, each with a special purpose. The primary criteria which dictate the configuration of the base are the physical size and shape of the lamp, and the electrical characteristics of the circuit. These will be discussed in detail later in this chapter. Typical bases are shown in Figure 3-3.

BULB – The bulb, or tube, as it is sometimes called, is the glass enclosure which contains the working parts of the lamp and provides a controlled environment inside the lamp.

CATHODE – Made of tungsten wire, the cathode resembles the filament used in an incandescent lamp. Its function, however, is greatly different. Cathodes, located at the ends of the tube, act as terminals for the electric arc. They are wound into a coil and filled with a special "emission" material, typically barium, strontium, and calcium oxide, which act as a reservoir of free electrons. These electrons are emitted to form the arc stream, and collide with electrons in the mercury atoms. In operation, the cathodes are heated, either by the current flow or by applying a voltage across them. This heating helps free the electrons from the cathode.

Several different types of cathodes are used, as shown in Figure 3-4; however, they typically take the form of a coiled coil or triple coil.

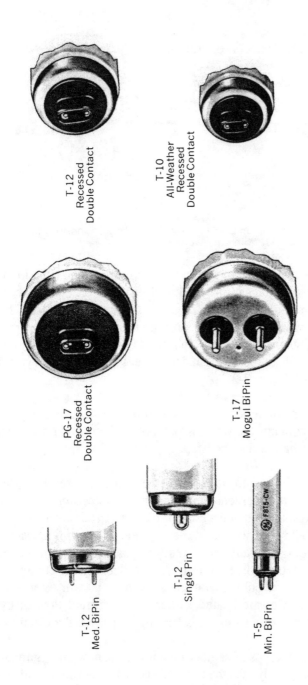

Figure 3-3
Common fluorescent lamp bases. (*Courtesy GE Lighting*)

F40 Mainlighter

High Output with Plate Anodes

F40 Staybright with Floating Shield

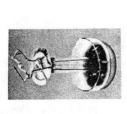

T-12 1500-Milliampere

Power Groove with Wire Anodes

Coiled Coil

Triple Coil

Stick Coil

Figure 3-4
Typical cathodes used in fluorescent lamps. *(Courtesy GE Lighting)*

FILL GAS – A variety of fill gases are used to facilitate the starting of the arc within the lamp. Standard lamps typically employ argon, neon, and sometimes xenon. Energy-saving fluorescent lamps also use krypton. The gases are at a near vacuum, on the order of 1 to 3 torr (.02 to .06 PSIA). They ionize quite readily when voltage is applied across the lamp, and provide a path for the arc.

MERCURY VAPOR – As discussed earlier under Theory of Operation, mercury atoms emit ultraviolet radiation which is converted to visible light.

PHOSPHOR – The phosphor is the light-producing element of the lamp. It is applied as a thin coating on the inside of the glass tube. Phosphors are chemicals which are excited by ultraviolet radiation and re-radiate the energy at different wavelengths. Several different chemicals are used in varying combinations to produce different colors. Calcium halophosphate is the most common and forms the base for phosphors used in white lamps. Other phosphors are added to produce other colors.

BALLASTS

Once an arc has been initiated in a fluorescent lamp, the lamp exhibits a negative electrical resistance characteristic: the resistance to current flow through the lamp decreases as the current increases. This "run-away" current condition would destroy the lamp in a few seconds if allowed to occur. An auxiliary device, a ballast, is required to limit current and prevent the condition. In addition to limiting current, the ballast also provides the proper voltage for lamp starting and operation. Thus the ballast stabilizes the power supply to the lamp. In fact, the ballast receives its name from the lead, iron, sand, or water "ballast" used in ships to stabilize them in heavy seas. Typical ballast construction is shown in Figure 3-5.

Fluorescent lamps, and thus ballasts, are produced in three basic circuit types. These are called preheat, instant start, and rapid start. Voltage and current characteristics vary between types, and lamps are not interchangable between ballasts.

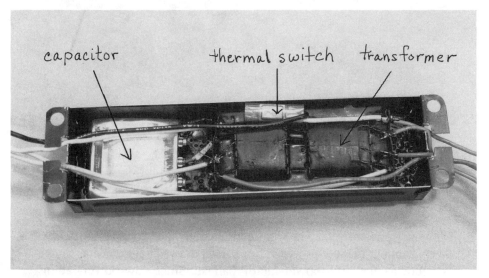

Figure 3-5
Typical transformer type ballast.

Most ballasts are current-limiting transformers. The heart of the ballast consists of coils of enameled aluminum or copper magnet wire, wound on a core of transformer-grade steel. This assembly is impregnated with a non-magnetic insulating material such as hot asphalt to aid in conducting heat away from the windings and provide electrical insulation from the case. A capacitor to correct power factor and assist in lamp starting may also be included.

Another type of ballast, which is used for most fluorescent lamps of 20 watts and less, is the "simple reactance ballast," commonly called a "choke coil." It is inexpensive, and its small physical size is well suited to the small fixtures typically used with small lamps. The choke consists of a long piece of enameled copper magnet wire wound on a steel core, similar to the core used for transformers. The high inductive reactance of the choke serves to limit current, and it is suitable for use on lamps whose operating voltage rating does not exceed the line voltage. A typical choke is shown in Figure 3-6.

BALLAST POWER FACTOR

Transformer-type ballasts are available in both high-power-factor and low-power-factor types. The low-power-factor type is also called

Figure 3-6
Choke coil ballast.

"normal" power factor. The watts consumed by the two types are nearly the same, but the low-power-factor ballast draws about twice as much current as the high-power-factor ballast. Since low-power-factor ballasts require more current, the power supply system must use more branch circuits, larger electrical distribution panels and switchgear, and the utility must frequently install a larger transformer to supply the load. The additional current means that line losses are higher, and the user will pay a higher power cost each month. The utility will also be impacted since losses in transformers will be higher, and this loss occurs ahead of the meter. As a result, many utilities have rules limiting the use of low-power-factor ballasts and/or have penalties in rate schedules for low-power-factor loads.

Low-power-factor ballasts are lower in first cost, but the added cost of wiring, plus the added power cost for line losses makes them a poor economic choice. The use of low-power-factor, transformer-type ballasts should be discouraged from a purely economic standpoint unless they are the only type manufactured for a particular lamp. In general, lamps of less than 20 watts are operated on choke coils, which are low-power-factor types. In these cases the choke is justified since high-power-factor ballasts are not normally available for small lamps.

THERMAL PROTECTION OF BALLASTS

If the ballast is to be used inside a building or in other locations where heat might cause a fire hazard, the ballast must also include an integral thermal protector. This is a heat-activated switch which will open when the internal coil temperature exceeds 105 degrees Celsius. The switch is normally activated when the ballast fails, causing the lamps controlled by the ballast to turn off. When the ballast has cooled, the switch closes and the lights are turned on for a short period of time, until the ballast heats up. Regular cycling of a fluorescent lamp is an indication that the ballast has failed. Thermally protected ballasts are known as "Class P," and are required by the National Electric Code when ballasts are installed inside buildings. Choke coils, since they are inherently current-limiting, are exempted from this requirement.

BALLAST FACTOR

Lumen output ratings for lamps are determined by actual tests performed by the lamp manufacturers, in accordance with ANSI Standards. As specified by the Standard, lamps are tested on a special ballast known as a linear reactor. Linear reactors are high-quality, high-cost transformers, which drive the lamp at a higher light output than the smaller, less expensive ballasts which are used in commercially available fixtures. BALLAST FACTOR is the term which specifies the percentage of rated lamp lumens that will be produced when the lamp is operated on a particular commercial ballast. A ballast factor of 95% means that the lamp can be expected to produce about 95% of its rated lumens when operated on the ballast.

The Certified Ballast Manufacturers Association (CBM) has established standards for ballast performance. A ballast carrying the CBM label has been certified as producing 95% ± 2-1/2% of the lamp manufacturer's rated lumen output. In practice, most CBM rated ballasts drive the lamp at about 93%-94% rated output. They will also not cause premature lamp failures; the lamp can be expected to comply with the lamp manufacturer's published lamp mortality curves.

Non-CBM rated ballasts can be expected to drive the lamp at less than 92-1/2% rated light output, and lamp life may also be shortened. Most ballasts for standard lamps of 40 watts and greater are CBM rated. Choke coils, typically used on small lamps, are not CBM certified, and typically drive the lamp at about 80% rated output.

It should be noted, however, that the CBM certification is for a lamp/ballast combination. A ballast may meet CBM Standards when driving a standard lamp, but fall far short in light output if a reduced-wattage, energy-saving lamp is installed. In fact, energy-saving lamps produce about 82%-88% rated light output when operated on ballasts which have been certified for standard lamps.

BALLAST LIFE

Average ballast life at 50% duty cycle and a hot spot temperature of 90°C is about 12 years. Raising or lowering ballast temperature will greatly affect this estimate. For instance, a 10-degree increase will cut life by 50%, while a 10-degree decrease can be expected to double the life. Ballasts installed in recessed troffers are usually hotter than 90 degrees and can be expected to last about 10 years. A well ventilated industrial fixture is generally much cooler, and a ballast life of 15-20 years is not uncommon. Energy-saving ballasts run cooler than standard ballasts and life expectancies are typically doubled.

BALLAST POWER LOSSES

Power losses in ballasts typically average 10% to 25% for standard ballasts. This figure can be reduced by 50% or more when using high-efficiency, energy-saving ballasts. In general, single lamp ballasts have the same or higher losses than 2-lamp ballasts, and losses increase as lamp size increases. An exception to this rule is the 30-watt, rapid-start system, which has higher losses than the 40-watt, rapid-start system. A table of typical lamp and ballast watts for common lamps is shown in Figure 3-7. Manufacturers' literature should be consulted for data on specific applications.

| | | TYPICAL BALLAST WATTS | | | |
| | | Standard Ballast | | Energy Saving Ballast | |
	Lamp Watts	Ballast Watts	System Watts	Ballast Watts	System Watts
1 Lamp/ballast					
F4T5	4	5	9	–	–
F20T12	20	5	25	–	–
2 Lamps/ballast					
F30T12	30	21	81	–	–
F40T12	40	16	96	6	86
F96T12	75	23	173	8	158
F96T12/HO	110	37	257	17	237

Figure 3-7
Lamp, ballast, and system watts for typical lamps and ballasts. Note that energy saving ballasts are much more efficient. Listed watts are based on ANSI tests and actual wattage will vary according to field conditions.

ELECTRONIC BALLASTS

Electronic ballasts differ from magnetic ballasts in both construction and operation. They use electronic components to alter voltage and limit current, and usually employ an oscillator circuit to generate high frequency, typically 20 to 30 kilohertz. Fluorescent lamps respond positively to frequency increases, as shown in Figure 3-8. As frequency increases, lamp efficacy also increases, so the system becomes more efficient.

Electronic ballasts weigh less than conventional transformer-type ballasts, and have lower internal power losses. Some manufacturers offer integral manual dimming capability at minimal added cost, and continuous photocell dimming is available as an option on some ballasts to reduce power consumption in daylighted buildings.

Electronic ballasts vary widely in ballast factor, with published ratings ranging from 70% to slightly over 100%. Care must be taken to assure that the proper ballast factor is included in lighting design calculations. This procedure will be discussed in detail in Chapter 6.

PREHEAT CIRCUITS

The first commercial fluorescent lamps were of the preheat type, and most of the tubular fluorescent lamps of 20 watts and less still use this circuit. In addition, some 30-watt, 40-watt, and 90-watt lamps are made in preheat types. The larger wattages, however, are old circuit designs and seldom used in modern fluorescent systems. The basic circuit is illustrated in Figure 3-9.

When the circuit is energized (turned on), the starter switch is in the closed position so current flows through both cathodes. This causes the cathodes to heat up, which in turn boils emission material off of the cathode. After a few seconds, sufficient emission material has evaporated to establish a conductive path through the lamp. The starter switch then opens, causing the magnetic field in the ballast to collapse. The collapsing magnetic field induces a high-voltage in the ballast windings, which initiates the arc across the lamp. The starter switch remains open while the lamp is in operation, so no external voltage is available to provide cathode heating, and the only cathode heat results from the current in the arc stream. The starting process takes several seconds so the lamp does not come on immediately when the wall switch is turned on. Typical starter types are shown in Figure 3-10.

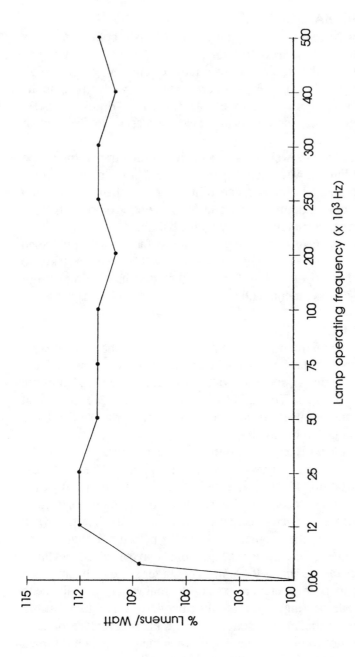

Figure 3-8

Efficacy increase for typical F40T12 lamp as a function of lamp operating frequency. Other lamp diameters also show increases, but magnitudes vary. (Plotted from data in the IES Lighting Handbook, and "High Frequency Characteristics of Fluorescent Lamps up to 500 kHz," E.E. Hammer, JIES Winter 1987)

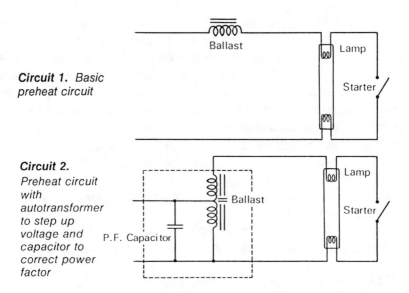

Circuit 1. *Basic preheat circuit*

Circuit 2.
Preheat circuit with autotransformer to step up voltage and capacitor to correct power factor

Figure 3-9
Typical preheat circuits. Top diagram is for choke coil ballasts, Bottom represents transformer type ballast. *(Courtesy GE Lighting)*

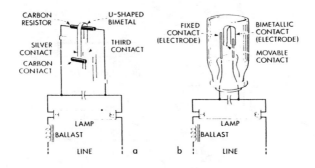

Figure 3-10
Starter switches for preheat circuits: (a) thermal type; (b) glow switch type. *(Courtesy Illuminating Engineering Society, IES Lighting Handbook 1981 Reference Volume)*

Starters may be   either manual or automatic. Manual starters are simply momentary contact switches which are manually held in the closed position. They are commonly found on small fluorescent desk lamps. Automatic starters usually employ bi-metallic strips as the active switching element.

A bi-metallic strip is composed of two layers of metal with different coefficients of expansion. When the strip is heated, the metals expand at different rates, which causes the strip to bend. When the strip is cold, it is straight, and the switch is in the closed position. As the strip heats, it bends, and pulls away from the electrical contact. This opens the switch. Automatic switches are available in two basic types: thermal and glow switch. Thermal switches are well suited for applications involving low temperature or fluctuating voltage. Glow switches are typically used for normal temperatures and stable voltage conditions.

Ballasts for preheat lamps take two forms: simple reactance ballasts (also known as choke coils), and preheat ballasts. Preheat ballasts are of the transformer type.

Preheat lamps use bi-pin bases, since two points of electrical connection are required to allow current to flow through the cathode.

INSTANT-START (SLIMLINE) CIRCUITS

The instant-start lamp was introduced in 1944 to overcome the time delay in starting that is common to preheat lamps. Instant-start lamps ignite immediately when power is applied. A typical instant-start circuit is shown in Figure 3-11.

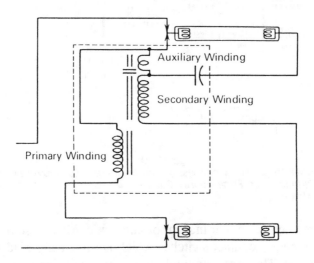

Figure 3-11
Typical instant start circuit. *(Courtesy GE Lighting)*

The instant-start system differs from the preheat in that arc initiation depends solely on high-voltage (400 to 1000 volts) across the lamp. The high voltage forces electrons through the lamp, causing an ionization of the fill gas and initiation of an arc. The high voltage is generated by the ballast and no external starter is required. Once an arc has been established, the current in the arc stream provides heating of the cathodes to facilitate the evaporation of emission material.

Instant-start circuits use transformer-type ballasts. These are larger than preheat transformer-type ballasts and have higher internal power losses. They can also be quite noisy and should not be used in areas such as offices, classrooms, and residences, where the noise may be objectionable.

Bases used on instant-start lamps are typically single-pin since external cathode heating is not required. Bi-pin bases are provided on two instant-start lamps, a 4-foot, 40-watt lamp, and a 5-foot, 40-watt lamp. These are very old circuit designs and are essentially obsolete; however, the 4-foot version is occasionally encountered in sign applications. The pins are internally shorted, effectively providing a single-pin cathode.

Electrical codes require the use of lampholders which interrupt power to the ballast when a lamp is removed (Figure 3-12). This is a safety precaution, since instant-start ballasts generate very high open-circuit voltages which can cause an electrical shock hazard. As a side benefit, instant-start ballasts equipped with circuit-interrupting lamp-holders will not consume energy if the fixture is delamped.

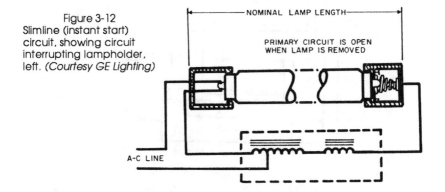

Figure 3-12
Slimline (instant start) circuit, showing circuit interrupting lampholder, left. (Courtesy GE Lighting)

NOMINAL LAMP LENGTH

PRIMARY CIRCUIT IS OPEN WHEN LAMP IS REMOVED

A-C LINE

The original instant-start lamps were of a smaller diameter than the preheat lamps , and were called "Slimline." The diameter has since been increased so the lamp is now the same diameter as most other lamps, but the "Slimline" name has been retained.

RAPID-START CIRCUITS

The rapid-start lamp was introduced in 1952 to combine the advantages of the preheat and instant-start circuits, and to overcome the drawbacks to each system. Rapid-start lamps start almost immediately, and at lower voltage. The cathodes are continuously heated while the lamp is in operation, thus stabilizing the rate of emission material evaporation from the cathode, and improving lamp life. The ballast is smaller than the instant-start ballast, and is less expensive. Rapid-start ballasts are also quiet in operation and seldom cause objectionable noise. They are the most popular system in use today, and are offered in a wide variety of wattages, lengths, and colors. A typical rapid-start circuit is shown in Figure 3-13.

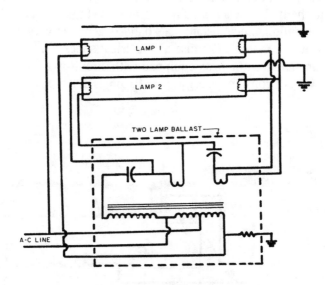

Figure 3-13
Typical two-lamp, series-sequence, rapid-start system. *(Courtesy GTE Sylvania)*

When the circuit is energized, separate windings in the ballast apply about 3.5 volts across each cathode. This voltage heats the cathode and evaporates emission material, which circulates through the lamp, and ionizes the fill gas. When sufficient ionization has occurred, typically less than 1 second, an arc strikes between the cathodes. The cathode heat is applied continuously while the lamp is operating. Before arc initiation the circuit is in an "open circuit" condition, and a high-voltage condition exists (about 277 volts for the 4-foot lamp), which facilitates arc ignition. Once the arc has been established, the voltage drops to a lower value for normal lamp operation.

Ballasts for rapid-start lamps are typically of the transformer type. Since the circuit is very similar to the preheat circuit, many 4-foot and smaller lamps will also operate on preheat ballasts. The small-wattage lamps will also operate on choke coils. Operation of rapid-start lamps on preheat of choke coil ballasts usually results in reduced lamp life since the cathodes are not continuously heated.

Since rapid-start lamps require external heating of the cathodes, they use bases with two electrical contacts. Bi-pin bases are the most common; however, a heavier, better insulated base, called the recessed double contact, is used on high-current lamps, explained in the next section.

VOLTAGE AND CURRENT RATINGS

Unlike incandescent lamps, fluorescent lamp voltages are not the same as the voltage rating of the electrical power supply. Each lamp type and wattage is designed to operate at a specific voltage, often different than the power supply voltage. For instance, the 4-foot, rapid-start lamp operates at a nominal 101 volts; the 8-foot slimline is designed for 197 volts; and the 6-inch, 4-watt lamp is rated at 29 volts, yet these lamps are normally powered by 120-volt or 277-volt electrical circuits. The ballast, as a transformer, steps the supply voltage up or down as required to provide the proper lamp operating voltage.

Fluorescent lamps are also designed to operate at a specific current level. This current, in conjunction with the proper voltage, will cause the lamp to consume the rated wattage. Most lamp currents are specified in milliamperes, or thousandths of an ampere. The standard 4-foot, 40-

watt, rapid-start lamp is rated for 430 milliamperes, or 0.43 amps. The 8-foot slimline lamp is rated at 425 milliamperes (.425 amps) and, at 197 volts, consumes 75 watts of power. Other common current ratings are 800 milliamps and 1500 milliamps, which are called high output and very high output, respectively. Voltage and current ratings for most fluorescent lamps are published by the Illuminating Engineering Society in the IES Lighting Handbook, or may be obtained from lamp manufacturers.

Reduced-current ballasts are manufactured for some lamps as an energy conservation feature. These ballasts drive the lamp at lower power input with a commensurate reduction in light output. They are recommended for use only in cases where the illuminance produced by lamps operated on conventional ballasts exceeds that required for the task, and should not be confused with energy-efficient, full-light-output ballasts, which are commonly used.

LAMP IDENTIFICATION

Fluorescent lamp codes consist of a series of letters and numbers. For most common lamps they may be interpreted as follows:

Tubular preheat lamps

F4T5/CW	F	4	T	5	CW
	↑	↑	↑	↑	↑
	fluorescent	watts	shape (tubular)	diameter in 1/8"	color

Tubular instant-start lamps

F96T12/CW	F	96	T	12	CW
	↑	↑	↑	↑	↑
	fluorescent	length* (inches)	shape (tubular)	diameter (in 1/8")	color

*except F40T12/IS and F40T17/IS, where the first number represents watts, and "IS" denotes instant start.

Tubular rapid-start lamps 40 watts and less
F30T12/RS/CW

F	30	T*	12*	RS*	CW
↑	↑	↑	↑	↑	↑
fluorescent	watts	shape	diameter in 1/8"	rapid start	color

*omitted for standard 40-watt lamp

Tubular 800-milliamp, rapid-start lamps
F96T12/HO/CW

F	96	T	12	HO	CW
↑	↑	↑	↑	↑	↑
fluorescent	length (inches)	shape	diameter in 1/8"	high output	color

Tubular 1500-milliamp, rapid-start lamps
F48T12/CW/1500

F	48	T	12	CW	1500
↑	↑	↑	↑	↑	↑
fluorescent	length (inches)	shape	diameter in 1/8"	color	current

These codes may vary slightly between manufacturers, and some specialty lamps will carry additional numbers or letters to denote special features. For example, the prefix "FC" indicates a circular lamp, and a suffix "/U/6" indicates a U-shaped lamp with a 6-inch spacing between the legs.

Perusal of several lamp manufacturers' catalogs is recommended as a means of becoming familiar with fluorescent lamp codes.

SHAPES

Fluorescent lamps are produced in a wide variety of shapes, as shown in Figure 3-14. The shapes are identified by one or more letters, which are included in the Lamp Ordering Code. Lamp manufacturers use the same codes for most common lamps, such as "F96T12CW," which denotes an 8-foot, cool white slimline lamp. There are exceptions, however, for some of the newer lamps and some less common older lamps. For example, the single ended compact lamp is called a "Biax," "PTT," or "PL" lamp by three major manufacturers, yet the lamps are essentially the same.

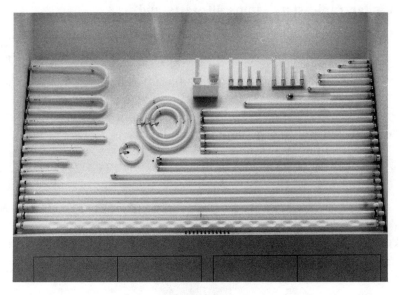

Figure 3-14
Typical fluorescent lamp shapes.

There are also minor differences between manufacturers in the location of the alpha shape identifier in the lamp-ordering code for some lamps. For example, a jacketed, 8-foot, 1500-milliamp lamp designed for use at temperatures of –20°F is identified by one manufacturer as an "FJ96T12," while another manufacturer calls the same lamp an "F96T12J." Note that the "J" is in the prefix in the first case, and follows the "Diameter" designator, 12, in the second case.

Common shape identifiers, and their meanings, are:

"C" – Circular lamp. The letter "C" typically appears in the prefix, immediately after the "F."

"T" – Tubular lamp, omitted from the F40 lamp code for rapid-start lamps.

"J" – Used in conjunction with a "T" designator, "J" denotes a lamp with a clear outer glass jacket over the regular "T" envelope, which acts as a thermal insulator and permits operation of the lamps at sub-zero temperatures.

"BX," "PL," and "PTT" – These identifiers denote a family of single ended compact lamps.

"U" – A lamp shaped like a letter "U."

LAMP LIFE

Rated life for fluorescent lamps is the point at which 50% of a large number of lamps will have failed, and the remaining 50% will still be in service. Some lamps will begin failing at about 40% rated life; however, this number will be small. As burning hours accumulate, the number of failures will increase until, at rated life, one half of the lamps have burned out. Figure 3-15 shows this curve. Note that the frequency of failure is relatively low until lamps have operated for about 70% of their rated life. At this point the failure rate increases dramatically until, at rated life, 50% failure has occurred.

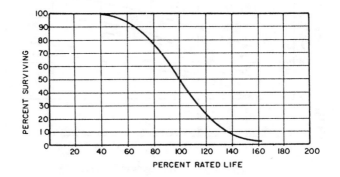

Figure 3-15
Mortality curve for fluorescent lamps. Some lamps fail before end of rated life, while some lamps operate beyond rated life. Rated life is the time at which 50% of a statistically large sample of lamps will have failed. *(Courtesy GTE Sylvania)*

Group relamping, the replacement of all lamps in a system regardless of the number of failures, is typically done at about 70% rated life to minimize the number of lamps which must be replaced on an individual basis.

The life of a fluorescent lamp will end when all of the emission material has left one cathode. Each time the lamp is started, emission material leaves the cathodes at a much faster rate than under normal operating conditions, thus shortening lamp life. Lamps which are operated for long hours with a minimal number of starts can be expected to last longer than lamps which are started frequently. The life of a fluorescent lamp is then a function of the number of times the lamp is started.

As can be seen in Figure 3-16, lamp life increases as burning hours per start increase. Most manufacturers rate fluorescent lamps for operation at 3 hours per start. This results in a rated life of 20,000 hours for an F40 lamp. A few manufacturers base their life ratings on 10 hours per start, and publish life ratings of 26,000 hours, which makes it appear that their lamp will last longer than a lamp which has been rated at 3 hours per start. Careful examination of Figure 3-16, however, shows that both lamps actually have the same rated life. Care must be taken when evaluating lamp life to assure that both lamps are compared on the same basis.

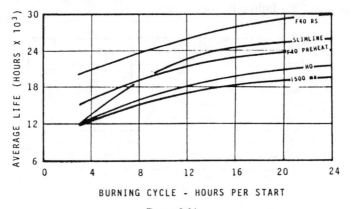

Figure 3-16
Fluorescent lamp life is shortened by frequent switching. The value of the energy saved by turning lights off for periods of only a few minutes is generally greater than the value of the lost lamp life unless relamping is difficult and expensive.

The shortening of lamp life as a function of the number of times the lamp is started raises an obvious question: is it more economical to leave lamps on or turn them off? In the 1950's and early 60's it was common practice to leave lamps on unless they were to be left off for long hours. Many buildings were built with no switches for lights— circuit breakers were used to control entire circuits. In those days, however, power costs were very low. Rates of less than 1 cent/kWh were common, and lamps had shorter rated lives.

The economics of turning fluorescent lamps off vary as a function of energy, lamp, and replacement labor cost. In general, lamps should be turned off whenever possible, and the breakeven point is usually only a few minutes. An often overlooked factor is the possibility of an increase in real time life even though the burning life of the lamp is shortened, since the lamp is not deteriorating when it is turned off.

PRECAUTION FOR SLIMLINE CIRCUITS

When one lamp in a two-lamp slimline circuit fails, the lamp may flicker, or try to restart. When this occurs the lamp is rectifying, and direct current (DC) is applied to the ballast windings. This may result in excessive ballast heating, with a subsequent reduction in ballast life. Failed slimline lamps should be replaced as soon as possible to prevent this condition.

EFFICACY

The efficacy of fluorescent lamps ranges from about 15 lumens per watt for the 4-watt lamp to 103 lumens per watt for a high-efficiency lamp operated on a high-frequency electronic ballast. These figures are based on total input watts of the lamp and ballast, and ballast losses should always be included when evaluating the efficacy of a ballasted system.

In general, efficacy tends to increase as lamp wattage increases; however, high-current rated lamps, such as the 1500-milliamp types, tend to exhibit lower efficacies than medium- and low-current density lamps.

LAMP LUMEN DEPRECIATION

The light output of fluorescent lamps decreases with accumulated burning hours. This is due primarily to a deterioration of the phosphors, and the deposition of evaporated cathode material on the inside surface of the bulb wall. The actual loss of light for a specific lamp depends upon the current loading of the lamp and the types of phosphors. Figure 3-17 shows the depreciation curves for typical standard cool white phosphors for F40 and slimline lamps, high-output lamps, and 1500-milliamp lamps. The current loading of these lamps is classified as light, medium, and heavy, respectively.

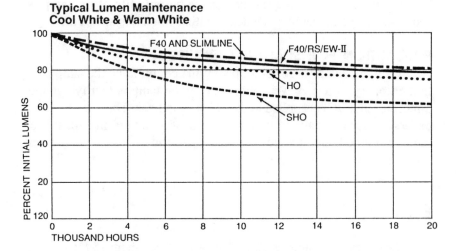

Figure 3-17
Typical lamp lumen depreciation as a function of accumulated burning hours
for cool and warm white lamps. *(Courtesy Philips Lighting Co.)*

These curves are generic and may not be representative of the lamp
which is actually used, so it is highly recommended that lamp manufacturers' data be consulted for specific applications. Deluxe and some
other color-improved phosphors generally deteriorate more rapidly than
standard phosphors. Note, however, that the newer high color
rendering "tri-phosphor" lamps provide excellent color and very good
lumen maintenance characteristics.

Lighting systems are generally designed to produce a specified
illuminance at some time in the future, typically the time at which lamps
will be replaced and/or the fixtures will be cleaned. This means that the
depreciation in light output over time must be considered when
designing a system.

The expected light output of the lamp at the time of relamping is
used in design calculations if the system is to be group relamped. If
lamps are to be replaced only when they fail, the light output at 70%
rated life is normally used. The "70% of life" factor should be used
instead of 100% life since lamps will be failing quite rapidly at this
point in life, and new lamps which replace them will produce full light
output. This compensates for the continued deterioration of older
lamps, and the average light output of the total number of lamps in the

installation will not, under normal conditions, drop below the output of the total number of lamps at 70% rated life. The method of including lamp lumen depreciation in design calculations will be discussed in detail in Chapter 6.

VOLTAGE VARIATIONS

The published lamp-operating data are based on operation of the lamp at rated voltage. This voltage will be supplied to the lamp only if the primary voltage to the ballast is maintained at the proper level. Variations in primary voltage will affect lamp volts, watts, current, and light output.

Decreases in voltage will result in reductions in light output, power consumption, and current. If voltage drops too low, the arc will extinguish and the lamp will go out. Prolonged operation at greatly reduced voltage can cause a reduction in cathode temperature, which causes rapid loss of cathode emission material, and a shortening of lamp life. Overvoltage operation overdrives the lamp, resulting in increases in light output, power consumption, and current, and also shortens life. Overvoltage also causes excessive ballast heat which shortens ballast life. The effects of voltage variations are shown in Figure 3-18. Note that the effects are less pronounced than the effects on incandescent lamps.

Figure 3-18
Effect of voltage
variations on
typical two-lamp
rapid-start systems.
(Courtesy GTE
Sylvania)

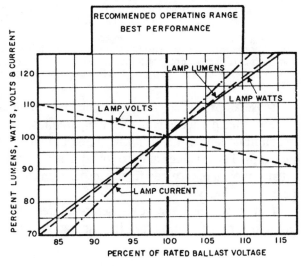

TEMPERATURE

Fluorescent lamps, unlike other types, are greatly affected by variations in temperature. Lamps are designed to produce full light output at a specific temperature, and variations above or below this design level will result in decreased light output. This is accompanied by a decrease in power consumed by the lamp. The variations are the result of changes in the pressure within the lamp, which affects mercury gas pressure. These variations are shown for common lamps in Figure 3-19.

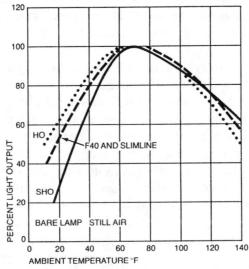

Figure 3-19
Most fluorescent lamps are designed for optimum performance when operated in a still-air, ambient temperature of 77°F. Operation in higher or lower temperatures affects performance. *(Courtesy Philips Lighting Co.)*

Most lamps are designed to operate in a still-air ambient temperature of 77°F. This produces bulb wall temperatures of about 100°F for F40 and slimline lamps, 122°F for high-output lamps, and 167°F for 1500 ma lamps. These values may be found in some well ventilated fixtures, but are atypical of recessed fixture installations unless the fixtures are of the air-handling type.

Lamps may not start at low temperatures, and special ballasts are required for some lamp types if temperatures drop below 50°F. Most

430 ma rapid start standard ballasts will start standard lamps at 50°F or higher. Special low-temperature ballasts are rated at 0°F. Instant-start and slimline ballasts typically start standard lamps at temperatures of 0°F and higher. Special low-temperature ballasts for high-output and 1500 ma lamps will provide reliable starting for standard and low-temperature lamps at –20°F.

Energy-saving lamps are an exception to the above ratings; they will not start reliably at temperatures below 60°F.

Special jacketed 1500 ma lamps are available for operation in ambient temperatures of –20°F. These lamps are equipped with a clear outer jacket which traps heat and provides a higher bulb-wall temperature. They should not be used in normal-temperature environments.

Manufacturers' literature should be consulted for specific applications involving abnormal temperatures.

DIMMING

Fluorescent lamps may be dimmed by a variety of means; special dimming ballasts and controls, variable transformers, and solid-state devices are typically used. The introduction of solid-state electronic technology to the field of fluorescent lamp dimming has brought about rapid changes in equipment, and the information presented here regarding these devices may soon be obsolete due to technological advances.

Conventional dimming ballasts and controls provide uniform, flicker-free dimming from full light output to almost zero. They are typically used when the dimming is for aesthetic purposes and large reductions in light are desired. Dimming ballasts are expensive when compared to standard ballasts. A dimming ballast is required for each pair of lamps, and a special electronic control is required. These systems maintain proper cathode voltage across the dimming range and should not affect lamp life.

Variable autotransformers, either motor-driven or manually operated may be used to provide a limited dimming range as an energy conservation measure. When equipped with photocell controls they may also be used to dim lights in conjunction with daylight, and can compensate for the gradual loss of light as a lighting system ages.

Transformer-type dimmers are limited to a dimming range of about 40%. Below this point lamps will typically extinguish due to insufficient voltage. Cathode voltage is also reduced and lamp life may be affected if excessive long-term dimming is used.

Transformers can provide an economically feasible method of achieving energy conservation, and are well suited to large lighting systems.

Fixed-reduction transformers provide a single or, in some cases, step-function dimming. They are typically used on large lighting systems which produce more light than is required, and a reduction of 15% to 30% in illuminance is acceptable. These systems typically start lamps at full voltage, and then reduce voltage in a single or series of steps to a predetermined level. They are less expensive than variable systems but do not provide the same level of control. As with variable systems, care must be taken to avoid excessive reductions in cathode voltage. Reductions in primary voltage of 30% of less will normally be acceptable, but lamp manufacturers should be consulted for recommendations on specific lamps.

Electronic dimmers for fluorescent lamps typically employ a wave-chopping method to reduce power to the lamp. They may be sized for a single fixture, several fixtures, or several circuits. These are relative newcomers to the lighting field, and some systems exhibit undesirable characteristics such as flickering, low power factor, waveform distortion, and high harmonic content. Electronic dimmers require careful evaluation by qualified personnel to assure their suitability.

Regardless of the dimming means employed, the relationships between power input and light output for fluorescent lamps follow a reasonably predictable path. Relationships between light and power are shown in Figure 3-20.

SPECTRAL POWER DISTRIBUTION AND COLOR

Fluorescent lamps are available in a wide variety of colors, and different colors are being introduced by lamp manufacturers at frequent intervals. In fact, many of the recent lamp developments involve new

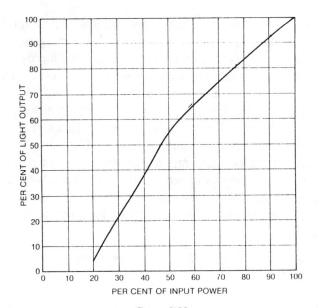

Figure 3-20

Percent light output vs input power for a typical F40 dimming system. *(Courtesy Illuminating Engineering Society, IES Lighting Handbook, 1981 Reference Volume)*

phosphors, or combinations of phosphors, to improve both efficacy and color rendering.

Many of the new lamps use combinations of rare earth phosphors which produce high concentrations of light at the peak sensitivity regions of the human eye: red, green, and blue. The eye then blends the colors to produce other colors, and the result is good color rendering and improved energy efficiency. These lamps are frequently called "Prime Color" or "Tri-phosphor" lamps since their spectral distribution is accentuated in the prime colors. Note, however, that an estimated 10% of the population, mostly male, suffer from some form of color-vision deficiency, commonly called "color blindness." Total color blindness, where individuals see no color and live in a "black and white" environment, is very rare. The vast majority of color blind people see colors, but in most cases do not see the full range of colors perceived by individuals with so-called "normal" color vision, since one or more of the color receptors in their eyes do not function normally. These individuals will typically perceive "Tri-phosphor" lamps as distorting colors, and will normally prefer standard color lamps.

Spectral Power Distribution Curves, also called Spectral Energy Distribution Curves, are produced by most lamp manufacturers for many of their lamps. These curves provide a graphic representation of the relative energy produced by their lamps as a function of wavelength. Several typical curves are shown in Figure 3-21. Note that, while the curves vary by lamp color, they all contain spikes in blue, green, and yellow regions. These spikes are called "Mercury Lines," and represent primary emissions from the mercury atoms. The variations on the smooth parts of the curve represent variations in color output of the lamps due to the phosphors which are used. Lamps with strong components of specific colors can be expected to accentuate those colors, while deficiencies in specific colors indicate that those colors will be muted.

Figure 3-21

Spectral power distributions for typical fluorescent lamp colors. *(Courtesy GE Lighting)*

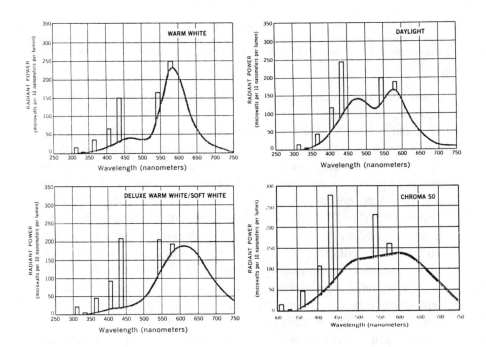

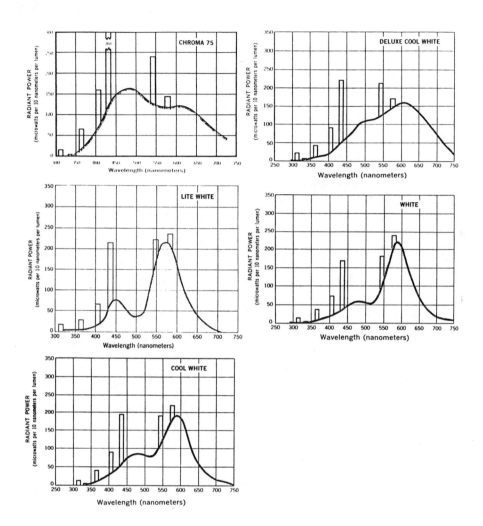

Chapter 4

LAMPS FOR GENERAL LIGHTING: HIGH-INTENSITY DISCHARGE

High-Intensity Discharge lamps are similar to fluorescent lamps in that they produce light by discharging an electric arc through a tube filled with gases and gaseous metals. They differ though, in that the tubes are considerable smaller than fluorescent tubes, and the pressures inside the tubes are considerably higher than the pressures inside the fluorescent tube. This results in a shift in the primary wavelengths from the ultraviolet region to the visible region, with only secondary ultraviolet emissions.

HID lamps are physically small light sources, which means that reflectors and refractors can be used to direct the light with good results. Some HID sources are also very efficient producers of light, especially in the larger lamp wattages. As with incandescent lamps, efficacy improves as lamp wattage increases.

There are three types of HID sources commonly used for lighting. These are mercury vapor, metal halide, and high-pressure sodium. A fourth source, low-pressure sodium, is classified as "miscellaneous discharge source." It is commonly used in some of the applications where HID sources might be used, and will be discussed in this chapter.

While HID lamps share many common characteristics, there may be dramatic differences between them in efficacy, color rendering, and life. These important differences will often be the deciding factors in lamp selection for a particular application.

OVERVIEW

MERCURY VAPOR LAMPS

The first high-pressure, high-intensity discharge lamp was of the mercury vapor type and was introduced in 1934. Like its predecessors, the Weintraub/Steinmetz and Cooper-Hewitt lamps, the mercury lamp produced a bluish light with poor color rendering characteristics. The first lamps were rated at 400 watts, produced about 16,000 initial lumens, and had a life of 4,000 to 6,000 hours. They were used in factories, for street lighting, and for other applications where color rendering was not important. These lamps required a large arc tube, about 7-3/4 inches long, and were designed for use in a vertical-burning position. If used horizontally, as in many street lighting luminaires, the arc would bow and burn through the glass arc tube. This problem was overcome by installing a magnet in the fixture to draw the arc and prevent contact with the wall of the arc tube.

In 1950 the use of phosphors to improve color rendering was introduced. These coatings, similar to the phosphors used in fluorescent lamps, converted some of the ultraviolet radiation to visible wave-lengths, primarily red, and improved their acceptance for industrial lighting. These improvements were not sufficient, however, to make the lamp acceptable for indoor use in merchandising or office environments. It was not until 1966 that Deluxe phosphors were introduced and the lamp was acceptable for these color-critical applications.

Unfortunately, mercury lamps are relatively inefficient producers of light, compared to fluorescent or other HID sources, and have limited utility in contemporary lighting design. Their use is generally confined to low wattage lamps for landscape lighting, where their blue-green color enhances many plants and trees, and to very low-cost outdoor luminaires used for some residential and light industrial applications.

METAL HALIDE LAMPS

During the early 1960's, in an attempt to improve the color rendering of the mercury lamp, different metals were added to the arc stream of a mercury lamp. The result was a lamp with improved color rendering and higher efficacy. Thus was born the metal halide lamp.

Early metal halide lamps suffered from short life and color variations between lamps, had special ballasting requirements, could be operated only in certain positions, and were expensive. After years of improvements, many of these limitations have been overcome. Life ratings now range from 5,000 hours to 20,000 hours, colors are fairly consistent, ballasts are readily available, and some lamps can be operated in any burning position. In addition, phosphor coatings have improved color rendering, and a 3000 Kelvin lamp which closely approximates the color of incandescent lamps is available.

Metal halide lamps are commonly used in industrial facilities where good color rendering is required, in offices and stores, and for many sports lighting applications. Metal halide is the preferred source for televised sports due to its excellent match with color television cameras.

HIGH-PRESSURE SODIUM LAMPS

One year before the introduction of metal halide lamps, our most efficient white light source was developed. The first high-pressure sodium lamp was rated at 400 watts, had an efficacy of 105 lumens per watt, and a life of 6,000 hours.

The lamp was made possible by the development of a translucent tube made of aluminum oxide, which could withstand the highly corrosive action of sodium, and stand up to the extremely high temperatures generated by the sodium arc.

In the quarter century since their introduction, high-pressure sodium lamps have been improved in both color and life, which averages 24,000+ hours for most sizes. Wattages now range from 35W to 1,000W, and special color-improved versions have found limited acceptance in office and school environments. When used in combination with metal halide, the color blend is pleasing, and this combination is becoming increasingly popular when used in indirect luminaire types.

High-pressure sodium lamps produce a slightly yellowish light, and are favored for industrial applications where color rendering is not critical. They are also widely used for outdoor area lighting and street lighting. The high efficacy of high-pressure sodium lamps makes them popular with consumers due to lower operating cost, and first cost of a high-pressure sodium system can be lower than mercury systems since fewer luminaires are required. For example, a typical industrial plant lighted with two hundred fifty 400-watt mercury fixtures can achieve the same illuminance with only one hundred 400-watt high-pressure sodium luminaires.

LOW-PRESSURE SODIUM LAMPS

Low-pressure sodium lamps were introduced in Europe in 1932, and in the United States during the following year. The early lamps were available in two wattages, 145W and 180W, and produced 6,000 and 10,000 lumens, respectively. Efficacies were 45-55 lumens per watt, and life was rated at 3,000 hours.

The lamp did not achieve popularity in this country since the light it produced was monochromatic yellow, and its efficacy was not sufficiently higher than fluorescent or mercury lamps to justify the poor color. The lamp did gain popularity in Europe, and the application of an indium oxide film to the outer glass envelope allowed visible light to pass out of the lamp yet reflect infrared energy back to the arc tube. This improved efficacy, and low-pressure sodium lamps currently produce up to 150 lumens per watt.

Low-pressure sodium lamps are typically used for outdoor lighting in applications where color rendering is unimportant. The monochromatic yellow emissions render only specific shades of yellow, and all other colors appear to be shades of gray. The wavelengths produced by low-pressure sodium lamps are near the peak sensitivity of most black and white television cameras, so the source is acceptable for most black and white TV surveillance systems. The lamps are physically large, about 4' for the 180-watt size, and optical control is difficult to achieve, so light distribution from luminaires is somewhat uncontrollable.

THEORY OF OPERATION

High-intensity discharge lamps produce light in a manner similar to fluorescent lamps. An electric arc is discharged through a small tube which is filled with gases and vaporized metals. The free electrons in the arc stream collide with electrons in the atoms of the metals and, as a result of the collision, give up energy. This energy causes the electron in the atom to jump to a higher energy level. The electrons return to their original energy level almost immediately and, in the process, give off the energy that was absorbed. This energy is in the form of particles called photons.

High-intensity discharge lamps operate at much higher internal pressures than fluorescent lamps, and this increase in pressure causes a shift in the primary emissions from the ultraviolet region to the visible region. Note that there are still some secondary ultraviolet emissions in mercury vapor and metal halide lamps that can be used to excite phosphors to produce additional visible light wavelengths other than the primary emissions. Both the efficiency with which electrical energy is converted to light, and the colors of the light produced, depend upon the composition of the metals in the arc stream and the pressure within the arc tube.

LAMP CONSTRUCTION AND OPERATION

MERCURY VAPOR LAMPS

The basic parts of a typical mercury vapor lamp are shown in Figure 4-1.

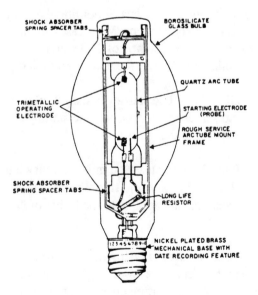

Figure 4-1
Typical mercury vapor lamp construction. (*Courtesy GTE Sylvania*)

BULB – The outer envelope is made of heat-resistant borosilicate glass and provides a controlled environment for the inner components. The bulb helps maintain the high-arc tube temperature, up to 1,000°C. The bulb also absorbs ultraviolet energy and prevents the emission of harmful wavelengths. An inert fill gas, generally nitrogen, is used to protect the inner components from oxidation, and regulates temperature. Typical bulb shapes are shown in Figure 4-2.

ARC TUBE – The arc tube contains mercury vapor, metallic mercury, and a small quantity of inert gas, usually argon. The tube is made of quartz glass to withstand the high operating temperatures.

ELECTRODES – Lamps contain three electrodes. Two operating electrodes, one located at each end of the arc tube, provide terminals for the arc. Operating electrodes are made of tungsten wire, and are impregnated with a rare earth oxide emission material. This material supplies free electrons for the arc. A starting electrode is located near one of the operating electrodes and assists in starting the lamp.

BASE – The base provides a means of holding the lamp in the socket and makes the electrical connections. Most bases are made of brass, and are frequently nickel plated. Lamps of 100 watts or less typically use medium bases, while mogul bases are employed on larger lamps.

STARTING RESISTOR – The starting resistor limits the current in the starting circuit, and serves to "disconnect" the starting electrode once an arc has been established in the lamp.

PHOSPHOR – Phosphors are used on most mercury lamps. These are similar to the phosphors used on fluorescent lamps, and serve to improve color. Phosphors are applied to the inside of the bulb.

Operation

A starting voltage is applied to the electrodes. The resistance between the operating electrodes is very high at this point in time and an arc cannot be initiated. The starting electrode, however, is located close enough to an operating electrode to allow an arc to pass between them. The arc ionizes the argon gas and vaporizes some of the mercury. The

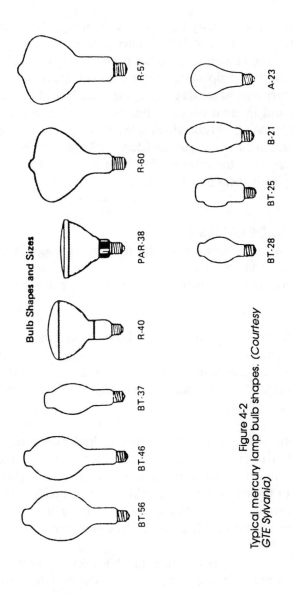

Figure 4-2
Typical mercury lamp bulb shapes. (Courtesy GTE Sylvania)

ionized argon and mercury vapor circulate through the tube and, being electrically conductive, reduce the resistance between the operating electrodes. When this resistance has been sufficiently reduced, an arc strikes between the operating electrodes. This arc heats and vaporizes more mercury and, as the quantity of vaporized metal in the arc stream increases and internal pressure builds, the lamp increases in light output. The starting resistor which is in series with the starting electrode limits current in the starting arc and acts to take the starting electrode out of the circuit once the main arc has been initiated. The entire starting process typically takes 5 to 7 minutes.

METAL HALIDE LAMPS

Metal halide lamps are similar to mercury vapor lamps in both construction and operation.

The basic parts of a typical metal halide lamp are shown in Figure 4-3.

BULB – Metal halide lamps have a borosilicate outer glass envelope which performs the same function as the envelope of mercury lamps. Bulb shapes are similar to mercury lamp shapes except that R, PAR, and A configurations are not typically used. A double-ended tubular lamp is also available.

ARC TUBE – The quartz glass arc tube is similar to the mercury vapor arc tube but is physically smaller. The tube contains mercury and combinations of metallic halides, which are the light-producing elements of the lamp. Typical combinations of halides are 1) sodium and scandium iodides, 2) sodium, thallium, and indium iodides, and 3) dysprosium and thallium iodides. Each of these combinations produces slight color differences.

Metal halide arc tubes operate at high temperatures and are subjected to severe thermal stresses. One or both ends of the tube is covered with a white reflective powder designed to equalize the temperature variations over the tube and reduce this stress. This white powder is used only on metal halide lamps and may be used to provide positive identification if white ends are observed on arc tubes.

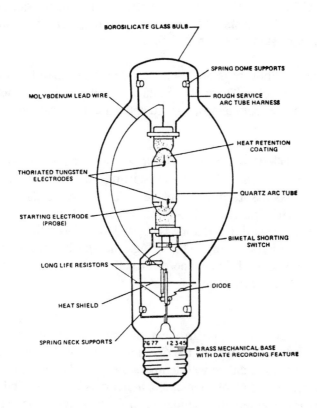

Figure 4-3.
Construction of typical metal halide lamp (*Coutesty GTE Sylvania*)

ELECTRODES – Most metal halide lamps contain three electrodes which are similar to the electrodes in mercury vapor lamps. Double-ended lamps typically have only the two operating electrodes, with the starting electrode omitted. Double-ended lamps may be used only with ballasts that include a special electronic igniter which provides a high voltage to start the lamp. This type of device is commonly used with high-pressure sodium lamps and will be discussed later in this section.

BASE – Screw shell bases are used on most metal halide lamps. Medium bases may be used on lamps on 100 watts or smaller, with mogul bases used on most higher wattage lamps. Double-ended lamps employ a special base which is attached to each end of the lamp.

STARTING RESISTOR – The starting resistor, like the resistor used in mercury lamps, limits starting current. It also helps eliminate current flow through the starting electrode after the lamp has started.

PHOSPHOR – Phosphors are used on a few metal halide lamps to impart a warmer color. These phosphors are similar to the ones used on fluorescent and mercury lamps, and are applied to the inside of the glass bulb.

SHORTING STRIP – If current is allowed to circulate between the operating electrode and the smaller starting electrode, the starting electrode will be dissolved by corrosive action of the metals inside the arc tube. The shorting switch is a bi-metallic strip which bends when heated so that the starting electrode is shorted to the operating electrode while the lamp is operating. This brings the starting and operating electrodes to essentially the same electrical potential and effectively eliminates circulating currents.

Operation

Metal halide lamps start in the same manner as mercury vapor lamps. Once started, however, the operation is somewhat different. In mercury lamps, the arc tube wall temperature is well above the boiling point of mercury, and all of the metallic mercury is vaporized and enters the arc stream. The metallic halides used in metal halide lamps have boiling points well in excess of the arc tube wall temperature, so not all of the metal is vaporized. Excess metal condenses at the coldest spot on the bulb wall, with the amount of metal condensed depending upon the cold spot temperature. The quantity of evaporated metal in the arc stream affects both the light output and spectral power distribution of the lamp.

Since this condition varies slightly between individual lamps, some color variation is normal. Some lamps are available with special arc tube configurations which increase the cold spot temperature and allow more metal to enter the arc stream, with an increase in light output. See Figure 4-4. Note that the horizontal configuration must be operated only with the tube bowed upward. To prevent operation in other positions the base is equipped with a small pin which permits its insertion only in special slotted sockets which are designed to maintain the proper orientation. This base is called a "position-oriented mogul base (POMB)," and fixtures must be ordered with an appropriate "position-oriented mogul socket (POMS)."

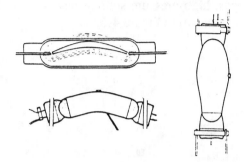

Figure 4-4
Typical metal halide arc tubes. Top left is a conventional tube used for standard and some high-output lamps. Lower left is special bowed configuration used for some high-output types designed for horizontal operation. Tube on right is used for some high-output types designed for vertical burning positions. (*Courtesy: GTE Sylvania*)

HIGH-PRESSURE SODIUM LAMPS

The construction and major components of a typical high-pressure sodium lamp are shown in Figure 4-5.

Figure 4-5
Typical high-pressure sodium lamp construction. (*Courtesy: GTE Sylvania*)

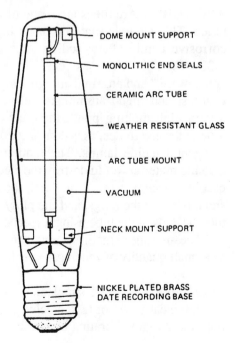

DOME MOUNT SUPPORT

MONOLITHIC END SEALS

CERAMIC ARC TUBE

WEATHER RESISTANT GLASS

ARC TUBE MOUNT

VACUUM

NECK MOUNT SUPPORT

NICKEL PLATED BRASS
DATE RECORDING BASE

BULB – The bulb is constructed of special heat-resistant borosilicate glass, and provides support and protection for the internal components. In contrast to mercury and metal halide lamps which use an inert fill gas, high-pressure sodium lamps are evacuated. Typical bulb shapes are shown in Figure 4-6.

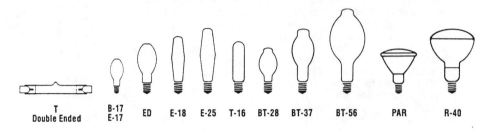

Figure 4-6
Typical bulb shapes used for high-pressure sodium lamps. (*Courtesy GTE Sylvania*)

ARC TUBE – Arc tubes are made of a special translucent aluminum oxide. This ceramic material has a high resistance to the extremely corrosive nature of hot sodium, and can withstand temperatures of 1,300°C. The tubes are cast, in a manner similar to ceramic "greenware," and are fired at high temperatures. The result is a strong, heat-resistant, high-transmittance tube of high purity. Since the tube has a very high melting point, the ends cannot be heated and crimped around the electrodes, as is done with quartz arc tubes used in mercury and metal halide lamps. The temperatures necessary to soften the ceramic material would destroy the electrodes. Instead, small plugs are cast of the same ceramic material, the electrodes are inserted through small holes in the plugs, and the plug/electrode assembly is cemented into the ends of the tube using a special process.

The arc tube contains an amalgam of mercury and sodium, as well as a small quantity of zenon gas which assists in starting the lamp.

ELECTRODES – The small diameter of high-pressure sodium arc tubes, typically 3/8" or less, precludes the use of a starting electrode, so only the 2 main operating electrodes are used. In place of a starting electrode, a special electronic starting aid is contained within the ballast to provide a high-voltage pulse to start the lamp.

BASE – Mogul bases are used for most high-pressure sodium lamp wattages except the 35W size, which uses a medium base. In addition, wattages up to 150W are also available with medium bases. Some manufacturers have recently introduced double-ended lamps which use a special double-ended base.

Operation

When the lamp is energized, a special electronic device in the ballast, called an "igniter," provides a high-voltage pulse to the lamp. This pulse typically ranges from 2,500 to 3,000 volts, and lasts for about 1 microsecond. The pulse consumes little energy due to its short time duration, but sufficient ionization of the zenon gas is achieved to lower the resistance to a point where an arc can be initiated between the operating electrodes. Several pulses are normally required to start the lamp. Once an arc has started, the heat which it generates causes more sodium to vaporize and enter the arc stream.

High-pressure sodium lamps go through three distinct color phases during startup. When the arc first starts, its color will be a bluish white, characteristic of zenon and mercury. As temperatures increase, more sodium enters the arc and the color shifts to a monochromatic yellow, characteristic of sodium at low pressure and temperature. As more heat is generated, the pressure inside the arc tube increases to 1-1.5 atmospheres, and the spectral emissions broaden to produce a warm, slightly yellowish-white light.

The entire starting process typically takes 2 to 3 minutes.

Low-Pressure Sodium Lamps

Low-pressure sodium lamps are not classified as high-intensity discharge lamps. They possess some characteristics of HID lamps, but also resemble fluorescent lamps in other areas. They are classified by the Illuminating Engineering Society as a "miscellaneous discharge source." A typical low-pressure sodium lamp is shown in Figure 4-7.

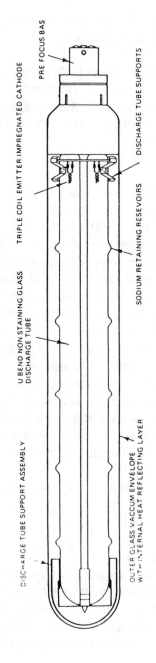

Figure 4-7

Construction of a typical low pressure sodium lamp. (*Courtesy Philips Lighting Co.*)

BULB – Low-pressure sodium envelopes are made of borosilicate glass and are coated on the inside with a heat-reflective film of indium oxide. The coating allows visible light to pass but reflects infrared energy back to the arc tube. This increases both arc tube temperature and light output. The tubes are quite large, nearly 4' for the 180-watt lamp, and frequently require mechanical support within the fixture.

ARC TUBE – The glass arc tube contains metallic sodium and a neon starting gas. The sodium is contained in numerous reservoirs, in the form of dimples in the glass, which are scattered throughout the arc tube. Tubes are typically bent into a U shape to increase the length of the arc, and are supported at both ends.

ELECTRODES – Electrodes provide a terminal for the arc and are located at opposite ends of the tube. They are similar to the electrodes used in high-intensity discharge lamps.

BASE – Low-pressure sodium lamps use a special pre-focus bayonet-type base.

Operation

Low-pressure sodium operation is similar to other arc discharge sources in that an arc is passed through a tube containing a metallic vapor. A mixture of neon and argon is used as a starting gas. When the lamp is energized, an arc strikes through the conductive gas, and the lamp produces a reddish-pink light, characteristic of neon. As current flows through the neon/argon mixture, heat is generated which vaporizes the metallic sodium. As the quantity of sodium in the arc stream increases, the lamp develops the characteristic monochromatic yellow color associated with sodium at low pressure and temperature. Typical operating pressures are about 0.005 torrs (about 0.00095 psi), with temperatures ranging from 250°C to 270°C.

BALLASTS

Like fluorescent lamps, HID and low-pressure sodium lamps present a negative resistance characteristic to the power supply circuit once the arc is established. A current-limiting transformer (a ballast) is therefore required. HID ballasts differ from typical fluorescent ballasts in that the component parts—the transformer, capacitor, and other auxiliaries—are not generally provided as a self-contained unit in a single enclosure. The individual components are mounted separately in the fixture and may be individually replaced as needed. The components of a typical ballast are shown in Figure 4-8.

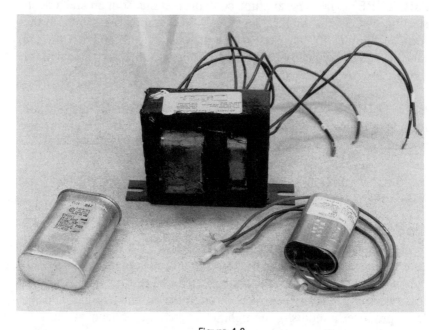

Figure 4-8
Typical HID ballast. Unlike fluorescent ballasts, the transformer (top) and capacitor (left) are installed individually in the fixture housing as separate components. The capacitor is omitted in low power factor ballasts. High pressure sodium fixtures also contain an electronic igniter, bottom. See text for explanation. *(Courtesy GE Lighting)*

Ballasts are available in several different types, each having advantages, disadvantages, and some specific applications. As a general rule, each ballast is manufactured for operation with a specific lamp type and wattage, and cannot be used with other lamp types, or with a lamp wattage other than the one for which it was designed. There are, of course, some exceptions to this rule. Any mercury lamp may be used in a metal halide ballast as long as the wattages match. In addition, several manufacturers produce metal halide and high-pressure sodium lamps designed for operation on some mercury ballast types.

Due to the wide variety of ballast circuits and relative obscurity of some types, only the more common ballasts will be discussed.

REACTOR ballasts, also called "high reactance," are the simplest type of ballast. Like choke coils used for small fluorescent lamps, they consist of a wire wound on an iron core which is placed in series with the lamp. Reactors are normally low power factor (50% to 60%); however, some types (commonly called high-reactance, high power factor) include a capacitor to raise power factor to 90% or higher. Reactors are the lowest-cost ballast and are typically used for small lamp wattages. Ballast losses are low, in the range of 5% to 10%. If the primary voltage varies, light output also varies, and regulation is poor. For example, variations in voltage of 3% can cause a 6% variation in wattage and light output.

Reactor ballasts are not recommended for installations where voltage variations of 5% or more are anticipated. Line current under starting conditions may be as much as 50% higher than normal operating current so circuit loading should be adjusted accordingly.

Reactor ballasts are typically used for mercury lamps of 175 watts or less and high-pressure sodium lamps of 150 watts and less, but may also be encountered in higher wattage mercury installations. They may also be used with metal halide lamps if the primary voltage is 480v, but this use is relatively uncommon.

CONSTANT WATTAGE AUTO TRANSFORMERS are also known as CWA, autoregulator, and autostabilized ballasts. They consist of a high-reactance autotransformer with a capacitor connected in series with the lamp. Regulation is good, with variations in line voltage of 13% resulting in only a 2% change in watts and light output. The CWA ballast is the most commonly used ballast for mercury vapor lamps.

LEAD PEAKED AUTOTRANSFORMERS are used to ballast metal halide lamps. They are similar to mercury CWA ballasts except that regulation provides a change in light output of about 7-10% for a 10% change in line voltage.

MAGNETIC REGULATORS, also called constant wattage ballasts, are used for high-pressure sodium lamps. They consist of a voltage-regulating transformer supplying the lamp through a reactor.

AUTOREGULATOR ballasts for high-pressure sodium lamps combine an autotransformer with a regulator circuit. A portion of the primary winding is common with the secondary, thus reducing the size of the ballast. Regulation varies depending on the amount of the primary winding which is coupled to the secondary, but is considered to be good. This ballast type may be used when line voltage variations do not exceed 10%.

LAMP LUMEN DEPRECIATION

Mercury vapor, metal halide, and high-pressure sodium lamps experience a gradual decrease in light output as burning hours accumulate, as shown in Figure 4-9. This decrease is caused by a buildup of evaporated electrode material on the inside of the arc tube wall, deterioration of electrodes, and shifts in the balance of chemicals within the arc tubes. Lumen ratings are based on light output after the first 100 hours of operation, during which time impurities inside the arc tube are burned off and the lamp stabilizes.

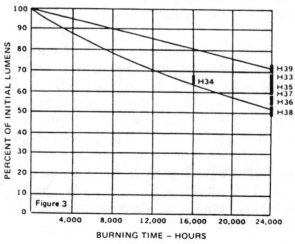

Approximate lumen maintenance of Color
Improved mercury lamps operating vertically.

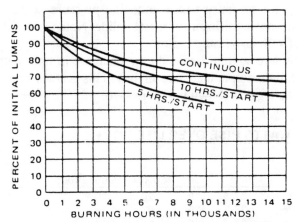

Approximate lumen maintenance of 400-Watt
Metalarc/C lamp at various burning cycles. Curves based on
vertical operation.

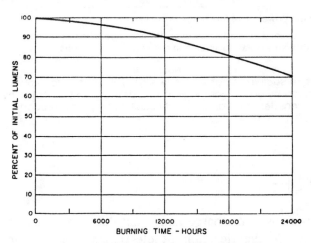

Lumen Maintenance of 150, 250, 400
and 1000 Watt Lumalux Lamps

Figure 4-9
Lamp lumen depreciation for typical HID lamps. Note that metal halide lamps
are greatly affected by the operating hours per state, and frequent switching
accelerates the deterioration in light output. Manufacturer's literature should be
consulted for data on specific lamps. (Courtesy GTE Sylvania)

The effects of frequent lamp starts on light output are considered insignificant for mercury vapor and high-pressure sodium lamps; however, they are significant for metal halide. Note the differences in light loss for various burning cycles.

Low-pressure sodium lamps do not decrease in light output over time. In fact, light output increases slightly. This is at the expense of an increase in wattage, as shown in Figure 4-10.

SOX WATTS RISE OVER LIFE HOURS

	100	2000	5000	10000	18000	Average Watts Over Life
SOX 18*	17	18	20	18	---	18.6
SOX 35	35	36	37	38	39	37.5
SOX 55	55	56	58	60	61	58.9
SOX 90	90	95	97	97	98	96.3
SOX 135	130	134	140	141	142	139.1
SOX 180	176	182	190	191	192	188.9

*Lamp power at 14,000 hours is 18 watts.

Representative watts rise over life characteristics for lamps with heat insulating shields at the end of the arc tube.

Figure 4-10
Low pressure sodium lamps increase slightly in light output over life. This is accompanied by an increase in wattage, as shown in the table. This increase should be considered for purposes of circuit loading and in economic studies. (*Courtesy Philips Lighting Co.*)

LAMP LIFE

High-intensity discharge lamps fail due to a variety of causes. These include deterioration of electrodes, and changes in the composition and characteristics of the various gases and metals used inside the arc tubes. Since the failure modes and causes of failure vary slightly between lamp types, they will be discussed individually. Note that lamp life is a function of operating hours per start, and when comparing different lamp manufacturers' ratings it is necessary to assure that comparisons are based on the same burning cycle. Comparisons based on unequal burning hours per start may be misleading since longer hours per start result in longer rated lives. Most lamp manufacturers base life ratings on operation for 10 hours per start.

MERCURY LAMPS

The typical failure mode of mercury lamps is simply an inability to start. The electrodes are coated with a mixture of metal oxides, called emission material, which evaporates slowly as the lamp operates. When the supply of material is exhausted, the lamp fails to start.

The rated life for most lamps is the point at which 50% of a large sample of lamps will have failed. The mercury lamp, however, is an exception to this rule. At the end of its rated 24,000-hour life, 20-40% of the lamps can be expected to have failed, depending upon the operating cycle. After 24,000 hours of operation, lamps have depreciated in light output to the point where further operation is considered to be uneconomical, thus the mercury lamp life rating is based on an assumed useful life as opposed to the conventional metric of "50% lamp failures." This life is based on 10 hours per start for most lamp sizes. When this cycle is altered the rated life changes. Extensions of the burning cycle increase rated life, while shortening the operating hours per start decreases life. Mortality curves for common mercury lamps are shown in Figure 4-11. Note the effect of different burning cycles on life.

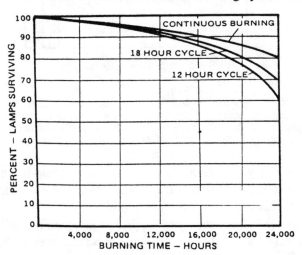

Figure 4-11
Mortality curves for typical mercury vapor lamps. Note that burning hours per start affect lamp life. (*Courtesy: GTE Sylvania*)

METAL HALIDE

The typical failure mode of metal halide lamps is an inability to start or to develop full light output, and is caused by a gradual deterioration of the electrodes. Electrode deterioration is most severe during the start-up cycle, and the life of the lamp is greatly impacted by the number of operating hours per start. Figure 4-12 shows typical mortality curves for a 400-watt lamp operated on various burning cycles. Note that frequent starts have a dramatic effect on life, and another light source should be considered for installations requiring frequent switching. Life ratings for metal halide lamps vary from 5,000 hours to 20,000 hours, and are subject to frequent revision based on lamp improvement. Manufacturers' literature should be consulted for ratings on specific lamps.

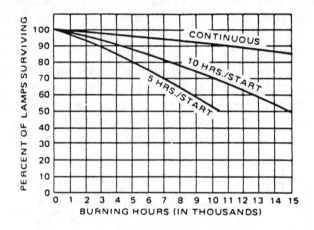

Figure 4-12

Mortality curves for typical 400-watt metal halide lamps. Lamp life is greatly affected by burning hours per start, and frequent switching of these lamps requires careful economic analysis. Manufacturer's literature should be consulted for data on specific lamps. *(Courtesy GTE Sylvania)*

Lamp life may also be affected by operating position, and some lamps operated in horizontal or tilted positions may have shorter lives than when operated in a vertical burning position.

Some metal halide lamps are subject to non-passive failure. The quartz glass arc tube may shatter, causing a scattering of hot glass. Lamps of 250 watts and smaller should be operated only in suitable enclosed fixtures. Other wattages, when operated in other than a vertical base-up of base-down ±15-degree configuration, should also be

*operated in suitable enclosed fixtures which are capable of containing
hot glass fragments at temperatures of up to 900 °C.*

*Lamps of all wattages, if operated continuously (24 hours per day,
7 days per week), should be turned off for at least 15 minutes once per
week to reduce the possibility of arc tube breakage. If this cannot be
done, lamps of all wattages should be used only in suitably enclosed
fixtures. Group relamping (the replacement of all lamps in a system
prior to a large number of failures) will reduce the probability of arc
tube breakage. Manufacturers' literature should be consulted for
specific applications.*

HIGH-PRESSURE SODIUM

The typical failure mode of high-pressure sodium lamps is an
inability to maintain light output. The lamp will start, build in light
output, and go out. This cycle will be repeated at 1- to 2-minute
intervals until the lamp is replaced or the igniter has failed. Since an
igniter normally produces about 50,000 pulses before failure, and
several pulses are required each time the lamp starts, failed lamps
should be replaced as soon as practical to prevent premature igniter
failure. Note that the current installed cost of an igniter ranges from
$100 to $150, including labor.

The cycling phenomenon is caused by a voltage rise across the arc
tube as it ages. As the lamp operates, emission material evaporates from
the electrodes and deposits on the arc tube wall in the form of a thin
black film. The darkened area absorbs heat which causes an increase in
temperature, with a commensurate increase in the amount of sodium
which enters the arc stream. This increases the pressure within the
lamp, and a higher voltage is required to maintain the arc. The voltage
requirement increases by about 1.2 to 2 volts per 1000 operating hours,
depending upon the lamp type. When the required voltage rises to a
level which the ballast cannot supply, the lamp fails and, after a cool-
down period of 1 minute or less, cycling begins. Most high-pressure
sodium lamps are rated for operation at 100 volts, and fail when the
voltage requirement exceeds 140 volts.

The life ratings of high-pressure sodium lamps are typically based
on operation for 10 hours per start, and are shown in Figure 4-13. Life
will be slightly extended if the burning cycle is increased, and shortened
slightly with more frequent switching.

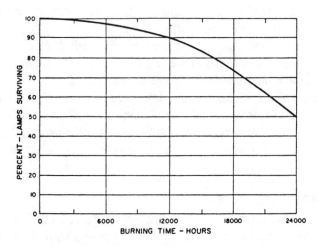

Figure 4-13
Mortality of typical high pressure sodium lamps. *(Courtesy GTE Sylvania)*

LOW-PRESSURE SODIUM

Low-pressure sodium lamps have a rated life of 10,000 to 18,000 hours, depending on wattage. Mortality curves are shown in Figure 4-14. Life is a function of operating hours per start, and will increase or decrease in a manner similar to HID lamps. Published curves are based on 10 hours per start.

The failure mode of low-pressure sodium lamps is similar to that of fluorescent lamps.

OPERATING POSITION

Mercury vapor lamps are rated for operation in any burning position from vertical to horizontal. When lumen ratings are published for a vertical position, they will be achieved only when the lamp is operated vertically. A loss of approximately 10% can be expected if the lamp is operated in a horizontal position, and while the loss is non-linear, it may be assumed to be so for estimating purposes.

Metal halide lamps are available in either "standard" or "high output" versions. The standard lamps are typically rated for operation in any burning position; however, operation in any position other than

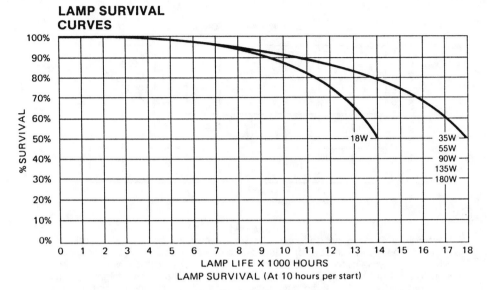

Figure 4-14
Mortality curves for typical low pressure sodium lamps. *(Courtesy Philips Lighting Co.)*

vertical will result in a decrease in light output. These reductions are shown in Figure 4-15. Some high output metal halide lamps are rated for operation in a specific orientation, e.g., vertical base up, horizontal, or vertical base down. A variation of ±15 degrees is normally acceptable, but no further tilt is advised.

High-pressure sodium lamps may be operated in any position with no noticeable effect on light output.

Low-pressure sodium lamps are designed for vertical ±110 degrees for sizes 18W through 55W, and horizontal ±20 degrees for 90W through 180W sizes.

STROBOSCOPIC EFFECT

All discharge lamps are "turned on and off" 120 times per second when operated on magnetic ballasts supplied by a 60 Hz power system. Each time the voltage passes through the zero point, the arc is extinguished and the lamp ceases to produce light. A rapidly rotating object at speeds which are multiples of 60, such as an 1800 rpm or 3600 rpm

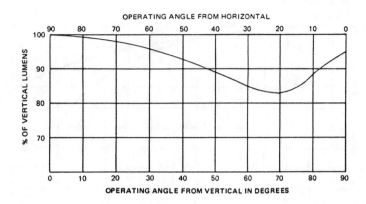

Figure 4-15
Light output of metal halide lamps decreases when lamps are operated in any position other than vertical. Some lamps are rated for operation in only specific orientations, while others are rated for operation in any position. Manufacturer's literature should be consulted on specific lamps, and the loss of light must be included in design calculations if lamps are to be tilted off vertical or horizontal positions. *(Courtesy GTE Sylvania)*

motor, can appear to be motionless. This condition could be hazardous, so the interaction of the light source with the rotating machinery may be important.

The phosphors used on most mercury vapor and some metal halide lamps have a decay time during which they produce light. Strobe effect with these lamps is seldom a problem. Clear metal halide lamps, due to the electrical characteristics of the ballast, also seldom cause excessive strobe; however, it is frequently noticeable and objectionable to some people.

High-pressure sodium and clear mercury lamps do strobe and may, under some conditions, be objectionable.

The effects of strobe may be minimized by using three-phase power distribution systems for lighting circuits, and alternating adjacent fixtures between phases so that two of any three adjacent fixtures will be producing light when the third fixture is not. This is discussed in greater detail in Chapter 11.

ULTRAVIOLET RADIATION

Ultraviolet radiation is produced when mercury atoms are subjected to electron bombardment. These emissions will pass through the quartz glass arc tubes used in mercury and metal halide lamps and, if phosphors are used, permit an improvement in the color rendering capabilities of the lamp. They also improve the lumen output of mercury lamps. In metal halide lamps the additional light produced by the phosphor offsets the loss of light of other wavelengths through absorption, and lumen output is not affected.

The ultraviolet emissions will not pass through the borosilicate glass which forms the outer bulb. This is fortunate, since the emissions fall into that region of ultraviolet energy which causes erythemal burns of the skin and conjunctiva (outer layer of the eye). It is possible for the outer envelope to break without affecting the operation of the arc tube. If this condition occurs, ultraviolet energy can escape and cause the aforementioned burns.

To minimize this possibility, lamp manufacturers produce special mercury and metal halide lamps which contain a protective switch which will disconnect the lamp from the electrical supply in case of bulb integrity failure. These lamps cost more than standard lamps, but are required in applications where bulb failure could result in exposure to the ultraviolet emissions. The special lamps are not normally required when lamps are operated in enclosed fixtures since the glass or plastic lens will normally block the harmful wavelengths.

Exposed mercury or metal halide lamps with the outer envelope broken should be turned off immediately and replaced with an intact lamp.

High-pressure sodium lamps are not subject to this condition since the ultraviolet wavelengths will not pass through the ceramic arc tube.

Low-pressure sodium lamps are also exempt since ultraviolet wavelengths are not produced by the sodium arc.

HOT RESTRIKE TIME

High-intensity discharge lamps will turn off if the power supply is interrupted, even momentarily. They will not restart until the lamp has cooled sufficiently to allow an arc to restrike. The time required for the lamp to cool down and restrike the arc is called "restrike time." Typical restrike times are:

Mercury vapor	3-5 minutes
Metal Halide	15-20 minutes
High-Pressure Sodium	Less than 1 minute
Low-Pressure Sodium	75% restrike instantly
	up to 20 minutes for remaining 25%.

No light will be produced during the restrike period, a condition which might be hazardous in many buildings. For this reason, HID lamps should not be used as the sole source of light in a building. Several options exist to ameliorate this condition. For example, a few fixtures might be equipped with an auxiliary quartz lamp which will provide some light in the event of a temporary power interruption. This option is available from most fixture manufacturers and is typically called a "quartz restrike system." Other options are the use of a few fluorescent or incandescent luminaires that will produce light immediately when power is restored, or battery-operated emergency lighting systems. High-pressure sodium lamps of 150 watts or smaller may be equipped with an "instant restrike system," which is simply an igniter which produces a 4,000v-5,000v pulse to provide instant restrike. Special high pressure sodium lamps containing a second "back up" arc tube that restrikes in a few seconds are also available.

SPECIAL POWER DISTRIBUTION AND COLOR RENDERING

HID lamps produce a wide variety of spectral power distributions (SPD's). These depend upon the lamp type and color, and will be discussed individually by type due to the dissimilarities. SPD's for several common lamps are shown in Figure 4-16.

MERCURY VAPOR

Mercury lamps are available in both clear and phosphor-coated versions. The clear lamp is very strong in blue and green, with substantial deficits in most other colors. Clear lamps are generally considered unacceptable for use indoors or in outdoor areas requiring even minimal color rendering.

The addition of phosphors to the inside of the bulb wall results in greatly improved color rendering with substantial increases in emissions in the red region. This type is suitable for most areas requiring reasonable color rendering, but is normally a poor choice due to the inherent inefficiency of mercury lamps.

Figure 4-16

Spectral power distributions for typical HID lamps. (a) is for clear mercury lamps, (b) deluxe mercury, (c) warm deluxe mercury, (d) clear metal halide, (e) phosphor coated metal halide, (f) high pressure sodium, (g) low pressure sodium. Note that there will be minor variations between manufacturers and between wattages. *(a through f courtesy GE Lighting)*

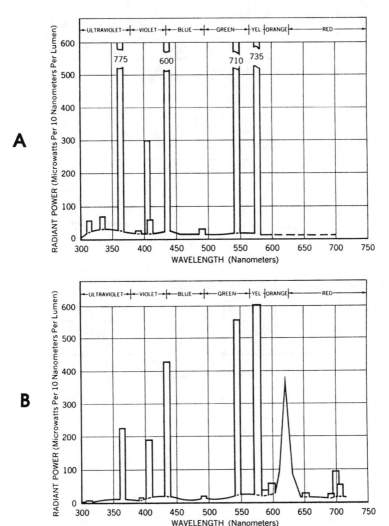

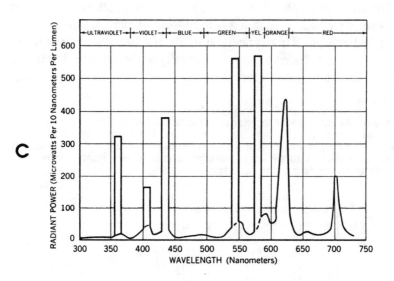

C

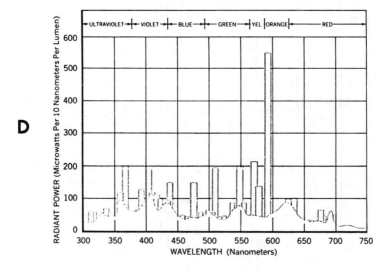

D

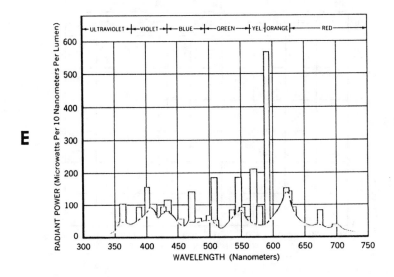

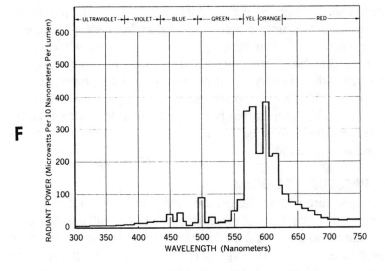

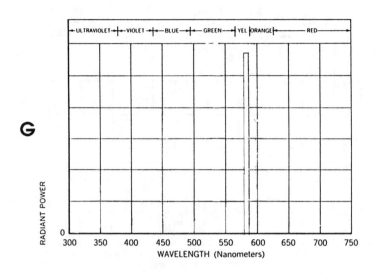

METAL HALIDE

Metal halide lamps are also available in both clear and phosphor-coated versions but, unlike mercury, both styles have good color-rendering properties. Clear lamps produce a cool, somewhat bluish-white light. Standard phosphor-coated versions produce a slightly warmer light, which is popular for offices. A special high-output lamp with a warm (3200K) phosphor produces a light very similar to incandescent. This lamp currently suffers from wide color variations between individual lamps, but should become very popular when this minor problem is solved.

HIGH-PRESSURE SODIUM

High-pressure sodium lamps are available in either standard or color-improved types. Both of these lamps are clear, with the difference being in the arc tube size and the internal arc tube pressure. The color-improved lamps operate at a slightly higher pressure than the standard lamps, which results in an improvement in color. This is at the expense of lamp life, which decreases from 24,000+ hours to 10,000 hours, and a decrease in light output of 10% to 15%. The improvement in color is quite dramatic, however, and justifies these reductions in some cases. Some manufacturers have recently introduced small lamps with apparent color temperatures of around 2500K, and CRI's in the 80's. The lamps produce a white light which is similar to incandescent, and may be suitable for merchandise lighting.

High-pressure sodium lamps are also produced in diffuse-coated versions which should not be confused with phosphor-coated mercury and metal halide lamps. High-pressure sodium lamps produce no ultraviolet emissions outside the arc tube, so phosphors cannot be excited. The diffuse coating is an inert compound and serves only to reduce direct glare from the lamp and to widen the light distribution patterns of fixtures in which they are used. Figure 4-17 shows a comparison between clear and diffuse-coated lamps when used in the same luminaire.

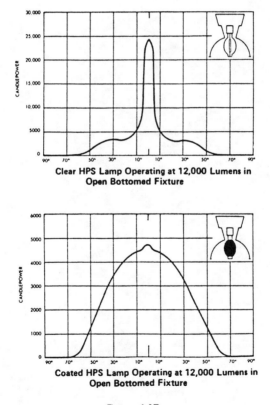

Clear HPS Lamp Operating at 12,000 Lumens in
Open Bottomed Fixture

Coated HPS Lamp Operating at 12,000 Lumens in
Open Bottomed Fixture

Figure 4-17
Typical candela distributions of clear and coated high pressure sodium lamps in the same fixture. Clear lamps tend to have a more concentrating distribution, while coated lamps spread the light. *(Courtesy GTE Sylvania)*

LOW-PRESSURE SODIUM

Low-pressure sodium lamps produce an essentially double-banded emission at 589.0 and 589.6 nanometers, with trace emissions in several other areas. These trace emissions are of such a low magnitude that the lamp can, for all practical purposes, be considered monochromatic. For this reason it is unsuitable for applications requiring any degree of color rendering since objects will appear to be either yellow or shades or gray. The SPD is near the peak sensitivity for some black and white TV cameras so it may be a viable source for applications involving TV surveillance systems.

Chapter 5

UNDERSTANDING AND USING THE PHOTOMETRIC TEST REPORT

Photometry is defined as the measurement of the intensity of light. When the intensity is measured at various angles from a luminaire, the process is called goniophotometry. Two types of goniometers, a rotating mirror and a fixed multiple cell, are shown in Figure 5-1. These precision light meters are used by photometric testing laboratories and some equipment manufacturers to develop photometric test reports. The reports provide information which allows the lighting engineer or designer to predict the performance of a lighting system, and to calculate the number of luminaires required to provide some specific design illuminance.

There are two types of photometric tests: Type A and Type B. Type A tests are by far the most common, and are used for all luminaires except aimable flood lights, which may be tested using Type B Photometry.

TYPE A PHOTOMETRY

THE PHOTOMETRIC TEST

The intensity of light distributed by the luminaire is measured at specified vertical angles around the fixture, as shown in Figure 5-2. The angles vary according to the type of luminaire under test, but are normally at 0°, 5°, 15°, 25°, ... 85°, and 90° for direct luminaires. Luminaires with uplight components are also measured at 95°, 105°, 115°, ... 175°, and 180°.

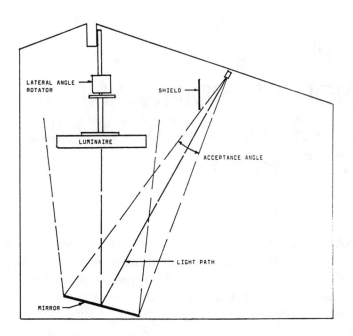

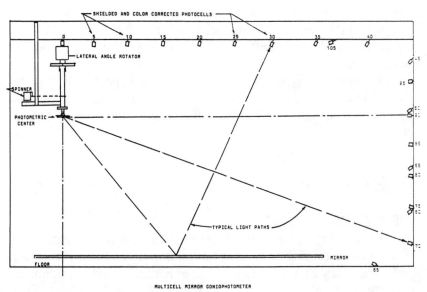

Figure 5-1 (a)
Schematic of a fixed multiple cell goniometer *(Courtesy Lighting Research Laboratory)*

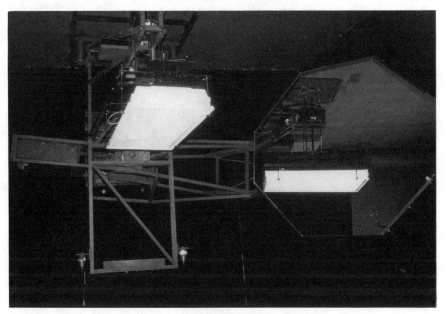

Figure 5-1 (b)
Rotating mirror Goniometer *(Courtesy Lighting Sciences, Inc.)*

Luminaires with symmetrical distribution patterns are normally photometered in one plane, as illustrated in Figure 5-2. Luminaires with asymmetrical distributions, such as fluorescent troffers and strips, are usually photometered in five planes. Some luminaires with wide variations in distribution between planes are photometered in 13 or more planes. The concept of these planes is illustrated in Figure 5-3. When measurements are required in more than one plane, the intensity is measured in each plane at each of the vertical angles described in the preceding paragraph. These measurements form the basis for the data contained in a photometric test report.

The format of the test report varies between laboratories, but they generally contain essentially the same information. Note, however, that some reports are more detailed than others, and that fixture manufacturers frequently print only portions of the test report in their sales and technical literature. Most manufacturers can provide copies of the actual test report upon request if additional data are needed. Test

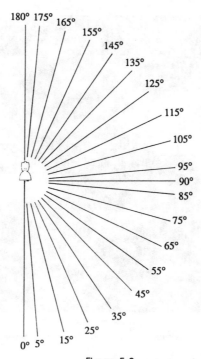

Figure 5-2
Vertical angles used in photometry. See test for explanation.

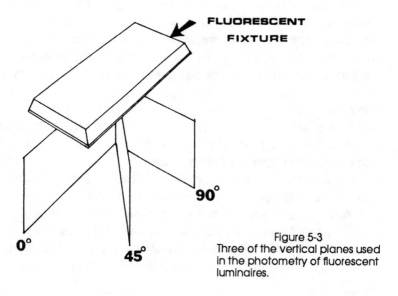

Figure 5-3
Three of the vertical planes used
in the photometry of fluorescent
luminaires.

reports generally contain, as a minimum, the following information, which is keyed to the typical test report shown in Figure 5-4:

1. Luminaire manufacturer's name, and catalog number of luminaire.
2. The lamp type and color used for the test.
3. A Candela Distribution Curve. This is a pictorial representation of the intensities of light at various angles. Curves for luminaires with symmetrical light distributions show only one line, while curves for fluorescent luminaires generally have two curves, one for intensities parallel to the fixture axis, and the other for intensities perpendicular to the axis.

   Note that the data are presented in polar format. The measured intensities are plotted, and the "best fit" line is drawn. The curve is precise only for the angles at which measurements were taken, and scaling further reduces accuracy, but interpolated values are of sufficient accuracy for lighting calculations. The primary use of the curve is for calculations of illuminance at a specific point in a space, as covered in Chapter 8.

   The curve also shows if light is concentrated directly below the fixture, as in Figure 5-5a, or widely spread, Figure 5-5b. If luminaires are mounted on a high ceiling and the intent of the lighting system is to provide horizontal illuminance on or near the floor, the fixture that produces the distribution in Figure 5-5a would be preferable since it directs light downward more effectively than the luminaire in Figure 5-5b, and less light will be lost on the walls.

   Conversely, if ceiling heights are low, Figure 5-5b provides a wider spread of light, and would generally be preferred.
4. A Candela Distribution Table. The table presents the data used to plot the "Candela Distribution Curve" in tabular form. The table will provide more accurate data than the curve when calculating the illuminance at a point if the angle is at or near an angle used in the test.
5. Zonal Flux Summary. This listing shows the lumens in each conic zone around the fixture. The data may be of use in assessing the suitability of the fixture when specific light distribution characteristics are required.

LIGHTING RESEARCH LABORATORY

P.O. BOX 6193, ORANGE, CALIF. 92667
BILL F. JONES, P.E. (714) 771-1312

REPORT PREPARED FOR: NO—LITE CORP
CATALOG NUMBER: 6X48-1F40
REPORT NO: LRL 687-3E   LAMPS: 1   F40/CW
IES SPACING CRITERION: 5
DESCRIP.: 48" X 6" INDIRECT UNIT
            W/INNER RIBBED ACRYLIC LENS

PAGE 1 OF 2
LUMENS: 3150
CU'S ARE BASED ON S/MH RATIO = 1

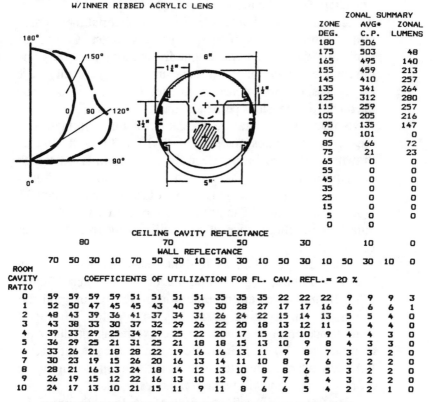

ZONAL SUMMARY

ZONE DEG.	AVG* C.P.	ZONAL LUMENS
180	506	
175	503	48
165	495	140
155	459	213
145	410	257
135	341	264
125	312	280
115	259	257
105	205	216
95	135	147
90	101	0
85	66	72
75	21	23
65	0	0
55	0	0
45	0	0
35	0	0
25	0	0
15	0	0
5	0	0
0	0	0

CEILING CAVITY REFLECTANCE																	
80				70				50			30			10			0
WALL REFLECTANCE																	
70	50	30	10	70	50	30	10	50	30	10	50	30	10	50	30	10	0

ROOM CAVITY RATIO — COEFFICIENTS OF UTILIZATION FOR FL. CAV. REFL.= 20 %

RCR	70	50	30	10	70	50	30	10	50	30	10	50	30	10	50	30	10	0
0	59	59	59	59	51	51	51	51	35	35	35	22	22	22	9	9	9	3
1	52	50	47	45	45	43	40	39	30	28	27	17	17	16	6	6	6	1
2	48	43	39	36	41	37	34	31	26	24	22	15	14	13	5	5	4	0
3	43	38	33	30	37	32	29	26	22	20	18	13	12	11	5	4	4	0
4	39	33	29	25	34	29	25	22	20	17	15	12	10	9	4	4	3	0
5	36	29	25	21	31	25	21	18	18	15	13	10	9	8	4	3	3	0
6	33	26	21	18	28	22	19	16	16	13	11	9	8	7	3	3	2	0
7	30	23	19	15	26	20	16	13	14	11	10	8	7	6	3	2	2	0
8	28	21	16	13	24	18	14	12	13	10	8	8	6	5	3	2	2	0
9	26	19	15	12	22	16	13	10	12	9	7	7	5	4	3	2	2	0
10	24	17	13	10	21	15	11	9	11	8	6	6	5	4	2	2	1	0

THESE COEFFICIENTS WERE COMPUTED BY THE ZONAL—CAVITY METHOD,
I.E.S. RECOMMENDED PRACTICE, AND PREPARED FROM THE CANDLEPOWER
DISTRIBUTION DATA IN PHOTOMETRIC REPORT LRL 687-3E

* AVERAGE OVER ZONE.   FOR SPECIFIC CANDELA VALUES, SEE P. 2.

COMPUTED BY ILLUMINATION COMPUTING SERVICE   CERTIFIED BY

Unless otherwise stated, all tests are performed in accordance with IES or other applicable standard procedures; are accurate within standard photometric tolerances; and are based on rated performance conditions. LRL assumes no responsibility for applicability of this data to field conditions or to any sample of the luminaire other than the specific unit tested.

LIGHTING RESEARCH LABORATORY

P.O. BOX 6193, ORANGE, CALIF. 92667
BILL F. JONES, P.E. (714) 771-1312

NO-LITE CORP   6X48-1F40   PAGE 2 OF 2
                           CANDLEPOWER   LRL 687-3E

ZONE DEG.	0	22	45	67	90
		CANDELAS			
180	506	506	506	506	506
175	506	503	504	500	500
165	488	486	492	506	504
155	434	443	466	475	472
145	375	389	415	429	436
135	267	297	337	393	403
125	179	234	312	398	428
115	100	159	279	361	376
105	43	105	240	298	308
95	10	63	159	204	215
90	5	42	120	157	167
85	6	24	77	103	111
75	5	6	25	34	36
65	0	0	0	0	0
55	0	0	0	0	0
45	0	0	0	0	0
35	0	0	0	0	0
25	0	0	0	0	0
15	0	0	0	0	0
5	0	0	0	0	0
0	0	0	0	0	0

ZONE	LUMENS	% BARELAMP	% LUMINAIRE
0 TO 30	0	0.0	0.0
0 TO 40	0	0.0	0.0
0 TO 60	0	0.0	0.0
0 TO 90	94	3.0	4.9
90 TO 180	1822	57.9	95.1
0 TO 180	1917	60.8	100.0

LUMINANCE VALUES

ANG.	AVG. PARA	AVG. PERP
85	146	412
75	41	105
65	0	0
55	0	0
45	0	0

Unless otherwise stated, all tests are performed in accordance with IES or other applicable standard procedures; are accurate within standard photo-metric tolerances; and are based on rated performance conditions. LRL assumes no responsibility for applicability of this data to field conditions or to any sample of the luminaire other than the specific unit tested.

Figure 5-4
Abbreviated Type "A" photometric report. Data may also include VCP Tables, discussed in Chapter 9. *(Courtesy Lighting Research Laboratory)*

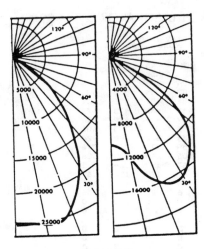

Figure 5-5
Typical candela distribution curves for concentrating (left) and wide (right) distributions.

6. Fixture Efficiency. Fixture efficiency is the percentage of lamp lumens which exit the fixture.

7. Coefficients of Utilization. These are the percentage of lamp lumens which reach the workplane, and are used to determine the number of luminaires required to light a room. They are a very powerful tool, and are frequently used by the lighting designer. Due to their great importance they will be discussed in detail later in this chapter.

8. Spacing Criteria, formerly called "Spacing to Mounting Height Ratio." This is the maximum recommended spacing of fixtures, as a function of their mounting height above the workplane, to achieve uniformity of illuminance. Its use will be discussed in detail in Chapter 9.

In addition, the following information may be provided:

9. Maximum Luminance and Luminance Ratio Table. This lists the luminance of the luminaire, typically in footlamberts, when viewed from several different angles and planes. The Illuminating Engineering Society has deprecated the footlambert, and recommends the use of candelas/foot^2, so

future test reports may use these units. Cd/ft^2 may be converted to footlamberts by multiplying Cd/Ft^2 by π (3.1416), and footlamberts may be converted to cd/ft^2 by multiplying footlamberts by 0.3183.

10. Test Distance. This is the distance, in feet or meters, from the luminaire to the photocell. The distance should be at least five times the maximum luminous dimension of the luminaire. Chapter 8 will explain the reasons for this constraint.

11. Visual Comfort Probability Table. Visual Comfort Probability (VCP) is a metric which is used to evaluate the effects of direct glare from the luminaire. It is the percentage of people which can be expected to find a given lighting installation comfortable when viewed from the least comfortable viewing position in the room. VCP will be discussed in detail in Chapter 10.

COEFFICIENTS OF UTILIZATION

APPLICATION

By definition, the average illuminance in a space, E_{ave}, is equal to the luminous flux per unit area, or:

$$E_{ave} = \frac{\Phi}{A}$$

Flux is expressed in lumens, and the area may be in square feet or square meters. If the square foot is the unit of area, the illuminance, E, will be in footcandles. If the SI system is used, the area will be in square meters and the unit of illuminance is the Lux.

For example, if 6000 lumens are evenly distributed over an area of 100 square feet, the average illuminance will be:

$$E_{ave} = \frac{6000 \text{ lumens}}{100 \text{ sq. ft.}} = \frac{60 \text{ lumens}}{\text{sq. ft.}} = 60 \text{ footcandles}$$

The example assumes that the 6000 lumens have reached the surface. A more practical approach recognizes that not all of the lumens produced by a lamp will reach the workplane. Since lighting design calculations are frequently employed to determine the number of luminaires required to produce a specified illuminance, the percentage of lamp lumens which actually reach the workplane (CU) must be factored into the calculation. If the CU is 0.50, the number of lamp lumens required will be:

$$\Phi = \frac{E \times A}{CU} = \frac{60 \text{ Fc} \times 100 \text{ sq. ft.}}{0.50} = 12,000 \text{ Lm}$$

In order to provide 60 Fc at the workplane, the number of lumens in the ceiling will need to be doubled.

The CU of a fixture is specified to that fixture, and will vary as a function of three factors: the physical characteristics of the luminaire, the room proportions (ratio of vertical wall surface area to horizontal surface area), and the percentage of light which is reflected by room surfaces.

The coefficient of utilization should not be confused with fixture efficiency, which is simply the percentage of lamp lumens which exit the fixture. Fixture efficiency is a fixed value and does not change as a function of room characteristics; coefficient of utilization is dynamic, and varies with room proportions and surface reflectances.

The physical characteristics of the luminaire influence the coefficient of utilization in several ways. In order for light to reach the workplane it must first be emitted by the fixture. Luminaires with dark interior surfaces or hidden corners will absorb more light than fixtures with smooth, highly reflective surfaces, and will typically have lower coefficients of utilization. Lenses and diffusers, which may be necessary to direct light in a specific direction or reduce glare, also reduce the amount of light which leaves the fixture. Since all mass within a fixture absorbs some light, the number of lamps in a fixture will also influence the coefficient of utilization: fixtures with large numbers of lamps will generally have lower CU's than fixtures with fewer lamps.

In any room some light travels directly from the luminaire to the workplane, while other light travels to the walls, floor, and in some cases, to the ceiling. These surfaces will reflect some light to the workplane. The remaining light will either be reflected to other room surfaces or absorbed by the surface upon which it fell. In either case, some of the light which strikes other surfaces in the room will not reach the workplane. Figure 5-6 shows a simplified representation of the direct and reflected components of light which reach the workplane in two rooms; one with a high ratio of wall area to floor area, and a second room with a low ratio of wall to floor area.

Note that in room A some of the light from each fixture strikes the walls and is lost through absorption. In room B, only the fixtures near the wall cast significant quantities of light on walls, while the fixtures in the center of the room project virtually all of their light directly to the workplane. On the average, the luminaires in room B direct a larger percentage of the total lamp lumens in the system to the workplane, and a lower percentage is lost through absorption by the walls. We can then conclude that, on the average, a luminaire will have a higher coefficient of utilization in a room with a low percentage of its total surface area in walls and, as this ratio increases, more wall area will exist to absorb light. This will result in a decrease in the coefficient of utilization.

As the reflectances of room surfaces increase, the percentage of interreflected light within the room increases, and a higher percentage of the light will eventually reach the workplane. Dark-colored room surfaces will absorb more light, with a resultant reduction in the coefficient of utilization.

A typical Table of Coefficients of Utilization is shown in Figure 5-7. The CU's are the decimal values contained in the body of the table. They are accessed by entering the table under the appropriate columns and rows which denote room surface reflectances and room size. Note that some manufacturers print CU tables in integer form; however, it is understood that CU's are decimal values.

Item 1 identifies the reflectance of the ceiling or, if the luminaire is suspended below the ceiling, the effective reflectance of the cavity above the luminaire. The concept of effective cavity reflectances will be discussed in detail in Chapter 7. At this time we will consider only the actual ceiling reflectance.

This area of the table is typically labeled ρ_{cc}. The Greek letter ρ (rho) denotes reflectance, and the subscript $_{cc}$ identifies the ceiling cavity.

Item 2 identifies the reflectance of the walls.

Item 3, RCR, stands for "Room Cavity Ratio," and represents a modified ratio of wall area to horizontal surface area in the room. The physical size of a room has no influence on the percentages of light

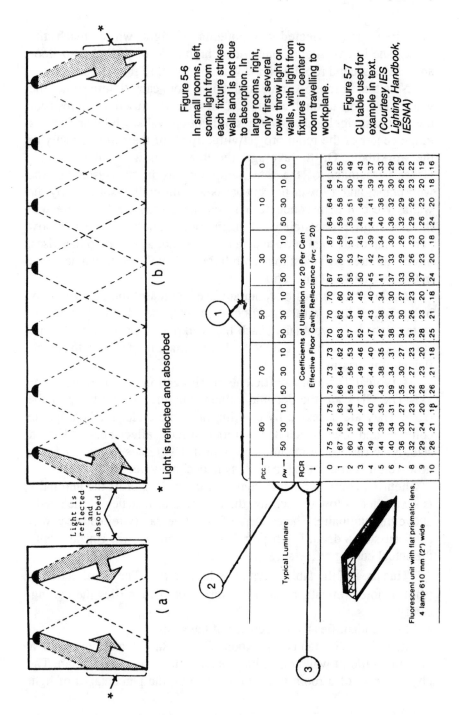

* Light is reflected and absorbed

(a)

(b)

Figure 5-6
In small rooms, left, some light from each fixture strikes walls and is lost due to absorption. In large rooms, right, only first several rows throw light on walls, with light from fixtures in center of room travelling to workplane.

Figure 5-7
CU table used for example in text.
(Courtesy IES Lighting Handbook, IESNA)

Typical Luminaire

Fluorescent unit with flat prismatic lens, 4 lamp 610 mm (2') wide

pcc →	80			70			50			30			10			0
pw →	50	30	10	50	30	10	50	30	10	50	30	10	50	30	10	0
RCR ↓	Coefficients of Utilization for 20 Per Cent Effective Floor Cavity Reflectance (ρ_{FC} = 20)															
0	.75	.75	.75	.73	.73	.73	.70	.70	.70	.67	.67	.67	.64	.64	.64	.63
1	.67	.65	.63	.66	.64	.62	.63	.62	.60	.61	.60	.58	.59	.58	.57	.55
2	.60	.57	.54	.59	.56	.53	.57	.54	.52	.55	.53	.51	.53	.51	.50	.49
3	.54	.50	.47	.53	.49	.46	.52	.48	.45	.50	.47	.45	.48	.46	.44	.43
4	.49	.44	.40	.48	.44	.40	.47	.43	.40	.45	.42	.39	.44	.41	.39	.37
5	.44	.39	.35	.43	.38	.35	.42	.38	.34	.41	.37	.34	.40	.36	.34	.33
6	.40	.34	.31	.39	.34	.31	.38	.34	.30	.37	.33	.30	.36	.32	.30	.29
7	.36	.30	.27	.35	.30	.27	.34	.30	.27	.33	.30	.26	.32	.29	.26	.25
8	.32	.27	.23	.32	.27	.23	.31	.26	.23	.30	.26	.23	.29	.26	.23	.22
9	.29	.24	.20	.28	.23	.20	.28	.23	.20	.27	.23	.20	.26	.23	.20	.19
10	.26	.21	.18	.26	.21	.18	.25	.21	.18	.24	.20	.18	.24	.20	.18	.16

1

2

3

which are directed to the walls and workplane; only the ratio of wall area to horizontal surface area is important. A room with dimensions of 10' x 10' x 10' will have the same proportions as a room measuring 20' x 20' x 20', and the percentages of light directed to walls and floor will be identical in either room. Using this fact, tables of coefficients are assembled based on these ratios rather than actual room sizes.

In a rectangular room the wall area equals the height of the room times the perimeter or,

wall area = hp

This may also be stated as:

wall area = 2h (L+W) where L = length and W = width

The horizontal surface area equals the ceiling area plus the floor area, or:

horizontal area = 2LW

The ratio of wall area to horizontal surface area, then, is:

$$\text{ratio} = \frac{2h(L+W)}{2LW} \text{ or } \frac{h(L+W)}{LW}$$

For most rooms, the ratio will be less than 1. When the Zonal Cavity System was developed, it was desired to have ratios ranging from 1 to 10 for most rooms, so a constant, 5, was placed in the numerator to meet this constraint.

The room cavity is defined as the distance from the workplane to the fixture, and the room cavity ratio (RCR) is:

$$RCR = \frac{5h(L+W)}{LW}$$

USING THE CU TABLE

Consider a room with dimensions and data as shown in Figure 5-8.

The room cavity ratio is calculated as:

$$RCR = \frac{5h(L+W)}{LW} = \frac{5(6)(10+10)}{(10)(10)} = 6$$

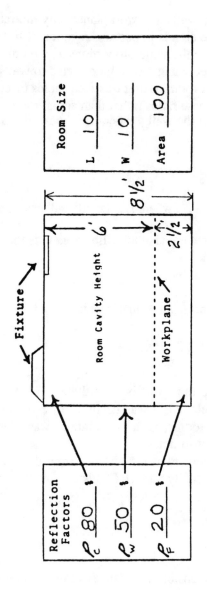

Figure 5-8
Physical characteristics of room used in example calculation.

To access the CU table:
1. The ceiling reflectance is 80% so locate the column under the ρ_{CC} heading of 80%.
2. The wall reflectance is 50% so locate the column under the ρ_W heading of 50% which is under the ρ_{CC} column of 80%.
3. Locate the RCR value of 6 under the RCR column.
4. Read the CU directly from the table. The example CU is 0.40, or 40%. This means that 40% of the lumens produced by the lamps will reach the workplane.

Returning to the original example, the lamp lumens required to light the 100-square-foot room with the example fixture can now be calculated:

$$\text{Lamp lumens required} = \frac{(Fc)\,(area)}{CU} = \frac{(60)\,(100)}{0.40} = 15{,}000\ Lm$$

INTERPOLATION

In the above example, the RCR conveniently worked out to 6, a whole number. Since the CU table lists CU's for integer values of RCR's from 1 to 10, it was a simple matter to read the CU directly from the table. RCR's in the real world seldom are integers, so a process for determining CU's for decimal values is required.

Inspection of the CU table indicates that the change in CU is not precisely linear. It is close enough, however, to permit a sufficiently accurate estimate of the CU to be made using linear interpolation. For example, assume that the RCR is 5.75 instead of 6. The CU can be determined by:
1. Determine the CU's for RCR = 5, and RCR = 6
    CU @ RCR 5 = .44
    CU @ RCR 6 = .40
2. Subtract the smaller CU from the larger
    .44 – .40 = .04
3. Multiply the result (.04) by the decimal portion of the RCR
    (.04) (.75) = .03
4. Subtract the product (.03) from the larger CU
    .44 – .03 = .41

The CU for RCR = 5.75 is 0.41

Since CU tables are published for a limited number of ceiling and wall reflectances it is usually necessary to interpolate between columns for reflectances not listed in the table. For example, assume that the ceiling reflectance is 55%, wall reflectance is 50%, floor reflectance is 20%, and RCR is 1. Since a 55% ceiling column is not generally published, it will be necessary to interpolate between the 70% and 50% ceiling columns.

1. Determine the CU's for ρ_{cc} = .70 and ρ_{cc} = .50

   CU @ ρ_{cc} = 70% is .73

   CU @ ρ_{cc} = 50% is .70
2. Subtract the smaller cu from the larger

   $(.73) - (.70) = .03$
3. The spread between the listed CU's is 20%, and the desired reduction is 15% below the higher value so multiply the difference between the CU's (.03) by 15/20

   $(.03)(15/20) = .02$
4. Subtract the product from the larger CU

   $.73 - .02 = .71$

It may also be necessary to interpolate between RCR's and wall reflectances. In these cases the procedure outlined above is repeated for each case until a final CU is obtained.

Note that while several different ceiling and wall reflectances are listed in the CU table, there is no choice of floor reflectance. Tables are typically developed using an assumed floor reflectance of 20%. For reflectances of other than 20%, a correction factor is applied to adjust the CU accordingly. These factors will be discussed in Chapter 7.

DEVELOPMENT OF CU TABLES

Tables of coefficients of utilization are developed using candela distribution data obtained from photometric tests, as previously discussed. The data are then mathematically reduced to tabular form. The actual calculations are typically performed by computer since they involve many redundant calculations. The number crunching is purely mechanical; however, the general concepts are important and merit discussion.

The calculation consists of two basic parts: the determination of the percentage of lamp lumens which travel directly downward to the workplane and; the determination of the percentage of lumens which will be reflected to the workplane by the room surfaces.

Assume that a symmetrical distribution fixture has been photometered and the intensities (in candelas) are as shown in Figure 5-9.

Angle (Degrees)	Candelas	Angle (Degrees)	Candelas
0	13456	95	25
5	12831	105	56
15	15170	115	129
25	17083	125	68
35	11174	135	149
45	6118	145	1650
55	1111	155	791
65	229	165	87
75	68	175	72
85	34	180	0
90	27		

Figure 5-9
Candela distribution table used for example calculation of coefficient of utilization.

The data are presented as candelas at various angles and, since coefficients of utilization represent percentages of lamp lumens which reach the workplane, the candela data must be converted to lumens. We are also concerned with the angles at which the lumens are emitted, since light at high angles has a higher probability of striking a wall than light which travels more nearly downward. Ideally, the flux at each angle surrounding the luminaire would be measured and converted to lumens falling on an infinitely small area at that angle. From a practical viewpoint this is not possible since there are an infinite number of angles which could be subtended from a luminaire, and the photosensor has a dimension which is much lager than the infinitely small point at which the flux could be measured. In practice, the area surrounding the luminaire is divided into a series of conic solid angles as shown in Figure 5-10a.

There are 18 conic solid angles, each creating a band with its mid-point corresponding to an angle at which a measurement was taken during the photometric test. Note that only four of these bands are shown in the Figure. The remaining angles progress upwards at 10-degree intervals, but are omitted from the diagram for clarity. The flux contained within each band (solid conic angle) can be determined from:

$$\Phi = 2\pi I (\cos\theta_1 - \cos\theta_2)$$

where θ_1 and θ_2 are the angles which define the zone, and I is the mid-zone intensity.

Figure 5-10b illustrates the angles used in the calculation.

Since most type-A photometry is based on 10° zones, the quantity "$2\pi (\cos \theta_1 - \theta_2)$" can be reduced to tabular form to reduce redundancy in the calculation process. This results in a table of factors, called "Zonal Constants" which, when multiplied by the mid-zone intensity, yield the lumens contained in the zone. Zonal Constants for 10° zones are shown in Figure 5-11, and the equation for determining zonal lumens becomes:

$$\Phi_N = I_N (\text{Zonal Constant}_N)$$

where

Φ = zonal lumens
I   = mid-zone intensity
N   = zone

Asymmetrical distribution fixtures may have different intensities in different planes even though the vertical angle, θ, is constant. The value of I in the equation is obtained by averaging the intensities for all planes in which measurements were taken. For example, assume that a fluorescent luminaire was photometered in 0°, 22.5°, 45°, 67.5°, and 90° planes. For purposes of calculation the luminaire is assumed to have both X axis symmetry and Y axis symmetry, so the intensity at 0° will equal the intensity at 180°; the

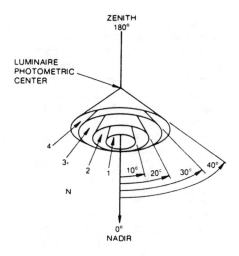

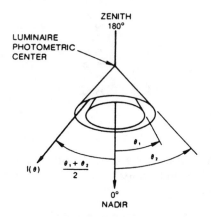

Figure 5-10
(a) Conic solid angle zones of 10-degree width used in calculation of zonal flux.
(b) angles used for zones of any width when calculating zonal flux. (*Courtesy IES Lighting Handbook, IESNA*)

intensity at 22.5° will be the same at 112.5°, 202.5°, and 292.5°. The intensities at 45° and 67.5° will have similar symmetry at corresponding angles in the other three quadrants. The average intensity for each vertical angle can be found from:

$$I_{AVE_N} = \frac{I_{0°} + 2I_{22.5°} + 2I_{45°} + 2I_{67.5°} + I_{90°}}{8}$$

Figure 5-11
Constants used in the Zonal Cavity Method of computing luminous flux from
candlepower data. *(Courtesy IES Lighting Handbook, IESNA)*

1-Degree Zones		2-Degree Zones		5-Degree		10-Degree Zones	
Zone Limits (degrees)	Zonal Constant	Zone Limits (degrees)	Zonal Constant	Zone Limits (degrees)	Zonal Constant	Zone Limits (degrees)	Zonal Constant
0-1	0.0009	0-2	0.0038	0-5	0.0239	0-10	0.095
1-2	.0029	2-4	.0115	5-10	.0715	10-20	.283
2-3	.0048	4-6	.0191	10-15	.1186	20-30	.463
3-4	.0067	6-8	.0267	15-20	.1649	30-40	.628
4-5	.0086	8-10	.0343	20-25	.2097	40-50	.774
5-6	.0105	10-12	.0418	25-30	.2531	50-60	.897
6-7	.0124	12-14	.0493	30-35	.2946	60-70	.993
7-8	.0143	14-16	.0568	35-40	.3337	70-80	1.058
8-9	.0162	16-18	.0641	40-45	.3703	80-90	1.091
9-10	.0181	18-20	.0714	45-50	.4041		
				50-55	.4349		
				55-60	.4623		
				60-65	.4862		
				65-70	.5064		
				70-75	.5228		
				75-80	.5351		
				80-85	.5434		
				85-90	.5476		

The downward flux, upward flux, and total flux from the
luminaire can then be found from:

$$\text{Downward flux} = \sum_{N=1}^{9} 2\pi I (\cos \theta_1 - \cos \theta_2)$$

or, more simply stated:

$$\text{Downward flux} = \sum_{N=1}^{9} I_N (\text{Zonal Constant}_N)$$

and,

$$\text{Upward flux} = \sum_{N=10}^{18} 2\pi I (\cos \theta_1 - \cos \theta_2) = \sum_{N=10}^{18} I_N (\text{Zonal Constant}_N)$$

The total luminaire lumens is the sum of the downward and upward lumens.

The fixture efficiency may now be calculated from:

$$\text{Efficiency} = \frac{\text{(Downward lumens + Upward lumens)}}{\text{Total lamp lumens}}$$

The fractional downward flux, Φ_D, and fractional upward flux, Φ_U, are now found from:

$$\Phi_D = \frac{1}{\Phi_T} \sum_{N=1}^{10} \Phi_N$$

$$\Phi_U = \frac{1}{\Phi_T} \sum_{N=10}^{18} \Phi_N$$

Where Φ_T is the total lamp lumens installed in the luminaire.

The equations may be more simply stated as:

Φ_D= (Total downward lumens)/(Total lamp lumens)

Φ_U= (Total upward lumens)/(Total lamp lumens)

Note that the terminology is somewhat misleading in that the terms Φ_D and Φ_U infer flux in lumens. This is not the case, however, and it is important to realize that they refer to the percentage of lamp lumens which are directed downwards and upwards.

It was shown in the beginning of this chapter that, in any room, some percentage of lamp lumens from each fixture travels directly to the workplane. This percentage is called the "direct ratio," D_G. The direct ratio is calculated from the data obtained from photometric testing of the luminaire by:

$$D_G = \frac{1}{\Phi_D \Phi_T} \sum_{N=1}^{9} (K_{GN} \Phi_N)$$

where

Φ_D = fractional downward flux

Φ_T = total lamp lumens

G   = the room cavity ratio
K_{GN} = the zonal multiplier

Note that the Direct ratio, D_G, is a function of the room cavity ratio, and that D_G represents the direct ratio at a specific RCR which is identified by the subscript. For example D_1 is the direct ratio for RCR 1, D_2 for RCR 2, etc.

K_{GN}, the Zonal Multiplier, is the percentage of flux contained in each zone, N, that is directly incident on the workplane. It is the average flux in the zone, per luminaire, from an installation consisting of luminaires which are spaced uniformly within the room. Zonal multipliers are functions of the room cavity ratio. In the past they were based on standardized spacing to mounting height ratios, and tables of the multipliers will be found in older versions of the IES Lighting Handbook and other publications. These tables have been replaced with a single equation, and zonal multipliers are now calculated from:

$$K_{GN} = e^{(-AG^B)}$$

where G is the room cavity ratio and N is the zone for which the multiplier is being calculated.

Values for A and B are found in the IES Lighting Handbook, 1983 Reference Volume, and are reproduced in Figure 5-12. Note that this value is not specific to any one luminaire in the installation; it is the average value for all of the luminaires in the installation. The development of Zonal Multipliers is complex and is beyond the scope of this text.

Figure 5-12
Constants for the Zonal Multiplier equation. *(Courtesy IES Lighting Handbook, IESNA)*

Zone	A	B
1	0.	0.
2	0.041	0.98
3	0.070	1.05
4	0.100	1.12
5	0.136	1.16
6	0.190	1.25
7	0.315	1.25
8	0.640	1.25
9	2.10	0.80

This completes the basis for calculating the direct component of the coefficient of utilization. The reflected component is based on flux transfer theory to determine the percentages of flux which will reach the workplane from the walls, ceiling, and floor. These ratios are called C_1, C_2, and C_3, respectively, and are functions of surface reflectances, room cavity ratio, and the form factor $f_{2\to3}$. They are calculated from:

$$C_1 = \frac{(1-\rho_1)(1 - f^2_{2\to3})\, G}{2.5\rho_1(1 - f^2_{2\to3}) + G\, f_{2\to3}(1-\rho_1)}$$

$$C_2 = \frac{(1 - \rho_2)(1 + f_{2\to3})}{1 + \rho_2 f_{2\to3}}$$

$$C_3 = \frac{(1 - \rho_3)(1 + f_{2\to3})}{1 + \rho_3 f_{2\to3}}$$

$$C_0 = C_1 + C_2 + C_3$$

where $f_{2\to3}$ is approximated from:

$$f_{2\to3} = 0.026 + 0.503e^{(-0.270RCR)} + 0.470e^{(-0.119RCR)}$$

The coefficient of utilization can than be calculated from:

$$CU = \frac{2.5\rho_1 C_1 C_3 (1 - D_G)\, \Phi_D}{G(1 - \rho_1)(1 - \rho_3)\, C_0} + \frac{\rho_2 C_2 C_3 \Phi_U}{(1 - \rho_2)(1 - \rho_3)\, C_0}$$

$$+ \left[1 - \frac{\rho_3 C_3 (C_1 + C_2)}{(1 - \rho_3)\, C_0}\right] \frac{D_G\, \Phi_D}{1 - \rho_3}$$

where ρ_1, ρ_2, and ρ_3 are the reflectances of the walls, floor, and ceiling, respectively.

The special case of RCR = 0 does not exist in the practical world since it indicates a room which has a floor and a ceiling, but is infinitely large so as to have no walls. It is of interest, however, since rooms are frequently larger than RCR 1 and a basis for interpolation to obtain a CU is desirable. In this special case there are no walls to absorb light and all downward flux from the luminaire travels directly to the workplane. There may be, however, a reflected component from the ceiling. The CU for RCR = 0 may be calculated from:

$$CU_{G=0} = \frac{\Phi_D + \rho_2\Phi_U}{1 - \rho_2\rho_3}$$

A sample calculation of a coefficient of utilization is shown at the end of this chapter for those who wish a more detailed explanation of the calculation process.

COEFFICIENTS FOR OTHER CALCULATIONS

The calculation of illuminance at a point will be discussed in detail in Chapter 8. It should be noted at this time, however, that the calculation of the reflected component of illuminance at the point requires the use of three additional coefficients: the Wall Exitance Coefficient (WEC), the Ceiling Cavity Exitance Coefficient (CCEC), and the Wall Direct Radiation Coefficient (WDRC). These coefficients are calculated in a manner similar to the Coefficient of Utilization, and require much of the same data. The equations are:

$$WEC = \frac{2.5}{G} \left\{ \frac{\rho_1(1 - D_G)\,\Phi_D}{(1 - \rho_1)} \times \left[1 - \frac{2.5\rho_1C_1(C_2 + C_3)}{G(1 - \rho_1)C_0}\right] \right.$$
$$\left. + \frac{\rho_1\rho_2C_1C_2\Phi_U}{(1 - \rho_1)(1 - \rho_2)\,C_0} + \frac{\rho_1\rho_3C_1C_3D_G\Phi_D}{(1 - \rho_1)(1 - \rho_3)C_0} \right\}$$

$$CCEC = \frac{2.5\rho_1\rho_2C_1C_2(1 - D_G)\,\Phi_D}{G(1 - \rho_1)(1 - \rho_2)C_0} + \frac{\rho_2\Phi_U}{(1 - \rho_2)}\left[1 - \frac{\rho_2C_2(C_1C_3)}{(1 - \rho_2)C_0}\right]$$
$$+ \frac{\rho_2\rho_3C_2C_3D_G\Phi_D}{(1 - \rho_2)(1 - \rho_3)C_0}$$

$$WDRC = \frac{2.5\Phi_D(1 - D_G)}{G}$$

These coefficients are typically not published by fixture manufacturers, and are seldom available for specific luminaires. The IES Lighting Handbook contains tables of these coefficients for typical luminaires, and may be used when the photometrics of the fixture to be used approximate the typical fixture.

TYPE B PHOTOMETRY

Type B photometry has largely been replaced by Type A photometry, and is seldom encountered in current practice. However, it has been used in the past for aimable floodlights and some manufacturers still use it, so Type B test reports may be encountered when designing systems using these luminaries.

In Type B photometry the luminaire is rotated about a fixed vertical axis, and a horizontal axis which corresponds to each vertical angle at which measurements are taken, as shown in the grid pattern in Figure 5-13. For example, the luminaire is aimed directly at the photocell so that the vertical and horizontal angles are both zero, and the intensity measured. While still on the 0° vertical axis, the luminaire is rotated 5° horizontally, and another intensity reading taken. The luminaire is then rotated horizontally in 10° steps, with readings taken at each step. The luminaire is then rotated 5° on the vertical axis, and each previously described horizontal angle measurement is repeated. The luminaire is then rotated vertically in 10° steps, and each horizontal measurement repeated for each vertical angle. Intensities for both positive (upward) and negative (downward) vertical angles are measures.

Figure 5-13
Method of obtaining
Type "B" photometric
data. Luminaire is
mounted on mov-
able base, "A," and
aimed at grid coor-
dinates shown in "b."
(Courtesy IES Lighting
Handbook, IESNA)

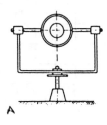

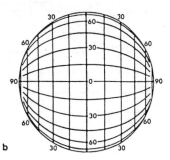

The results are then tabulated, and a table of lumens in each zone of the grid is calculated, as shown in Figure 5-14.

These test reports are used for some outdoor lighting calculations for floodlights, as explained in Chapter 14.

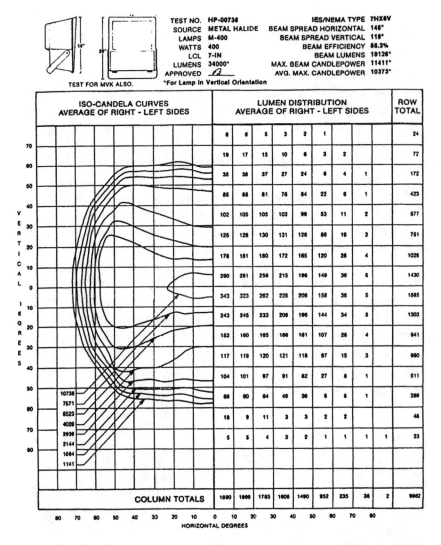

TEST NO.	HP-00738		
SOURCE	METAL HALIDE	IES/NEMA TYPE	7HX6V
LAMPS	M-400	BEAM SPREAD HORIZONTAL	146°
WATTS	400	BEAM SPREAD VERTICAL	119°
LCL	7-IN	BEAM EFFICIENCY	56.3%
LUMENS	34000*	BEAM LUMENS	19126*
APPROVED	_B_	MAX. BEAM CANDLEPOWER	11411*
		AVG. MAX. CANDLEPOWER	10373*

TEST FOR MVK ALSO.   *For Lamp in Vertical Orientation

ISO-CANDELA CURVES — AVERAGE OF RIGHT - LEFT SIDES

LUMEN DISTRIBUTION — AVERAGE OF RIGHT - LEFT SIDES

6	6	5	3	2	1				ROW TOTAL
6	6	5	3	2	1				24
19	17	15	10	6	3	2			72
35	38	37	27	24	6	4	1		172
85	88	81	76	64	22	6	1		423
102	105	105	103	96	53	11	2		577
125	126	130	131	129	86	19	3		751
178	181	180	172	165	120	28	4		1026
290	281	256	215	196	149	36	5		1430
343	323	282	228	206	158	38	5		1585
243	245	233	206	196	144	34	5		1303
152	160	165	166	161	107	28	4		941
117	119	120	121	118	67	15	3		680
104	101	97	91	82	27	8	1		511
88	60	64	49	36	6	6	1		289
18	9	11	3	3	2	2			48
5	5	4	3	2	1	1	1	1	23

Vertical scale (degrees): 70, 60, 50, 40, VERTICAL 30, 20, 10, 0, DEGREES 10, 20, 30, 40, 50, 60, 70, 80

ISO-candela values: 10738, 7671, 6523, 4028, 2930, 2144, 1684, 1141

COLUMN TOTALS

1890	1866	1785	1606	1490	952	235	36	2	9662

80 70 60 50 40 33 20 10 0 10 20 30 40 50 60 70 80
HORIZONTAL DEGREES

Figure 5-14.
Typical Type "B" photometric report. *(Courtesy Hubbell Lighting)*

EXAMPLE CALCULATION OF COEFFICIENT OF UTILIZATION

Calculate the CU for RCR 1, ρ = 80-50-20, for a luminaire with the candela distribution in Figure 5-15. Lamp is a 400-watt high-pressure sodium rated at 50,000 lumens.

θ (Degrees)	I	θ (Degrees)	I
5	28800	95	25
15	25950	105	25
25	21300	115	225
35	13700	125	525
45	6350	135	825
55	3475	145	1025
65	1000	155	1600
75	125	165	1650
85	75	175	825

Figure 5-15

Step 1. Calculate the flux in each zone.

Zone	Mid-Zone Angle	cd	Zonal Constant	Lumens
1	5°	28800	.095	2736
2	15°	25950	.283	7344
3	25°	21300	.463	9862
4	35°	13700	.628	8604
5	45°	6350	.774	4915
6	55°	3475	.897	3117
7	65°	1000	.993	993
8	75°	125	1.058	132
9	85°	75	1.091	82
			Total Downward Flux	37785

<u>Step 1.</u> *(Continued)*

Zone	Mid-Zone Angle	cd	Zonal Constant	Lumens
10	95°	25	1.091	27
11	105°	25	1.058	26
12	115°	225	.993	223
13	125°	525	.897	471
14	135°	825	.774	639
15	145°	1025	.628	644
16	155°	1600	.463	741
17	165°	1650	.283	467
18	175°	825	.095	78
			Total Upward Flux	3316

Total flux = 37785 + 3316 = 41101

Efficiency = $\dfrac{41101}{50000}$ = 82%

<u>Step 2.</u> Determine K_{gn} & Flux in each zone which is directly incident on the workplane

$$K_{gn} = e^{-(AG^B)}$$

Flux directly incident on workplane = $K_{gn} \Phi_n$
First, determine k_{gn} for each zone

Zone

1	$K_{gn} = e^{-0}$	$= 1$
2	$= e^{(-.041)}$	$= .96$
3	$= e^{(-.070)}$	$= .93$
4	$= e^{(-.100)}$	$= .90$
5	$= e^{(-.136)}$	$= .87$
6	$= e^{(-.190)}$	$= .83$
7	$= e^{(-.315)}$	$= .73$
8	$= e^{(-.640)}$	$= .53$
9	$= e^{(-2.10)}$	$= .12$

Then, determine the flux in each zone that is directly incident on the workplane

1 $K_{gN_1}\Phi_1$ = (1) (2736) = 2736
2 $K_{gN_2}\Phi_2$ = (.96) (7344) = 7050
3 $K_{gN_3}\Phi_3$ = (.93) (9862) = 9172
4 $K_{gN_4}\Phi_4$ = (.90) (8604) = 7744
5 $K_{gN_5}\Phi_5$ = (.87) (4915) = 4276
6 $K_{gN_6}\Phi_6$ = (.83) (3117) = 2587
7 $K_{gN_7}\Phi_7$ = (.73) (993) =   725
8 $K_{gN_8}\Phi_8$ = (.53 (132) =   70
9 $K_{gN_9}\Phi_9$ = (.12) (82) =   10
                              $\overline{34370}$

Step 3. Determine the fractional downward flux, Φ_D

$$\Phi_D = \frac{1}{\text{Lamp Lumens}} \sum_{N=1}^{9} \Phi_N$$

$$= \frac{1}{50000 \text{ Lm}} (37785 \text{ Lm}) = .76 \qquad (37785 \text{ Lm from Step 1})$$

Determine the fractional upward flux, Φ_U

$$\Phi_U = \frac{1}{\text{Lamp Lumens}} \sum_{N=10}^{18} \Phi_N$$

$$= \frac{1}{50000 \text{ Lm}} (3316 \text{ Lm}) = .07 \qquad (3316 \text{ Lm from Step 1})$$

Step 4. Determine the direct ration, D_G

$$D_G = \frac{1}{\text{Downward Flux}} \sum_{N=1}^{9} (K_{gN} \Phi_N)$$

$$D_G = \frac{1}{37785 \text{ Lm}} (34370 \text{ Lm}) = .91$$

Step 5. Calculate C_1, C_2, C_3, and C_0

$$C_1 = \frac{(1-\rho_1)(1 - f^2_{2\to3})\,G}{2.5\rho_1(1 - f^2_{2\to3}) + Gf_{2\to3}(1-\rho_1)}$$

$$= \frac{(1 - .5)(1 - .827^2)(1)}{(2.5)(.5)(1 - .827^2) + (1)(.827)(1 - .5)} = .195$$

$$C_2 = \frac{(1-\rho_2)(1 + f_{2\to3})}{1+\rho_2 f_{2\to3}} = \frac{(1 - .8)(1 + .827)}{1 + (.8)(.827)} = .220$$

$$C_3 = \frac{(1-\rho_3)(1 + f_{2\to3})}{1+\rho_3 f_{2\to8}} = \frac{(1 - .2)(1 + .827)}{1 + (.2)(.827)} = 1.254$$

$C_0 = C_1 + C_2 + C_3 + .195 + .220 + 1.254 = 1.669$

Note that $f_{2\to3}$ may be found in the text, IES Lighting Handbook, or calculated from:
$$f_{2\to3} \doteq 0.026 + 0.503\,e^{-0.270RCR} + 0.470e^{-0.119RCR}$$

Step 6. Calculate the CU

$$CU = \frac{2.5\rho_1 C_1 C_3 (1 - D_G)(\Phi_D)}{G(1-\rho_1)(1-\rho_3)\,C_0} + \frac{\rho_2 C_2 C_3 \Phi_U}{(1-\rho_2)(1-\rho_3)\,C_0} + \left[1 - \frac{(\rho_3 C_3)(C_1 + C_3)}{(1-\rho_3)\,C_0}\right]\frac{D_G\,\Phi_D}{(1-\rho_3)}$$

$$= \frac{(.80)(.220)(1.254)(.07)}{(1 - .80)(1 - .20)(1.669)} + \left[1 - \frac{(.20)(1.254)(.195 + .220)}{(1 - .20)(1.669)}\right]\frac{(.91)(.76)}{(1 - .20)}$$

$$= .03 + .06 + .80 = .89$$

Chapter 6

LIGHT LOSS FACTORS

In Chapter 1 it was shown that the average illuminance produced by a lighting system is equal to the flux per unit area, or $E = \Phi/A$. The concept of coefficient of utilization was introduced in the preceding chapter to recognize that only a portion of the lamp lumens actually reach the workplane and, if the average illuminance produced by a luminaire or system of luminaires is to be determined, the coefficient of utilization must be included in the calculation. The basic equation then took the form:

$$E = \frac{\Phi \times CU}{A}$$

This assumes, however, that the lighting system is operating under ideal conditions and that all components are performing as rated. This is seldom the case, and the reductions from rated performance of system components must be included in lighting design calculations. These reductions, called "light loss factors," are found to varying degrees in all lighting systems.

Light loss factors have typically been relegated to a position of relative unimportance in the past. While there are 10 officially recognized individual factors and one new factor which is undergoing additional investigation at the time of this writing, it has been common practice for many designers and engineers to lump them together in a single number, usually ranging from .75 to .85, and apply this factor to all lighting designs. This value recognizes only two or three of the factors and, as will be seen, can result in substandard designs. It is unfortunate that many designers are not even aware of the other eight or nine factors.

The proper application of light loss factors to the design process is increasing in importance as a result of two major changes in our industry: the trend towards lower design illuminance, and the introduction of a vast array of new equipment with operating characteristics which vary considerably from the more traditional types of equipment. A thorough knowledge of the fundamentals of equipment-operating characteristics is mandatory if the lighting professional is to perform effectively.

Light loss factors are divided into two categories: nonrecoverable and recoverable.

NONRECOVERABLE FACTORS

Nonrecoverable factors are ones for which no ameliorative actions can reasonably be taken. There are 6 officially recognized nonrecoverable factors, 5 of which may exist from the time the system is first energized: 1) temperature factor, 2) voltage factor, 3) ballast factor, 4) lamp tilt factor, 5) equipment operating factor, & 6) luminaire surface depreciation factor. The 6th factor, luminaire surface depreciation, occurs over time. A 7th factor, alternately called "Luminaire Thermal Factor," "Application Factor," or "Ballast/Lamp Photometric Factor," was presented at the 1983 IES Annual Conference, and expanded upon in a paper presented at the 1985 Conference. We will use the term "Luminaire Thermal/Application factor" in this discussion; however, it should be noted that an official name has not yet been adopted.

TEMPERATURE FACTOR
(Fluorescent Systems Only)
The production of light by a fluorescent lamp is a function of the pressure of the mercury vapor contained in the lamp. If this pressure increases or decreases from the design level, both light output and power consumption will be affected. The lamp contains more liquid mercury than can be vaporized, and the quantity of vaporized metal is determined by the gas pressure and temperature. The excess mercury will condense at the coldest spot on the bulb wall. The temperature of this cold spot regulates the quantity of mercury which is vaporized, and thus affects light output. Temperatures above the design level increase the pressure, while lower temperatures reduce pressure. In either case, light output drops, as shown in Figure 6-1.

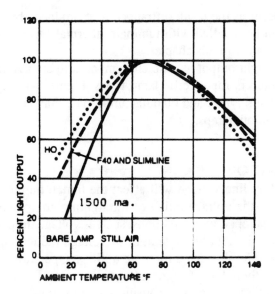

Figure 6-1
Effect of temperature on the light output of typical fluorescent lamps.
Manufacturers' data should be consulted for data on specific lamps. (*Courtesy Philips Lighting Co.*)

F40 and F96 slimline lamps are designed to produce rated output at bulb-wall temperatures of about 100°F. High-output lamps are designed to operate at about 120°F, and 1500 ma. lamps at about 165°F. These temperatures will be attained when lamps are operated in still air at a temperature of 77°F (25°C). Fluorescent luminaires are typically photometered at this temperature, so changes in bulb-wall temperature due to heat which is trapped in enclosed fixtures will be incorporated into coefficients of utilization as long as the ambient temperature is about 77°F. Note that this is only true when the luminaire is surrounded by 77°F air. When troffers are recessed into ceilings, or other luminaire types are surface mounted, the temperature above the fixture is frequently well above 77°F, which will result in elevated temperatures within the fixture.

The primary application of the temperature factor is in cases where fluorescent strip, industrial, and other well-ventilated open types of luminaires are employed in abnormal temperature environments such as foundries or refrigerated rooms and cases. Compensation for temperature increases in troffers and surface-mounted luminaries which occur as the result of trapped heat produced by lamps and ballasts should be handled with luminaire thermal factors, which are discussed separately in this chapter.

Graphs of light output vs ambient temperature are available from lamp manufacturers for specific lamp types. Figure 6-1 may be used for the listed lamp types, and manufacturers' data may be obtained for other, less common, lamps.

VOLTAGE FACTOR

Variations in line voltage will affect the lumen output of lamps. Variations from the voltage rating of the lighting equipment may be the result of local operating conditions, such as voltage drop within the building power distribution system, or of voltage reduction programs which have been implemented by some utilities. Operation of lighting equipment at other than design voltage may also be intentionally implemented to reduce power consumption, increase lamp life, or increase light output. For example, dimming of lamps by means of primary voltage reduction is occasionally done to reduce operating costs for fluorescent and HID systems.

Generalized data for overvoltage operation of discharge lamps and both over and under voltage operation of incandescent lamps may be found in the IES Handbook for typical lamps, and specific information may be obtained from lamp manufacturers.

BALLAST FACTOR

When discharge lamps are tested by lamp manufacturers, they are operated on a special "reference ballast." Commercially available ballasts typically drive the lamp at lower light output, and this reduction must be considered in the design process. Ballast factor is simply the percentage of rated lumens which the lamp can be expected to produce when operated on the specific ballast which will be used.

Most core and coil type ballasts for fluorescent lamps meet performance standards established by the Certified Ballast Manufacturers Association. These standards require that the ballast drive the lamp at 95% ± 2-1/2% of the rated lumen output, and that the life of the lamp meet the lamp manufacturer's rating. Note that the rating is specific not only to the ballast, but to the lamp. Since different lamp types have different electrical characteristics it is possible for a ballast to be CBM rated for one lamp type but not for another. This is the case with CBM rated ballasts for many common lamps such as the F40 and F96, where standard lamps will be driven at about 94% rated light output, but reduced-wattage energy-saving lamps, which have different electrical characteristics and thus provide an impedance mismatch, will be driven at anywhere from 82% to 88%, depending upon the specific lamp/ballast combination. In this case the ballast is CBM rated for standard lamps but is not rated for energy-saving lamps.

Ballast factors are subject to periodic change due to improvements in lamp and ballast design, so manufacturers' test data should be periodically consulted to assure that current factors are used. Figure 6-2 is representative of typical factors at the time of this writing.

	Ballast Type		
	Standard	Energy Saving	High Efficiency
4' rapid start system			
standard lamp	.94	.94	.96
reduced wattage lamp	.88	.86	.94
8' slimline system			
standard lamp	.94	.93	.97
reduced wattage lamp	.87	.85	.96
8' high output system			
standard lamp	.98	—	1.03
reduced wattage lamp	.93	—	.98

Figure 6-2
Typical ballast factors for common lamp/ballast systems. Note that energy-saving, reduced-wattage lamps have lower ballast factors than standard lamps when operated on the same ballast.

Ballast factors for solid-state (electronic) ballasts vary between manufacturers and types; however, ranges of .75 to .95 are not uncommon. The higher factors are typically accompanied by higher power consumption. Manufacturers should be consulted for appropriate factors.

Non-CBM rated ballasts can be expected to drive lamps at lower output. While no reliable data are available, outputs of 50% to 80% of rating are common. Be aware that lamp life may also be adversely affected. Choke coils which are commonly used to ballast small fluorescent lamps, typically 20 watts and less, are not CBM rated.

There are no standards for HID ballast factors, so reliable data are not available. A factor of 0.90 is commonly used, but may not be accurate for all lamp/ballast combinations.

LAMP TILT FACTOR
(HID Lamps Only)

This factor is applied primarily to metal halide lamps when the lamp is operated in other than a vertical or horizontal burning position, as in some sports lighting and similar applications where the lamp is tilted to achieve proper luminaire aiming. Tilting the lamp off its design axis will cause a shift in the bulb cold spot with a resulting change in the amount of metal which enters the arc stream. This results in a decrease in the light output of the lamp. Typical tilt factors for metal halide lamps are shown in Figure 4-15 (Chapter 4). Mercury vapor and high-pressure sodium lamps are relatively unaffected by tilting the lamp off of the rated design axis. Specific information is available from lamp manufacturers.

EQUIPMENT OPERATING FACTOR
(HID Lamps Only)

The equipment operating factor applies only to high intensity discharge luminaires and is comprised of three individual components: ballast factor, lamp tilt factor, and the effects of power which is reflected from the reflector back to the lamp.

As previously mentioned, there are no standards for HID ballast factors but a value of 0.90 is commonly applied.

Lamp tilt factors are used only for metal halide lamps and were discussed in the preceding section. The effects of reflected power are contained in the coefficient of utilization. Thus it appears that there is no rationale for the application of a separate "Equipment Operating Factor" since it is composed of factors which are individually applied or included in coefficients of utilization.

Current practice is to simply apply a ballast factor, and use the lamp tilt factor for metal halide installations where the lamp is tilted off vertical or horizontal. The factor for reflected power is normally included in the coefficient of utilization and is not applied separately.

LUMINAIRE SURFACE DEPRECIATION FACTOR

This factor is applied in cases where luminaires are subjected to corrosive contaminants which are allowed to remain on the fixture surfaces for periods sufficient to cause a permanent deterioration of the reflecting surfaces.

These cases are the exception, not the rule, and no reliable factors exist. The loss should be estimated and included in design calculations when corrosive contaminants are known to exist and the conditions of maintenance are known to be inadequate to prevent a permanent loss of reflectance.

One test has been performed on a 10-year-old troffer used in a clean office environment. The fixture was not washed during the test period, and was found to have experienced a 3% reduction in coefficient of utilization. Based on this example it might be concluded that deterioration of luminaire surfaces is not a major concern in clean office environments, but considerably more study of this factor is needed.

LUMINAIRE THERMAL/APPLICATION FACTOR
(Fluorescent Systems Only)

The seventh, and unofficial, nonrecoverable light loss factor is still undergoing considerable discussion within the lighting community. Several excellent papers have been presented which outline the components of this factor and present the results of a number of tests which may be used to arrive at factors which may be included in design calculations. As of this writing the factor has not yet been named, nor are sufficient data available to provide absolute factors which can be used by designers. The existence of the factor is accepted, and by careful application of available data, the effects can frequently be estimated.

The Luminaire Thermal/Application Factor is composed of two different, yet interactive components. The first is the result of the use of lamps and/or ballasts which differ from the components used when the luminaire was photometered. Most photometry is performed using standard lamps and ballasts. When energy-saving lamps and ballasts, solid-state ballasts, or hybrid lamp/ballast systems are used, the resulting change in power dissipated within the fixture causes changes in the ambient temperature environment in which the lamp operates. Most enclosed fixtures maintain internal temperatures in excess of 77°F, thus the lumen output of the lamp will be reduced. The use of lower energy-consuming components reduces this temperature and brings the ambient temperature closer to the temperature for which the lamp was designed, thus increasing the light output of the lamp. Figure 6-3, excerpted from "Comparative Performance of 4-Foot Fluorescent Systems," Q.D. Dobras and A.L. Hart, General Electric Co., and presented at the 1985 IES Annual Conference, shows the results of studies conducted to determine luminaire thermal factors for a three-lamp troffer using a variety of lamps and ballasts. Note that even small changes in luminaire design can produce large changes in thermal factors. Much more testing is required to develop additional factors, and the inclusion of these data in photometric reports is also needed.

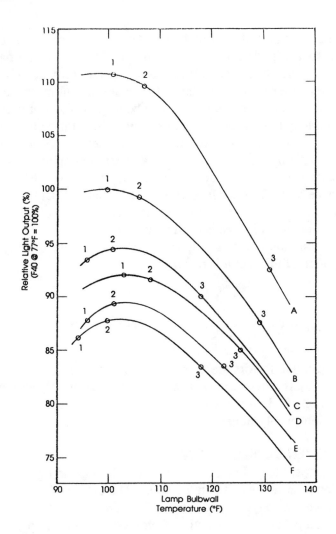

KEY
A. Maxi-Miser II Ballast, High Lumen F40 Lamp
B. Standard Ballast, Standard F40 Lamp
C. Maxi-Miser II Ballast, 34-Watt, Energy-Saving Lamp
D. Efficient Magnetic Ballast, T-8 Lamp
E. Standard Ballast, 34-Watt, Energy-Saving Lamp
F. Energy-Saving Ballast, 34-Watt, Energy-Saving Lamp

o 1 Rating point at 77 °F
o 2 Two-Lamp Strip
o 3 Four-Lamp
   Enclosed Troffer

Figure 6-3
Effects of temperature on light output of several lamp/ballast systems. Relative light output of each system is shown for operation at ANSI test conditions, in a typical two-lamp strip fixture, and in a four-lamp enclosed troffer. *(Replotted from GE Lighting test data)*

The second part of the Luminaire Thermal/Applications Factor is specific to the physical conditions under which the luminaire is installed. As previously mentioned, luminaires are surrounded by 77°F still air when they are photometered. In actual installations the temperature of the room under the fixture may be maintained at or near 77°F but the temperature of the plenum or cavity above the ceiling may be considerably higher. The movement of air through the plenum will also affect the operating characteristics of the luminaire. This factor was examined in a paper presented at the 1983 IES Annual Conference by Dr. Ian Lewin, entitled "Performance Characteristics of Fluorescent Lamp and Ballast Combinations." The conclusion, based on a small number of tests, is that reductions of up to 10% in performance are common due to the applications factor. Considerably more investigation of this factor is required.

RECOVERABLE FACTORS

Recoverable factors are those which occur over time and can be recovered by routine maintenance. These factors are 1) lamp lumen depreciation, 2) luminaire dirt depreciation, 3) room surface dirt depreciation, and 4) lamp burnouts factor.

LAMP LUMEN DEPRECIATION

All lamps, except low-pressure sodium, decrease in light output as a function of accumulated burning hours. The causes of this reduction vary between lamp types, as discussed in Chapters 2, 3, and 4.

The lamp lumen depreciation factor which will be applied to a lighting design depends not only upon the specific lamp type, but also upon the time interval at which lamps will be replaced. Group relamping of fluorescent lamps at 50% rated life, about 3 years in most offices, may reduce the required number of luminaires by about 5%, with commensurate reductions in first cost and power cost, when compared to replacement of lamps only upon failure. Systems using HID lamps are typically relamped when lamps fail, but careful analysis is required to determine the economic optimum relamping interval, and thus the lamp lumen depreciation factor.

If a system is to be group relamped after a specified number of burning hours, the lamp lumen depreciation factor for that number of accumulated burning hours is applied. For example, assume that an installation of F40CW lamps operates 3,500 hours per year and is to be group relamped after 3 years. The accumulated burning hours at the end of the 3-year period are 10,500, so a factor of 0.84 should be applied, as shown in Figure 6-4.

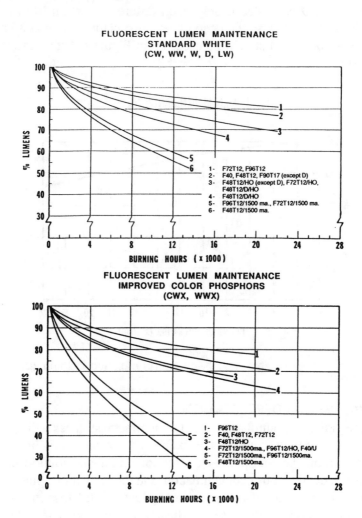

Figure 6-4
Depreciation curves for typical fluorescent lamps. *(Courtesy GE Lighting)*

When systems are relamped only upon lamp failure, the lumen maintenance factor at 70% lamp life is used. This is done since lamps typically start to fail at 40%-45% rated life in most systems and the lamps which replace them are producing near rated output. This results in a minimum average light output in large installations of the rated light output at 70% of rated life. For example, assume that burned-out lamps in an installation using 400-watt high-pressure sodium luminaires will be replaced only upon failure. The rated life of the lamp is 24,000 hours, so the lamp lumen depreciation factor at 16,800 hours (24,000 x .70) should be used. From Figure 6-5, this factor is about 0.83.

LUMEN MAINTENANCE

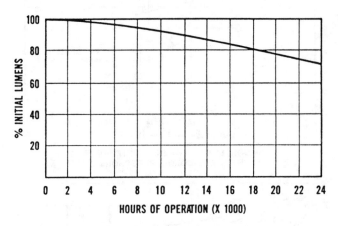

Figure 6-5
Depreciation curve for a typical 400-watt, high pressure sodium lamp. *(Courtesy GE Lighting)*

LUMINAIRE DIRT DEPRECIATION

The gradual accumulation of dirt on luminaire surfaces will reduce the quantity of light that exits the fixture. The extent of this light loss depends upon the luminaire construction, light distribution characteristics, degree of dirt contamination, and the type of dirt present.

The construction of the luminaire influences this loss in several ways. Luminaires with lenses, diffusers, or clear glass or plastic-bottom enclosures provide more surfaces for dirt buildup than open-bottomed fixtures, and typically have greater light loss. Fixtures with solid-top enclosures tend to trap dirt, while ventilated fixtures provide a means for airborne dirt to escape.

Similar fixture types tend to exhibit similar light loss characteristics due to dirt accumulation, and this makes it possible to classify luminaires into six different categories, based on construction. These categories are then referenced to six corresponding sets of dirt depreciation curves on page 9-8 of the IES Lighting Handbook, Reference Volume, which are reproduced here as Figure 6-6. To determine the applicable Category, divide the luminaire into two enclosures, top and bottom, by drawing an imaginary horizontal line through the fixture, as shown in Figure 6-7. Note that the line passes through the center of the lamp. Then locate the characteristics of each of the enclosures in Figure 6-8, reproduced from the IES Handbook. Some luminaires will fit into more than one category. If this occurs, use the lowest numbered Category.

Note that Figure 6-6 shows 5 different curves for each luminaire Category. These curves represent dirt conditions ranging from very clean to very dirty. Examples of these conditions are:

Very Clean	Class "A" office buildings, industrial clean rooms, etc.
Clean	Older office buildings, or ones in which the air conditioning system is not well maintained; industrial operations in which little or no dirt is generated, such as assembly of clean parts; inspection areas.
Medium	Offices adjacent to dirty areas such as older machine shops or foundries where dirt from the industrial operation may enter the space in limited quantities; light machining operations.
Dirty	Heavy industrial areas with a high degree of internally generated dirt and little or no dirt removal, such as foundries, heat treating, etc.
Very Dirty	Similar to dirty except no exhausting of dirty air, and an influx of outside dirt.

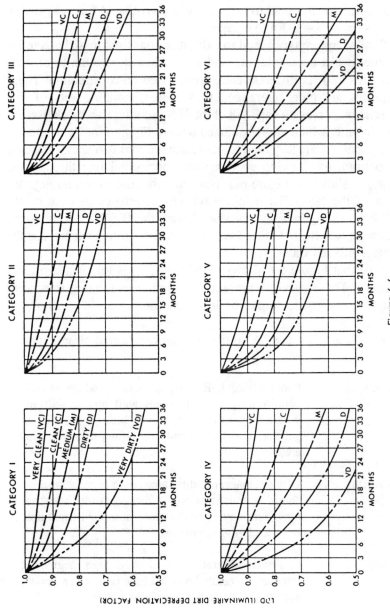

Figure 6-6
Luminaire dirt depreciation (LDD) curves for various luminaire types and dirt conditions (*Courtesy IES Lighting Handbook. IESNA*)

Maintenance Category	Top Enclosure	Bottom Enclosure
I	1. None.	1. None
II	1. None 2. Transparent with 15 per cent or more uplight through apertures. 3. Translucent with 15 per cent or more uplight through apertures. 4. Opaque with 15 per cent or more uplight through apertures.	1. None 2. Louvers or baffles
III	1. Transparent with less than 15 per cent upward light through apertures. 2. Translucent with less than 15 per cent upward light through apertures. 3. Opaque with less than 15 per cent uplight through apertures.	1. None 2. Louvers or baffles
IV	1. Transparent unapertured. 2. Translucent unapertured. 3. Opaque unapertured.	1. None 2. Louvers
V	1. Transparent unapertured. 2. Translucent unapertured. 3. Opaque unapertured.	1. Transparent unapertured 2. Translucent unapertured
VI	1. None 2. Transparent unapertured. 3. Translucent unapertured. 4. Opaque unapertured.	1. Transparent unapertured 2. Translucent unapertured 3. Opaque unapertured

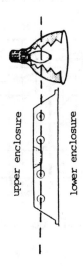

upper enclosure

lower enclosure

Figure 6-7
Division of luminaires into upper and lower enclosures for determination of maintenance category.

Figure 6-8
Characteristics of luminaire maintenance categories. (*Courtesy Illuminating Engineering Society of North America, IES Lighting Handbook.*)

To determine the luminaire dirt depreciation factor use the follow-ing steps:

1. Determine the appropriate luminaire category from Fig. 6-8.
2. Determine the dirt condition.
3. Determine the time interval, in months, at which luminaires will be washed.
4. Locate the set of curves for the appropriate luminaire category.
5. Locate the time interval between washing on the horizontal (X) axis, and project a line upward to intersect the appropriate dirt condition curve.
6. Project a horizontal line to the left to intersect the vertical (Y) axis and read the expected dirt depreciation factor.

For example, assume that a Category I luminaire is to be installed in a clean environment and washed at 12-month intervals. The expected dirt depreciation factor is 0.93 as illustrated in Figure 6-9. For luminaires which do not fit the classifications of Categories I thru VI, manufacturers' literature should be consulted. These include fix-tures with gasketed optical chambers and filters to prevent the entrance of dirt, and some luminaires with smooth glass reflectors.

Figure 6-9
Determination of a luminaire dirt depreciation factor for the example used in the text. *(Courtesy IES Lighting Handbook, IESNA)*

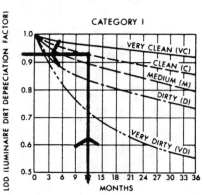

Luminaire Dirt Depreciation Factors may also be calculated from:

$$LDD = e^{-A(t^B)}$$

where t is the time, in decimal years, between fixture washings, and values for A and B are obtained from Figure 6-10.

Figure 6-10
The luminaire dirt depreciation factors shown in Figure 6-6 cover a 3-year period. If maintenance intervals exceed 3 years or the depreciation factors are to be programmed into a computer, the curve fitting equation in the text is used. This table lists the variables A and B used in the equation. (Courtesy IES Lighting Handbook, IESNA)

Luminaire Maintenance Category	B	A				
		Very Clean	Clean	Medium	Dirty	Very Dirty
I	.69	.038	.071	.111	.162	.301
II	.62	.033	.068	.102	.147	.188
III	.70	.079	.106	.143	.184	.236
IV	.72	.070	.131	.216	.314	.452
V	.53	.078	.128	.190	.249	.321
VI	.88	.076	.145	.218	.284	.396

ROOM SURFACE DIRT DEPRECIATION

Coefficients of utilization are normally selected based on reflectance values of clean or freshly painted room surfaces. As dirt accumulates on room surfaces the reflectance of light to the workplane from these surfaces decreases. This loss of light may be recovered by washing or repainting the room surfaces, and should be included in light loss calculations as a recoverable factor.

The determination of the room surface dirt depreciation factor is a two-part process, and is best explained with an example.

Assume that an office with a room cavity ratio of 2 is to be lighted with fluorescent troffers. The environment is very clean, and the room will be repainted every 18 months.

First, the "percent expected dirt depreciation" is determined from Figure 6-11 in the same manner that the luminaire dirt depreciation factor was determined. The "months" on the X axis is the expected time interval between washing or repainting of room surfaces. The expected dirt depreciation for the example is 10%, as shown on the graph.

Then, using Figure 6-12, determine the room surface dirt depreciation factor in the following manner:

(A) Locate the column heading which describes the luminaire distribution type. Note that luminaires are classified by the

manner in which they distribute light above and below
horizontal, as shown in Figure 6-13.

A troffer is a direct distribution luminaire.

(B) Locate the "10%" expected dirt depreciation column.

(C) Locate the "RCR 2" row.

(D) Read the RSDD factor directly from the table. The factor for
the example is 0.98.

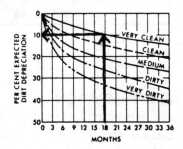

Figure 6-11
Dirt buildup on room surfaces absorbs light and affects the performance of a
lighting system. These curves are used to access the Table in Figure 6-12 to
estimate the loss of light due to dirt on walls and other room surfaces. See text for
full explanation. *(Courtesy IES Lighting Handbook, IESNA)*

Room Surface Dirt Depreciation Factors

Per Cent Expected Dirt Depreciation	Direct				Semi-Direct				Direct-Indirect				Semi-Indirect				Indirect			
Room Cavity Ratio	10	20	30	40	10	20	30	40	10	20	30	40	10	20	30	40	10	20	30	40
1	.98			.92	.97	.92	.89	.84	.94	.87	.80	.76	.94	.87	.80	.73	.90	.80	.70	.60
2	.98			.92	.96	.92	.88	.83	.94	.87	.80	.75	.94	.87	.79	.72	.90	.80	.69	.59
3	.98			.90	.96	.91	.87	.82	.94	.86	.79	.74	.94	.86	.78	.71	.90	.79	.68	.58
4	.97	.95	.92	.90	.95	.90	.85	.80	.94	.86	.79	.73	.94	.86	.78	.70	.89	.78	.67	.56
5	.97	.94	.91	.89	.94	.90	.84	.79	.93	.86	.78	.72	.93	.86	.77	.69	.89	.78	.66	.55
6	.97	.94	.91	.88	.94	.89	.83	.78	.93	.85	.78	.71	.93	.85	.76	.68	.89	.77	.66	.54
7	.97	.94	.90	.87	.93	.88	.82	.77	.93	.84	.77	.70	.93	.84	.76	.68	.89	.76	.65	.53
8	.96	.93	.89	.86	.93	.87	.81	.75	.93	.84	.76	.69	.93	.84	.76	.68	.88	.76	.64	.52
9	.96	.92	.88	.85	.93	.87	.80	.74	.93	.84	.76	.68	.93	.84	.75	.67	.88	.75	.63	.51
10	.96	.92	.87	.83	.93	.86	.79	.72	.93	.84	.75	.67	.92	.83	.75	.67	.88	.75	.62	.50

Figure 6-12
Room Surface Dirt Depreciation Factors. See text for explanation. *(Courtesy IES
Lighting Handbook, IESNA)*

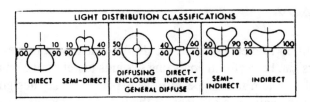

Figure 6-13
Classification of luminaires into categories based on their light distribution characteristics.

LAMP BURNOUTS FACTOR

On rare occasions the difficulty of replacing burned-out lamps will be extreme, and economics will dictate group relamping after some number of lamps have failed, without spot replacement of early failures. This is most common on large sports lighting or similar installations where spot replacement of individual lamps is not economically justifiable, but the loss of light from early failures is not acceptable. If, after careful analysis, it is determined that it is economically feasible to overlight initially in order to accept a fixed number of unreplaced early failures, this factor should be included in the design calculations. The factor is (1–% unreplaced failures). For example, if a 10% burnout is acceptable, the factor is 1–.10, or 0.90.

TOTAL LIGHT LOSS FACTOR

The total light loss factor is the product of the applicable individual factors.

If the individual factors for an installation are:

Ballast Factor = 0.94
Lamp lumen depreciation factor = 0.84
Luminaire dirt depreciation factor = 0.90
Room surface dirt depreciation factor = 0.98

then the total light loss factor is

(BF) (LLD) (LDD) (RSDD) = (0.94) (0.84) (0.90) (0.98) = 0.70

SUMMARY

In addition to their utility in the design calculation process, light loss factors can be useful in analyzing systems to determine what went wrong if the system fails to perform as expected. Many designers are unaware of the existence of many of the individual factors, especially thermal and applications factors. Clients may not be aware of the necessity of maintaining a lighting system in accordance with the specifications of the illuminating engineer, or may dismiss the recommended maintenance schedule as being costly and unnecessary. Any of these situations can result in substandard performance of a lighting system.

The recognition of light-loss factors, and their inclusion in the design calculation process, is vital if the lighting system is to meet the design specification. In the past this important fact has frequently been ignored, and the tendency to over design has, to a great extent, mitigated the need to apply the factors. The shortcomings of poor design practice were overpowered by brute force; if 10 fixtures were adequate, install 12 or 14 to be sure.

The complexities of our profession have increased dramatically in recent years due to the introduction of new equipment with operating characteristics which differ from traditional equipment, and the trend towards lower design illuminance. The application of realistic light loss factors by the illuminating engineer can be a powerful tool in achieving the goal of providing sufficient illumination at a reasonable cost.

Chapter 7

AVERAGE ILLUMINANCE CALCULATIONS – ZONAL CAVITY METHOD

In Chapter 1 the footcandle was defined as lumens of light per square foot of surface area, or $E = \Phi/A$, and was illustrated with the example of 6000 lumens evenly distributed over an area of 100 square feet. The average illuminance, E_{ave}, was found from:

$$E_{ave} = \frac{\Phi}{A} = \frac{6000 \text{ lumens}}{100 \text{ Sq. Ft.}} = 60 \text{ footcandles}$$

This example, however, assumes that the lumens have reached the workplane, and does not tell us the illuminance which can be expected if the 6000 lumens have been installed in a luminaire.

Chapter 5 introduced the concept of the coefficient of utilization as the percentage of lamp lumens which actually reach the workplane in a given room. If the fixture has a coefficient of utilization of 0.50 in the 100-square-foot room, the illuminance will be:

$$E = \frac{(\Phi)(CU)}{AREA} = \frac{(6000 \text{ lumens})(.50)}{100 \text{ Sq. Ft.}} = 30 \text{ fc}$$

This is a much more realistic estimate of the average illuminance in the room since it considers only the lumens which actually reach the workplane, yet it is still overly optimistic in most cases since it assumes that all lighting system components are in new condition and lamps are producing full-rated light output. Since this is seldom the case, it is necessary to include the light loss factors which were

171

discussed in the preceding chapter. In most cases we are concerned with the maintained illuminance (the minimum illuminance produced by the lighting system) at a time 5, or even 10 years, in the future, so all of the light-loss factors must be included. In some cases the initial illuminance will also be of interest, and this calculation will use only the initial light-loss factors: Voltage to Luminaires, Ballast Factor, and any other applicable non-recoverable factor..

When the light-loss factors are included, the equation becomes:

$$E = \frac{(\Phi)\ (CU)\ (LLF)}{AREA}$$

and if a light-loss factor of 0.68 is applied to the previous example, the average illuminance drops to:

$$E = \frac{(6000\ lumens)\ (.50)\ (.68)}{100\ Sq.\ Ft.} = 20\ fc$$

This calculation provides an accurate means of predicting the average maintained illuminance produced by a lighting system if the total number of installed lamp lumens is known.

In the design process, however, we are normally concerned with determining the number of fixtures required to produce a desired illuminance, so some rearranging of the equation is required, and it takes the form:

$$\#\ fixtures = \frac{(FC)\ (AREA)}{(rated\ lamp\ lumens\ per\ luminaire)\ (CU)\ (LLF)}$$

Note that the numerator, (FC) (AREA), is the total number of lumens required on the workplane to provide the desired illuminance, and the denominator, (lumens per luminaire) (CU) (LLF), is the lumens per fixture which will reach the workplane on a maintained basis. When the total number of required lumens is divided by the maintained lumens per luminaire which reach the workplane, the result is the total number of fixtures required.

For example, assume that the number of fixtures required to light a 50,000-square-foot industrial building is to be calculated for a maintained design illuminance of 50 footcandles. Lamps will be 400-watt, high-pressure sodium, rated at 50,000 lumens. Let us also assume that the following information has been obtained using methods and information presented in preceding chapters:

Coefficient of utilization = .76   Light-loss factor = .65

The required number of luminaires is found from:

$$\# \text{Fixt} = \frac{(FC)\,(AREA)}{(Lms/fixt)\,(CU)\,(LLF)} = \frac{(50)\,(50000)}{(50000)\,(.76)\,(.65)} = 101.2$$

The balance of this chapter will blend the information presented in preceding chapters and develop a method of easily determining the number of luminaires required to light a space. It will also introduce methods of including additional variables in the design calculation process.

THE ZONAL CAVITY METHOD

Life would be very simple for the lighting designer if all rooms were rectangular and had flat ceilings into which the luminaires were recessed. Unfortunately this is not always the case, since fixtures are frequently suspended from ceilings, and irregular room shapes and non-horizontal ceilings are relatively common.

The Zonal Cavity Method, more accurately entitled "The Zonal Cavity Method of Determining Intercavity Reflectances," provides a means of including the effects of these nonstandard conditions on coefficients of utilization. When luminaries are recessed into the ceiling, as illustrated in Chapter 5, the Room Cavity Height is the distance from the ceiling to the workplane. When luminaires are suspended from the ceiling, the room cavity height is based on the distance from the horizontal centerline of the lamps in the fixture, even though the actual ceiling is some distance above the luminaire (see Figure 7-1). This creates a cavity between the fixture and the ceiling; a cavity which may absorb light and reduce the *effective* reflectance of the ceiling.

If coefficients of utilization are based on the reflectance of the actual ceiling surface they will generally be optimistically high, since the cavity above the fixture typically absorbs light which otherwise might be assumed to be reflected to the workplane. Likewise, work-planes which are located some distance above the floor create a cavity which may also absorb light.

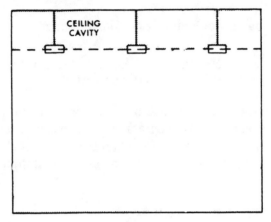

Figure 7-1
When luminaires are suspended below the ceiling, the ceiling cavity height is the distance from the ceiling to the centerline of the fixtures. The ceiling cavity height is 0 (zero) when luminaires are recessed or surface mounted. See text for explanation.

There are exceptions, however, when the cavity will reflect more light than the actual ceiling or floor surface, such as ceilings or floors which have very low reflectances, combined with walls of very high reflectance. These cases are rare, but they do occasionally exist and should be recognized since they actually result in coefficients of utilization which are higher than might be expected based on the actual ceiling reflectance. Likewise, workplanes which are located some distance above the floor create a cavity between the workplane and the floor which may absorb light, and reduce the effective reflect-ance of the floor surface.

The primary purpose of the Zonal Cavity Method is simply to provide a means of determining the *effective* reflectances of ceiling and floor cavities to allow the selection of the proper coefficient of utilization based on actual room characteristics. It is a very powerful design tool.

In the Zonal Cavity Method the room is divided into three vertical zones, called cavities. These are the ceiling cavity, the room cavity, and the floor cavity, as shown in Figure 7-2. The ceiling and floor cavity ratios are calculated, in the same manner as the room cavity ratio, and used to determine the effective reflectances of their respective surfaces. This procedure will be explained later in this chapter. The room cavity ratio is used to determine the coefficient of utilization based on the effective cavity reflectances.

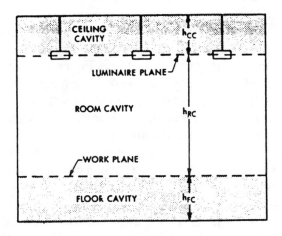

Figure 7-2
Rooms are divided into ceiling, room, and floor cavities, as shown, for Zonal Cavity calculations. *(Courtesy IES Lighting Handbook, IESNA)*

Remember that the Zonal Cavity Method is concerned only with determining the effective reflectances of the ceiling and floor cavities. The ceiling cavity reflectance represents the effective reflectance of the plane which separates the ceiling cavity from the room cavity, and is the "ceiling reflectance" used to access the coefficient of utilization table. Once the effective ceiling cavity reflectance has been determined, the actual ceiling reflectance plays no further part in the determination of the coefficient of utilization.

Coefficient of utilization tables are developed based on an assumed floor reflectance of 20%. Once the effective floor cavity reflectance has been determined, it will be used to modify the coefficient of utilization obtained from the table, as will be explained later in this chapter.

DETERMINATION OF
EFFECTIVE CAVITY REFLECTANCES

Effective cavity reflectances are based on radiative flux transfer; the transfer of light from a surface to another surface or to a plane. The theory behind flux transfer is complex, and is beyond the scope of this text. The application, however, is straightforward, and is of use to the lighting practitioner.

Effective cavity reflectances for simple cavities consisting of horizontal ceilings and uniform suspension lengths for fixtures, as shown in Figure 7-1, may be obtained from Figure 7-3, which is reproduced from the IES Lighting Handbook. Note the similarity between the Effective Cavity Reflectance Table and a coefficient of utilization table. The method of accessing the two tables is the same, but notice that some of the headings are slightly different.

A   "Percent Base Reflectance" refers to the actual reflectance of the surface behind the cavity. When determining the effective reflectance of the ceiling cavity, the actual reflectance of the ceiling surface is considered to be the "Base Reflectance." The actual reflectance of the floor will be used as the "Base Reflectance" when determining the effective floor cavity reflectance.

B   "Percent Wall Reflectance" is the actual reflectance of the *wall of the cavity* . For example, if the wall above the level of the fixtures is 80% reflectance, the wall below the level of the workplane is 20% reflectance, and the wall between the workplane and the fixtures is 50% reflectance, the effective ceiling cavity reflectance will be based on an 80% wall reflectance, the effective floor cavity reflectance will be based on a wall reflectance of 20%, and the coefficient of utilization will be based on a wall reflectance of 50%. See Figure 7-4.

C   "Cavity Ratio" is determined using the same equation used to calculate the Room Cavity Ratio, except that the cavity height is the height of the cavity being calculated.

Example

Consider a room with dimensions and reflectances as shown in Figure 7-5.

Per Cent Base* Reflectance	90										80										70										60										50									
Per Cent Wall Reflectance	90	80	70	60	50	40	30	20	10	0	90	80	70	60	50	40	30	20	10	0	90	80	70	60	50	40	30	20	10	0	90	80	70	60	50	40	30	20	10	0	90	80	70	60	50	40	30	20	10	0
Cavity Ratio																																																		
0.2	89	88	88	87	87	86	85	85	84	82	79	78	78	77	77	76	76	75	74	72	70	69	68	68	67	67	66	66	65	64	60	59	59	58	57	56	56	55	55	53	50	49	49	48	48	48	47	46	46	44
0.4	88	87	86	85	84	83	81	80	79	76	79	77	76	75	73	72	71	71	68	66	70	68	67	65	64	63	62	61	59	58	60	58	58	57	55	53	52	51	50	50	49	48	47	46	45	44	43	42	41	38
0.6	87	86	84	83	82	80	78	77	76	73	78	76	75	73	71	70	68	66	65	63	69	67	65	63	61	59	58	56	58	58	60	58	57	55	53	51	50	48	60	43	49	48	45	45	44	42	41	40	39	36
0.8	87	85	82	80	79	77	76	73	71	69	78	75	73	71	69	67	65	63	61	57	68	66	64	62	60	58	56	55	53	50	59	57	56	54	51	48	46	45	48	43	48	47	44	43	42	40	39	38	36	34
1.0	86	83	80	77	75	72	69	66	64	62	78	74	72	69	65	62	60	57	55	55	66	65	62	60	58	55	52	50	50	47	59	57	55	53	51	47	45	44	45	41	50	46	44	43	41	40	37	36	36	34
1.2	85	82	78	75	72	69	66	63	60	57	76	73	70	67	64	61	58	55	53	51	67	64	61	60	57	54	50	48	46	41	59	56	54	51	49	47	44	42	44	38	47	45	43	41	39	35	34	34	32	29
1.4	85	80	77	73	69	65	62	59	57	52	76	72	68	65	62	59	55	53	50	48	63	63	60	58	55	51	47	45	44	41	57	53	52	49	47	44	41	39	41	36	47	45	40	39	38	34	32	33	30	27
1.6	84	79	75	71	67	63	59	56	53	50	75	71	67	63	60	56	53	50	47	44	67	62	59	56	54	49	46	43	41	38	57	55	51	48	45	43	39	37	39	33	46	43	41	38	36	33	33	30	28	25
1.8	83	78	73	69	64	60	56	53	50	48	75	70	66	62	58	54	50	47	43	41	66	61	58	54	51	46	42	40	38	35	56	52	50	47	44	40	37	35	38	31	46	43	40	38	34	32	30	28	26	24
2.0	83	77	72	67	62	58	53	50	47	43	74	69	65	60	56	52	48	45	41	38	66	60	56	52	49	45	40	38	35	29	56	53	49	46	42	37	35	33	37	31	46	40	37	34	33	28	28	26	24	22
2.2	82	76	70	65	59	54	50	46	44	40	74	68	63	58	54	49	45	42	38	35	66	60	55	51	48	43	38	36	34	32	58	53	49	45	42	39	34	31	29	28	46	42	38	36	33	30	27	24	22	22
2.4	82	75	69	64	58	53	48	45	41	37	73	67	61	56	52	47	43	40	36	33	65	60	54	50	46	41	37	35	32	30	58	53	48	44	41	36	32	29	27	26	46	41	37	35	33	30	27	25	23	21
2.6	81	74	67	62	56	51	46	42	38	35	73	66	60	55	50	45	41	38	34	31	65	59	53	49	45	40	35	33	30	28	57	52	48	43	39	35	31	28	26	24	46	41	36	34	33	29	26	22	21	20
2.8	81	73	66	61	54	49	44	40	36	34	73	66	59	54	49	43	39	36	32	29	65	59	53	48	44	37	33	30	28	25	58	52	47	43	38	34	29	26	24	22	45	40	36	33	32	28	25	22	20	19
3.0	80	72	64	58	52	47	42	38	34	30	72	65	58	52	47	42	37	34	30	27	64	58	52	47	42	37	29	26	24	20	57	51	46	42	37	29	28	25	23	20	45	40	36	32	28	26	24	21	19	17
3.2	79	71	63	56	50	45	40	36	32	28	72	65	57	51	45	40	35	33	28	25	64	58	51	46	36	31	28	25	24	23	57	51	45	40	36	32	27	23	22	18	50	36	31	27	23	20	20	17	16	12
3.4	79	70	62	54	48	43	38	34	30	27	71	64	56	49	44	39	34	32	27	24	63	57	50	45	35	29	26	24	22	22	57	51	44	36	35	31	26	22	19	17	50	35	31	26	22	19	17	15	14	11
3.6	78	69	61	53	47	42	36	32	28	25	71	64	55	48	43	38	33	30	26	23	63	56	49	43	38	32	24	23	21	20	57	44	39	38	33	29	25	22	18	16	50	33	30	26	22	18	16	15	13	10
3.8	78	69	60	52	45	40	35	31	27	23	70	62	53	47	41	36	31	28	24	22	62	56	49	43	37	32	23	20	16	15	58	43	38	37	33	29	23	18	16	13	48	33	29	25	21	18	15	14	12	09
4.0	77	69	58	51	44	39	33	29	25	22	70	61	53	46	35	34	30	26	22	20	61	55	48	42	36	31	26	23	20	17	57	49	42	37	31	28	23	17	15	12	47	37	36	33	28	24	17	14	12	08
4.2	77	62	57	50	43	37	32	28	24	21	69	60	52	45	39	34	29	25	21	18	62	55	47	41	35	30	25	22	19	16	56	49	42	37	32	22	19	17	14	11	50	43	37	32	27	24	17	14	12	07
4.4	76	61	56	49	42	36	31	27	23	20	69	60	51	44	38	33	28	24	20	17	61	54	46	40	34	29	24	21	18	15	56	49	42	36	31	22	18	16	13	10	50	43	37	31	26	23	18	14	11	06
4.6	76	60	55	47	40	35	30	26	22	19	68	50	50	43	37	32	27	23	19	16	62	53	45	39	33	29	23	21	17	13	56	49	41	36	31	21	17	15	11	09	49	40	36	31	26	22	18	13	10	05
4.8	75	59	54	46	39	34	28	24	20	18	68	58	49	42	36	31	26	22	18	14	53	53	45	38	32	27	23	20	16	13	58	48	40	34	29	20	16	12	11	07	48	39	31	26	22	18	15	12	09	04
5.0	75	59	53	45	38	33	28	24	16	16	67	58	48	41	35	30	25	21	18	14	61	52	44	38	31	27	22	19	16	12	56	48	38	34	28	19	15	11	11	07	47	39	31	26	22	18	15	12	09	03
6.0	73	61	49	41	34	29	24	20	16	11	66	55	44	38	31	27	22	19	15	10	60	51	41	35	28	26	19	16	13	09	55	45	37	31	25	17	14	11	08	05	42	34	29	23	19	15	13	10	08	06
7.0	70	58	45	38	30	27	21	18	14	08	64	53	42	35	26	24	17	16	12	07	58	48	38	32	26	17	14	13	11	08	54	43	35	30	24	15	12	09	07	04	41	32	26	21	18	14	11	09	07	05
8.0	68	55	42	35	27	23	18	15	12	06	62	50	38	32	25	17	14	11	11	05	57	46	35	29	23	13	10	10	09	04	53	42	33	28	22	14	11	08	06	03	40	30	25	21	16	11	10	07	07	03
9.0	66	52	38	31	25	21	16	14	11	05	61	49	36	30	23	14	11	10	04	04	55	45	33	27	21	11	09	08	06	03	52	40	31	26	20	12	10	08	05	02	39	30	23	18	15	11	09	07	06	02
10.0	65	51	36	29	22	19	15	11	09	04	59	46	33	27	21	13	11	08	08	03	55	43	31	25	19	10	08	06	03	02	51	39	29	24	18	12	09	07	02	07	47	37	27	22	17	14	10	08	06	02

* Ceiling, floor or floor of cavity.

Figure 7-3

Table of effective cavity reflectances for Zonal Cavity calculations. *(Courtesy IES Lighting Handbook. IESNA)*

Per Cent Base* Reflectance	0										10										20										30										40									
Per Cent Wall Reflectance	90	80	70	60	50	40	30	20	10	0	90	80	70	60	50	40	30	20	10	0	90	80	70	60	50	40	30	20	10	0	90	80	70	60	50	40	30	20	10	0	90	80	70	60	50	40	30	20	10	0
Cavity Ratio																																																		
0.2	02	02	01	01	01	01	00	00	00	0	09	09	09	09	09	09	09	09	09	09	21	20	20	20	20	19	19	19	18	17	31	30	30	29	29	29	28	28	28	27	40	40	39	39	39	39	38	38	37	36
0.4	04	03	03	02	02	01	01	01	00	0	10	09	09	09	09	09	08	08	08	07	22	21	21	20	20	19	18	18	17	16	31	31	30	29	28	28	27	26	25	25	41	40	39	39	38	37	37	36	35	34
0.6	05	05	04	03	03	02	02	01	01	0	10	10	10	09	09	09	08	08	08	06	23	22	21	20	19	19	18	17	16	15	32	31	30	29	28	27	26	25	24	23	41	40	39	38	37	36	35	34	33	31
0.8	07	05	04	04	03	03	02	02	01	0	11	10	10	10	10	09	09	08	08	06	24	23	22	20	19	18	17	16	15	13	32	31	30	29	27	26	25	23	22	21	41	40	38	37	36	35	33	32	31	29
1.0	08	07	05	04	04	03	02	02	01	0	11	11	10	10	10	09	09	08	07	05	25	23	22	20	19	18	17	15	13	12	33	32	30	29	27	26	24	23	21	20	42	40	38	37	36	35	33	31	29	27
1.2	10	08	07	06	05	04	03	03	02	01	11	11	11	11	10	10	09	08	07	05	25	24	22	21	20	18	17	15	13	10	34	32	30	29	27	25	23	22	20	18	42	40	39	37	35	34	32	30	28	25
1.4	11	09	08	07	06	05	04	03	02	01	12	12	11	11	11	10	09	08	07	04	26	24	22	21	20	17	16	15	12	10	34	32	31	29	27	25	23	21	19	17	42	40	39	37	35	33	31	29	26	23
1.6	12	10	09	07	06	05	04	03	02	01	13	12	12	11	11	10	09	08	07	04	26	24	23	21	20	17	15	14	11	09	34	33	31	28	26	24	22	20	17	16	43	41	38	36	35	32	30	27	25	22
1.8	13	11	09	08	07	06	05	03	02	01	13	13	12	12	11	10	09	08	06	04	27	25	23	21	20	17	15	13	11	08	35	33	29	27	25	23	21	19	16	14	43	41	39	36	35	32	29	26	24	21
2.0	14	12	10	09	07	06	05	04	03	01	14	13	13	12	12	11	09	08	06	04	28	25	23	20	19	18	15	13	11	09	35	33	29	27	24	22	20	18	16	14	42	40	38	36	34	31	28	25	23	20
2.2	15	13	11	09	08	06	05	04	03	02	15	14	13	13	13	11	11	09	08	05	28	26	23	20	18	16	14	12	11	09	36	32	29	26	24	22	19	17	14	13	44	39	36	33	32	29	27	24	21	18
2.4	16	13	11	09	08	07	06	04	03	02	16	14	14	13	12	11	10	09	08	05	29	26	23	21	18	16	14	12	11	09	36	32	29	26	24	21	19	16	13	11	43	39	37	34	32	29	26	23	20	17
2.6	17	14	12	10	09	07	06	05	03	02	17	15	14	13	12	11	10	09	07	05	29	26	23	20	18	15	13	12	10	08	36	32	29	25	23	20	18	15	13	11	44	40	35	33	30	27	25	22	19	16
2.8	17	15	13	11	09	08	06	05	04	02	17	16	15	14	13	12	11	09	07	04	30	27	23	20	17	15	13	11	10	07	35	33	29	24	22	19	17	14	12	10	44	39	35	32	29	26	24	21	18	15
3.0	18	16	13	11	10	08	07	05	04	02	18	16	15	14	13	11	11	09	07	04	30	27	23	20	17	15	12	10	09	07	37	33	29	24	22	18	16	13	11	10	43	39	35	31	29	25	23	20	18	13
3.2	19	16	14	12	10	09	07	06	04	03	18	16	15	14	13	11	10	09	07	03	31	27	23	20	17	14	12	10	09	06	37	33	29	25	22	18	16	14	12	10	43	30	21	17	15	12	10	71	—	15
3.4	20	17	14	12	11	09	07	06	05	03	19	17	16	14	13	11	10	09	07	03	31	27	22	19	16	14	12	10	08	05	37	33	29	25	22	17	14	11	10	08	43	30	26	23	20	17	14	12	10	07
3.6	20	17	15	13	11	10	08	06	05	03	20	17	16	15	14	11	10	08	07	02	38	33	29	24	21	18	15	13	10	06	38	33	29	24	21	16	15	12	11	09	44	30	26	22	19	16	13	11	09	06
3.8	21	18	15	13	12	10	08	07	05	03	20	18	16	15	14	11	10	08	07	02	38	33	29	24	21	18	14	12	09	05	38	33	28	24	21	16	14	12	10	08	44	30	26	21	18	15	13	10	08	05
4.0	22	18	16	14	12	10	09	07	05	03	21	18	17	15	14	11	09	08	07	02	33	29	25	21	18	14	12	10	09	07	38	33	28	24	21	16	14	12	10	07	44	29	25	21	18	15	12	10	08	05
4.2	22	19	16	14	13	11	09	07	06	04	22	19	17	16	14	11	09	08	07	02	33	28	24	20	17	14	12	10	09	07	39	33	28	24	20	17	14	12	10	09	44	33	29	24	21	17	15	13	11	08
4.4	23	19	17	15	13	11	09	08	06	04	22	20	17	16	14	11	09	08	07	02	34	28	24	20	17	13	11	10	08	06	38	32	28	24	20	16	14	11	09	08	44	33	28	24	20	16	14	11	10	08
4.6	23	20	17	15	13	11	10	08	06	04	23	20	17	16	13	11	09	08	06	01	34	28	24	20	17	13	11	10	08	06	38	32	27	23	19	15	13	11	09	07	44	32	28	23	19	15	13	11	09	06
4.8	24	20	17	15	14	12	10	08	07	04	24	20	17	17	13	11	09	08	06	01	35	28	24	19	16	13	10	09	07	05	38	32	27	23	19	15	12	10	07	06	44	32	27	22	19	15	12	10	08	05
5.0	25	21	18	16	14	12	10	09	07	04	25	21	17	17	13	11	09	08	06	01	35	28	24	19	16	13	10	08	06	04	40	32	27	23	19	14	12	10	07	05	43	31	27	22	19	15	13	10	07	05
6.0	27	23	20	18	15	12	11	09	08	05	26	21	18	18	13	11	08	08	05	01	36	30	24	20	16	13	10	08	05	02	44	37	30	25	18	09	08	06	04	05	44	37	30	25	18	09	08	06	04	05
7.0	28	24	19	15	12	10	09	07	06	04	27	21	18	18	13	10	08	07	05	01	36	30	24	20	15	12	09	07	04	02	44	36	29	22	17	08	07	04	03	04	44	24	19	16	12	08	07	04	04	02
8.0	30	25	20	15	12	08	07	04	02	—	28	23	18	17	13	10	07	07	05	01	37	29	23	19	15	12	09	06	04	01	44	35	28	22	16	07	06	04	03	03	44	35	28	21	15	11	09	06	04	02
9.0	31	25	20	15	12	09	06	05	02	—	29	24	19	17	14	10	07	06	04	—	37	29	22	18	14	11	07	05	03	01	44	35	26	20	15	07	05	03	03	02	43	34	25	20	13	08	07	05	02	02
10.0	—	25	20	15	12	08	07	05	02	—	31	25	20	17	14	11	07	05	02	—	37	28	21	17	13	10	07	05	03	01	43	34	25	20	13	08	06	03	01	02	43	34	25	20	12	08	07	05	02	02

* Ceiling, floor or cavity.

Figure 7-3 (Continued)
Table of effective cavity reflectances for Zonal Cavity calculations. (Courtesy IES Lighting Handbook, IESNA)

Note that h_{cc} is the height of the ceiling cavity, as measured from the centerline of the lamps to the ceiling, h_{rc} is the room cavity height, the distance from the centerline of the lamps to the workplane, and h_{fc} is the height of the floor cavity, the distance from the workplane to the floor.

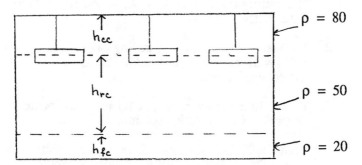

Figure 7-4

For the example, an 80% "wall" reflectance is used to determine the effective ceiling cavity reflectance, and a 20% "wall" reflectance is used in determining the effective floor cavity reflectance. The effective ceiling reflectance and a 50% wall reflectance are used to access the CU table.

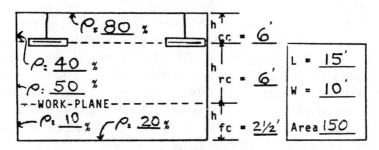

Figure 7-5

Room characteristics used for the example calculation.

To determine the effective ceiling cavity reflectance (ρ_{cc}):

1. Calculate the Ceiling Cavity Ratio (CCR)

$$CCR = \frac{5h_{cc}\ (Room\ Length + Room\ Width)}{(Room\ Length)\ (Room\ Width)}$$

$$CCR = \frac{5\,(6)\,(15 + 10)}{(15)\,(10)} = 5$$

2. Read the Effective Ceiling Cavity Reflectance from the table, Figure 7-6, in the following manner:
   A The ceiling reflectance is 80%, so locate the 80% "Base Reflectance" column.

   B The wall reflectance of the ceiling cavity is 40%, so locate 40% wall reflectance column under the "80% Base Reflectance" column.

   C The ceiling cavity ratio (CCR) is 5, so locate the "5" under the "Cavity Ratio" column.

   D Read the Effective Cavity Reflectance directly from the chart. The effective ceiling cavity reflectance for the example is 30%.

The coefficient of utilization can now be determined for a specific luminaire, using an effective ceiling cavity reflectance (ρ_{cc}) of 30%, rather than the actual ceiling reflectance of 80%.

Using the CU Table shown in Figure 7-7, the CU is 0.41, based on an effective ceiling cavity reflectance of 30%, a wall reflectance of 50%, and a floor reflectance of 20%. Note that if the actual ceiling reflectance of 80% had been used the CU would be read as 0.44, an overstatement of nearly 7%. If this CU were used in the design calculation, the illuminance would be about 7% lower than anticipated.

Effective floor cavity reflectances are determined by the same method, except the actual floor reflectance is used as the "Base Reflectance" and the reflectance of the walls between the floor and workplane is used as the Wall Reflectance. If the effective floor cavity reflectance varies greatly from 20% it will be necessary to adjust the CU to account for the difference in effective floor reflectance.

In the example room, the floor cavity height is 2-1/2 feet and the floor cavity ratio (FCR) is:

$$FCR = \frac{5(h_{fc})\,(\text{Room Length} + \text{Room Width})}{(\text{Room Length})\,(\text{Room Width})} = \frac{5(2.5)\,(15 + 10)}{(15)\,(10)} = 2.1$$

Per Cent Effective Ceiling or Floor Cavity Reflectances for Various Reflectance Combinations

Per Cent Base* Reflectance	90										80										70									
Per Cent Wall Reflectance \ Cavity Ratio	90	80	70	60	50	40	30	20	10	0	90	80	70	60	50	40	30	20	10	0	90	80	70	60	50	40	30	20	10	0
0.2	89	88	88	87	86	85	85	84	84	82	79	78	78	77	77	76	75	74	74	72	70	69	69	68	68	67	67	66	66	64
0.4	88	87	86	85	84	83	81	80	79	76	79	77	76	75	74	73	72	71	70	68	69	68	67	66	65	64	63	62	61	58
0.6	87	86	86	84	82	81	79	77	76	73	78	76	75	73	71	70	68	66	65	63	69	67	65	64	63	61	59	58	57	54
0.8	87	85	82	80	77	75	73	71	69	67	78	75	73	71	69	67	65	63	61	57	68	66	64	63	60	58	56	55	53	50
1.0	86	83	80	77	75	72	69	66	64	62	77	74	72	69	67	65	62	60	57	55	68	65	62	60	58	55	53	52	50	47
1.2	85	82	78	75	72	69	68	65	62	57	76	73	70	67	64	61	58	55	53	51	67	64	61	59	57	54	50	48	46	44
1.4	85	80	77	73	69	65	62	59	56	52	76	72	68	65	62	59	55	53	50	48	67	63	60	58	55	51	47	45	44	41
1.6	84	79	75	71	67	63	59	56	53	50	75	71	67	63	61	57	53	50	47	44	66	62	59	56	53	47	45	43	41	38
1.8	83	78	73	69	64	60	56	53	50	48	75	70	66	62	58	54	50	47	44	41	66	61	58	54	51	46	42	40	38	35
2.0	83	77	72	67	62	58	53	50	47	43	74	69	64	60	56	52	48	45	41	38	66	60	56	52	49	45	40	38	36	33
2.2	82	76	70	65	59	54	50	47	44	40	74	68	63	58	54	49	45	42	38	35	66	60	55	51	48	43	38	36	34	32
2.4	82	75	69	64	58	53	48	45	42	37	73	67	61	56	52	47	43	40	36	33	65	59	54	50	46	41	37	35	32	30
2.6	81	74	67	62	56	51	46	42	38	35	73	66	60	55	50	45	41	38	35	31	65	59	53	49	45	40	35	33	30	28
2.8	81	73	66	60	54	49	44	40	36	34	73	65	59	53	48	43	39	36	32	29	65	59	53	48	43	38	33	30	28	26
3.0	80	72	64	58	52	47	42	38	34	30	72	65	58	52	47	42	37	34	30	27	64	58	52	47	42	37	32	29	27	24
3.2	79	71	63	56	50	45	40	36	32	28	72	65	57	51	45	40	35	33	28	25	64	58	51	46	40	36	31	28	25	23
3.4	79	70	62	54	48	43	38	34	30	27	71	64	56	49	44	39	34	32	27	24	64	57	50	45	39	35	29	27	24	22
3.6	78	69	61	53	47	42	36	32	28	25	71	63	54	48	43	38	33	30	25	23	63	56	49	44	38	33	28	24	21	20
3.8	78	69	60	51	45	40	35	31	27	23	70	62	53	47	42	37	32	28	24	22	63	56	49	43	37	32	27	24	21	17
4.0	77	69	58	51	44	39	33	29	25	22	70	61	53	46	41	36	31	28	22	20	63	55	48	42	36	31	26	23	20	17
4.2	77	62	57	50	43	37	32	28	24	21	69	60	52	45	39	34	29	25	21	18	62	55	47	41	35	30	25	22	19	16
4.4	76	61	56	49	42	36	31	27	23	20	69	60	51	44	38	33	28	24	20	17	62	54	46	40	34	29	24	21	18	15
4.6	76	60	55	47	40	35	30	26	22	19	69	59	50	43	37	32	28	24	21	15	62	53	45	39	33	28	24	21	17	14
4.8	75	59	54	46	39	34	28	25	21	18	68	58	49	42	36	31	26	22	18	14	62	53	45	38	32	27	23	20	16	13
5.0	75	59	53	45	38	33	28	24	20	16	68	58	48	41	35	30	25	30	?	05	61	52	44	36	31	26	22	19	16	12
6.0	73	61	49	41	34	29	24	20	16	11	66	55	44	38	31	25	22	19	14	08	60	51	41	35	28	24	19	16	13	09
7.0	70	58	45	38	30	27	21	18	14	08	64	53	42	35	29	23	19	17	13	04	58	48	38	32	26	22	17	14	11	06
8.0	68	55	42	35	27	23	18	15	12	06	62	50	38	32	25	21	17	15	10	04	57	46	35	29	23	19	15	13	10	05
9.0	66	52	38	31	25	21	16	14	11	05	61	49	36	30	23	19	15	13	10	04	56	45	33	27	21	18	14	12	09	04
10.0	65	51	36	29	22	19	15	11	09	04	59	46	33	27	21	18	14	11	08	03	55	43	31	25	19	16	12	10	08	03

*Ceiling, floor or floor of cavity

Figure 7-6
Method of accessing the Effective Cavity Reflectance Table. See text for explanation. *(Table courtesy IES Lighting Handbook, IESNA)*

ρcc →	80			70			50			30			10			0
ρw →	50	30	10	50	30	10	50	30	10	50	30	10	50	30	10	0
RCR ↓	Coefficients of Utilization for 20 Per Cent Effective Floor Cavity Reflectance (ρfc = 20)															
0	.75	.75	.75	.73	.73	.73	.70	.70	.70	.67	.67	.67	.64	.64	.64	.63
1	.67	.65	.63	.66	.64	.62	.63	.62	.60	.61	.60	.58	.59	.58	.57	.55
2	.60	.57	.54	.59	.56	.53	.57	.54	.52	.55	.53	.51	.53	.51	.50	.49
3	.54	.50	.47	.53	.49	.46	.52	.48	.45	.50	.47	.45	.48	.46	.44	.43
4	.49	.44	.40	.48	.44	.40	.47	.43	.40	.45	.42	.39	.44	.41	.39	.37
5	.44	.39	.35	.43	.38	.35	.42	.38	.34	.41	.37	.34	.40	.36	.34	.33
6	.40	.34	.31	.39	.34	.31	.38	.34	.30	.37	.33	.30	.36	.32	.30	.29
7	.36	.30	.27	.35	.30	.27	.34	.30	.27	.33	.29	.26	.32	.29	.26	.25
8	.32	.27	.23	.32	.27	.23	.31	.26	.23	.30	.26	.23	.29	.26	.23	.22
9	.29	.24	.20	.28	.23	.20	.28	.23	.20	.27	.23	.20	.26	.23	.20	.19
10	.26	.21	.18	.26	.21	.18	.25	.21	.18	.24	.20	.18	.24	.20	.18	.16

ρ_{cc} = 30

ρ_w = 50   (Note that this valuebased on the reflectance of the wall of the room cavity and is not necessarily the same as

ρ_w used for the effective ceiling cavity reflectance.)

$$RCR = \frac{5 \text{ (cavity height) (length} \div \text{ width)}}{\text{area}} = \frac{5 \text{ (6) (15 + 10)}}{150} = 5$$

And the CU is .41

Figure 7-7
CU Table for example used in text. (*Table Courtesy IES Lighting Handbook, IESNA*)

From Figure 7-6, the effective floor cavity reflectance is about 10%. A correction factor should be applied since the effective floor cavity reflectance is not 20%, the assumed floor cavity reflectance used to calculate the CU Table. Tables of correction factors are published by the Illuminating Engineering Society in the IES Lighting Handbook and other technical publications. The table is reproduced here as Figure 7-8. Again, note the similarity to a CU table. The effective ceiling cavity reflectance, 30% for the example, is used to access the table. The wall reflectance is 50% (note that the wall of the room cavity is used), and the floor cavity ratio is 2.1. The correction factor, based on this information, is 0.981.

To determine the adjusted CU, simply multiply the previously determined CU, 0.41, by the correction factor. The adjusted CU for the example is:

$$(0.41) (0.981) = 0.40$$

This value, 0.40, is the CU to be used in the design calculation.

Multiplying Factors for Other than 20 Per Cent Effective Floor Cavity Reflectance

For 10 Per Cent Effective Floor Cavity Reflectance (20 Per Cent = 1.00)

% Effective Ceiling Cavity Reflectance, ρ_{cc}	80				70				50			30			10		
% Wall Reflectance, ρ_w	70	50	30	10	70	50	30	10	50	30	10	50	30	10	50	30	10
Room Cavity Ratio																	
1	.923	.929	.935	.940	.933	.939	.943	.948	.956	.960	.963	.973	.976	.979	.989	.991	.993
2	.931	.942	.950	.958	.940	.949	.957	.963	.962	.968	.974	.976	.980	.985	.988	.991	.995
3	.939	.951	.961	.969	.945	.957	.966	.973	.967	.975	.981	.978	.983	.988	.988	.992	.996
4	.944	.958	.969	.978	.950	.963	.973	.980	.972	.980	.986	.980	.986	.991	.987	.992	.996
5	.949	.964	.976	.983	.954	.968	.978	.985	.975	.983	.989	.981	.988	.993	.987	.992	.997
6	.953	.969	.980	.986	.958	.972	.982	.989	.977	.985	.992	.983	.989	.995	.987	.993	.997
7	.957	.973	.983	.991	.961	.975	.985	.991	.979	.987	.994	.984	.990	.996	.987	.993	.998
8	.960	.976	.986	.993	.963	.977	.987	.993	.981	.988	.995	.985	.991	.997	.987	.994	.998
9	.963	.978	.987	.994	.965	.979	.989	.994	.983	.990	.996	.986	.992	.998	.988	.994	.999
10	.965	.980	.989	.995	.967	.981	.990	.995	.984	.991	.997	.986	.993	.998	.988	.994	.999

Figure 7-8

Correction factors for use when effective floor cavity reflectance is other than 20%. (Courtesy IES Lighting Handbook. IESNA)

NON-HORIZONTAL CEILINGS

The effective cavity reflectance of non-horizontal ceilings such as hemispherical domes, barrel vaulted, and sawtooth roofs, may be calculated directly from:

$$\rho_{cc} = \frac{A_o \rho_s}{A_s - A_s \rho_s + A_o \rho_s}$$

where

A_o = Area of ceiling opening
A_s = Area of ceiling surface
ρ_s = Reflectance of ceiling surface

For ceilings which contain segments of varying reflectance, such as sawtooth roofs, a rearranged form of the equation will be easier to use:

$$\rho_{cc} = \frac{A_o}{A_{s_1}\left(\dfrac{1-\rho_s}{\rho_{s_1}}\right) + A_{s_2}\left(\dfrac{1-\rho_s}{\rho_{s_2}}\right) + \ldots A_{s_n}\left(\dfrac{1-\rho_n}{\rho_{s_n}}\right) + A_o}$$

where the subscripts $1, 2, \ldots n$ denote individual segments.

First consider the hemispherical dome, which is handled quite nicely by substituting into the first equation:

$$\rho_{cc} = \frac{A_o \rho_s}{A_s - A_s \rho_s + A_o \rho_s}$$

For a hemisphere:
$$A_o = \pi r^2$$
$$A_s = 2\pi r^2$$
so, by substitution

$$\rho_{cc} = \frac{\pi r^2 \rho_s}{2\pi r^2 - 2\pi r^2 \rho_s + \pi r^2 \rho_s} = \frac{\rho_s}{2 - \rho_s}$$

For a hemisphere with a surface reflectance of 0.90, the effective cavity reflectance is:

$$\rho_{cc} = \frac{\rho_s}{2 - \rho_s} = \frac{.90}{2 - .90} = .82$$

The barrel vaulted ceiling, shown in Figure 7-9, is slightly more complex since the wall segments which form the ends must be included. These walls typically have a different reflectance than the barrel-shaped part of the ceiling, so the second form of the equation will be used.

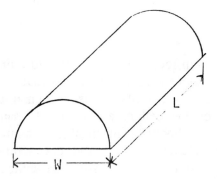

Figure 7-9
Dimensions used for direct calculation of effective reflectance of barrel vaulted ceilings.

Assume that the ceiling on Figure 7-9 has a length of 40', a width of 20', and reflectances are 80% for the ceiling surface and 50% for the wall surfaces which are the ends of the ceiling.

The areas of the surfaces are:
Let s_1 = barrel part of ceiling surface
s_2 = ends of ceiling surface

$$A_{S_1} = \frac{\pi W}{2} (L) = \frac{\pi(20 \text{ ft.})}{2} (40 \text{ ft.}) = 400\pi = 1257 \text{ ft.}^2$$

$$A_{S_2} = \frac{\pi W^2}{4} = \frac{\pi(20 \text{ ft.}^2)}{4} = 100\pi = 314 \text{ ft.}^2$$

and, by substitution into the equation:

$$\rho_{cc} = \frac{A_0}{A_{S_1}\left(\frac{1-\rho_{s1}}{\rho_{s1}}\right) + A_{S_2}\left(\frac{1-\rho_{s2}}{\rho_{s2}}\right) + A_0}$$

$$\rho_{cc} = \frac{800 \text{ ft.}^2}{1257 \text{ ft.}^2\left(\frac{1-.80}{.80}\right) + 314 \text{ ft.}^2\left(\frac{1-.50}{.50}\right) + 800 \text{ ft.}^2}$$

$\rho_{cc} = .56$

The sawtooth roof, found in many older industrial buildings, is handled in the same manner as the barrel vault, except it has one additional surface. These roofs consist of numerous identical sections which resemble the teeth on a saw, thus the name "sawtooth." Since each segment, or "tooth," is identical to each other segment, it is only necessary to calculate the effective cavity reflectance for a single segment, as it will be representative of the entire ceiling.

Consider the sawtooth roof shown in Figure 7-10. S_2 is normally glass, and is assumed to have a reflectance of 10%. For the example, S_1 and S_3 will be assumed to have reflectances of 50%.

First, calculate the areas of the opening and surfaces.

$A_0 = (L)(W) = (100')(20') = 2000 \text{ ft.}^2$

$A_{S1} = (L)(W_1) = (100')(20') = 2000 \text{ ft.}^2$

A_{S3} consists of two identical triangles. The area of one triangle is calculated from:

$$A = \sqrt{S(S-W)(S-W_1)(S-W_2)} \qquad \text{(Heron's Formula)}$$

where

$S = 1/2(W + W_1 + W_2) = 1/2(20' + 20' + 11') = 25.5'$

so,

$$A = \sqrt{25.5'(25.5'-20')(25.5'-20')(25.5'-11')} = 106\text{ft.}^2$$

and,

$A_{S3} = 106 \text{ ft.}^2 + 106 \text{ ft.}^2 = 212 \text{ ft.}$

Then, by substitution:

$$\rho_{cc} = \frac{A_O}{A_{s1}\left(\dfrac{1-\rho_{s1}}{\rho_{s1}}\right) + A_{s2}\left(\dfrac{1-\rho_{s2}}{\rho_{s2}}\right) + A_{s3}\left(\dfrac{1-\rho_{s3}}{\rho_{s3}}\right) + A_O}$$

$$\rho_{cc} = \frac{2000 \text{ ft.}^2}{2000 \text{ ft.}^2\left(\dfrac{1-.50}{.50}\right) + 1100 \text{ ft.}^2\left(\dfrac{1-.10}{.10}\right) + 212 \text{ ft.}^2\left(\dfrac{1-.50}{.50}\right) + 2000 \text{ ft.}^2}$$

$$\rho_{cc} = .14$$

As can be seen, the cavities formed by sawtooth roof sections may have effective reflectances far below the reflectance of the ceiling surface.

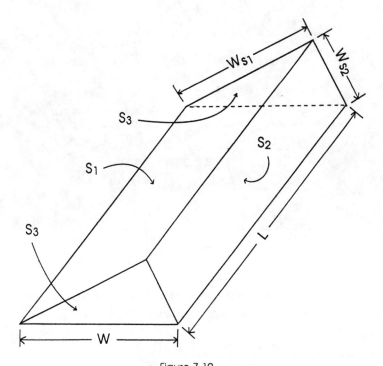

Figure 7-10
Dimensions used for direct calculation of effective reflectance of sawtoothed roof sections.

COFFERED CEILINGS AND
CEILINGS WITH BEAMS

Coffered ceilings and ceilings with beams are handled by simply determining the weighted average reflectance of the effective ceiling plane.

Figure 7-11 shows a typical coffered ceiling. Assume that each cavity has a length of 3', a width of 3', and is 1' deep. Coffers are separated by 1-foot-wide band so each coffered "section" measures 4' x 4'. All ceiling surfaces have reflectances of 70%.

Figure 7-11
Typical coffered ceiling.

First, the effective reflectance of the plane at the bottom of the coffer is determined in the same manner as a ceiling cavity.

$$\text{Cavity ratio} = \frac{5(h)\,(L+W)}{LW} = \frac{5(1')\,(3'+3')}{(3')\,(3')} = 3.33$$

From Figure 7-3, the effective reflectance is 50%.

The area of the typical ceiling section is 16ft.^2 and the area of the coffer is 9 ft.^2, so the area surrounding each coffer is 7 square feet, and the weighted average ceiling reflectance is:

$$\rho_{cc} = \frac{A_{s_1}\rho_{s_1} + A_{s_2}\rho_{s_2}}{A_{s_1} + A_{s_2}} = \frac{(9\text{ft.}^2)\,(.50) + (7\text{ft.}^2)\,(.70)}{16\text{ft.}^2} = .59$$

KEEPING TRACK OF THE DATA

At this point it is obvious that a great deal of information must be obtained for use in the design calculation process. The "Average Illuminance Calculation Sheet," Figure 7-12, has been developed for this purpose.

The following instructions, keyed to the form, will facilitate its use:

1. Enter the identification of the space, e.g., "Office Tower, Room 102."
2. Enter the Design Illuminance–see Chapter 10.
3. Enter the name of the luminaire manufacturer.
4. Enter the catalog number of the luminaire to be specified. (Note that entries 4 and 5 will become part of the job specifications. It is common practice in the construction industry to substitute fixtures. Since substitute fixtures may not provide the same performance as the specified fixture, it is sound practice to document the specified equipment in both calculations and specifications.
5. Enter the specified lamp type, e.g., F96T12CW.
6. Enter the rated *initial* lumen output of the lamp. Obtain from lamp manufacturer's data.
7. Enter the number of lamps per luminaire.
8. Enter the total rated lamp lumens per luminaire. Entry 6 x Entry 7.
9a. Enter the reflectance of the actual ceiling surface. For new construction or remodels when paint manufacturers and colors are specified by the architect, interior designer, or others, this information may be obtained from the paint manufacturer. For existing surfaces, the reflectances may be estimated, or may be obtained through measurement or comparison with color samples of known reflectance. See the section of Field Photometry, Chapter 10, for measurement procedures.

AVERAGE ILLUMINATION CALCULATION SHEET
ZONAL CAVITY METHOD

Project Name _____

Design Illuminance _____

Luminaire: Manufacturer: _____   Catalog No. _____

Lamps: Type & Color: _____   Lumen Rating: _____

Lamps Per Luminaire: _____   Total Lumens Per Luminaire _____

Fill in Sketch:

Determine Cavity Ratios from formula:

Ratio: = $\dfrac{5 \text{ (Cav. Ht.) (Length + Width)}}{\text{(Length) (Width)}}$

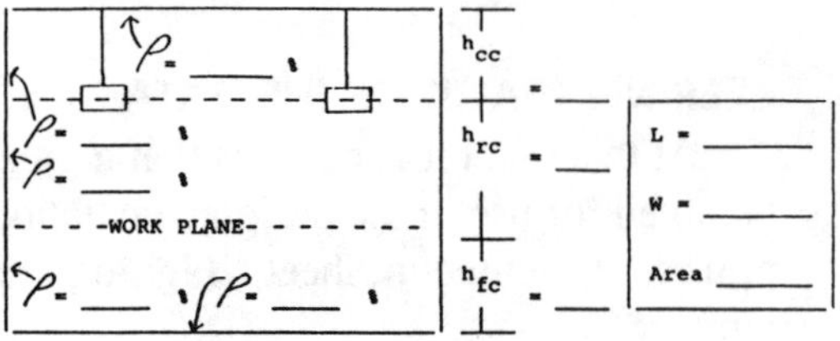

Ceiling Cavity Ratio _____   Room Cavity Ratio _____   Floor Cavity Ratio _____

Determine Effective Cavity Reflectances ρ_{cc} = ____ ρ_{fc} = ____

Obtain Coefficient of Utilization (CU) from Manufacturers data _____

Adjust CU if ρ_{fc} is other than 20%   CU _____ X Adj Factor _____ = _____ Adj. CU

Determine Light Loss Factors:

Non-Recoverable   Recoverable

Luminaire Ambient Temp. _____   Lamp Lumen Depreciation _____

Voltage to Luminaire _____   Luminaire Dirt Depreciation _____

Ballast Factor _____   Room Surface Dirt Depreciation _____

Luminaire Surface Depreciation _____   Lamp Burnouts Factor _____

          Total Light Loss Factor (LLF) Product of Above Factors _____

Calculate Number of Fixtures:

No. Luminaires = $\dfrac{\text{(Footcandles) x (Area)}}{\text{Lumens Per Luminaire X (Coefficient of Utilization X (Light Loss Factor)}}$

= $\dfrac{(\quad\quad) \text{ X } (\quad\quad)}{(\quad\quad) \text{ X } (\quad\quad) \text{ X } (\quad\quad)}$ = _____

Final Footcandle Level = $\dfrac{\text{(Number of Luminaires) X (Lumens Per Luminaire) X (CU) (LLF)}}{\text{AREA}}$

= $\dfrac{(\quad\quad) \text{ X } (\quad\quad) \text{ X } (\quad\quad) \text{ X } (\quad\quad)}{(\quad\quad)}$ = _____

Figure 7-12
Zonal Cavity calculation sheet.

AVERAGE ILLUMINATION CALCULATION SHEET
ZONAL CAVITY METHOD

Project Name ___(1)_____

Design Illuminance (2)_____

Luminaire: Manufacturer: (3)_____   Catalog No. ___(4)_____

Lamps: Type & Color: ___(5)_____   Lumen Rating: (6)_____

Lamps Per Luminaire: ___(7)_____   Total Lumens Per Luminaire __(8)___

Fill in Sketch:

Determine Cavity Ratios from formula:

Ratio: = 5 (Cav. Ht.) (Length + Width)
        ──────────────────────────────
            (Length) (Width)

P = (9a) P = (9b) P = (9c) P = (9d) P = (9e)

h_{cc} = (9f) h_{rc} = (9g) h_{fc} = (9h)

- - - -WORK PLANE- - - - - -

L = (9i)   W = (9j)   Area (9k)

Ceiling Cavity Ratio (10)___   Room Cavity Ratio (11)___   Floor Cavity Ratio (12)___

Determine Effective Cavity Reflectances P_{cc} = (13) P_{fc} = (14)

Obtain Coefficient of Utilization (CU) from Manufacturers data (15)___

Adjust CU if P_{fc} is other than 20%   CU (16)___ X Adj Factor (17)___ = (18)___ Adj. CU

Determine Light Loss Factors:

Non-Recoverable	Recoverable
Luminaire Ambient Temp. (19)_____	Lamp Lumen Depreciation (23)_____
Voltage to Luminaire ___(20)_____	Luminaire Dirt Depreciation (24)_____
Ballast Factor (21)_____	Room Surface Dirt Depreciation (25)_____
Luminaire Surface Depreciation (22)_____	Lamp Burnouts Factor ___(26)_____

Total Light Loss Factor (LLF) Product of Above Factors ___(27)_____

Calculate Number of Fixtures:

No. Luminaires = (Footcandles) x (Area)
                 ───
                 Lumens Per Luminaire X (Coefficient of Utilization X (Light Loss Factor)

$$= \frac{((28)) \times ((29))}{((30)) \times ((31)) \times ((32))} = (33)$$

Final Footcandle Level = (Number of Luminaires) X (Lumens Per Luminaire) X (CU) (LLF)
                         ──
                                                   AREA

$$= \frac{((34)) \times ((35)) \times ((36)) \times ((37))}{((38))} = (39)$$

Figure 7-12
Zonal Cavity calculation sheet.

9b. Enter the reflectance of the wall surface ABOVE THE PLANE OF THE LUMINAIRES. OMIT THIS STEP FOR RECESSED OR SURFACE-MOUNTED LUMINAIRES.

9c. Enter the reflectance of the walls of the room cavity. When multiple colors which vary widely are used on walls, or walls contain large inserts such as chalk boards or white boards, use the weighted average reflectance, which may be found from:

$$\rho_{ave} = \frac{A_{S_1}\rho_{S_1} + A_{S_2}\rho_{S_2} + A_{S_3}\rho_{S_3} + \ldots A_{S_n}\rho_{S_n}}{A_{S_1} + A_{S_2} + A_{S_3} + \ldots A_{S_n}}$$

Note that inserts of small area, such as a door or small window in a long wall, or small variations in paint reflectances, will not materially effect the coefficient of utilization and need not be calculated.

9d. Enter the reflectance of the walls of the floor cavity.

9e. Enter the reflectance of the actual floor surface.

9f. Enter the height of the ceiling cavity, the distance from the ceiling to the centerline of the lamps in the luminaire. Omit this entry for recessed or surface-mounted luminaires.

9g. Enter the height of the room cavity.

9h. Enter the height of the floor cavity.

9i. Enter the length of the room.

9j. Enter the width of the room

9k. Enter the room area.

10. Calculate and enter the ceiling cavity ratio. Omit this entry for recessed or surface-mounted fixtures, or for non-horizontal ceilings for which the effective ceiling cavity reflectance is directly calculated.

11. Calculate and enter the room cavity ratio.

12. Calculate and enter the floor cavity ratio when the workplane is located above the floor. Omit when the floor is the workplane.

13. Enter the effective ceiling cavity reflectance. Obtain from the "Effective Cavity Reflectance" table for systems with luminaires suspended below horizontal ceilings. Calculate directly, using equations given previously in this chapter for non-horizontal

ceilings. For recessed or surface-mounted luminaires enter the reflectance of the actual ceiling.

14. Enter the effective floor cavity reflectance. Obtain from the "Effective Cavity Reflectance" table. Enter the actual floor reflectance for rooms where the workplane is the floor.

15. Enter the coefficient of utilization for the luminaire. Obtain from manufacturers' data. When accessing the CU table, be sure to use the effective ceiling cavity reflectance, entry 13, when locating the proper cc column in the CU table.

16. Enter the coefficient of utilization from entry 15. Omit this step when the effective floor cavity reflectance is close to 20%, and proceed to entry 19.

17. Enter the correction factor from the table for ρ_{fc} of other than 20%.

18. Enter the adjusted CU. Multiply the CU from entry 16 by the correction factor, entry 17.

Note: Entries 19 through 27 relate to light loss factors. See Chapter 6.

19. Enter the Luminaire Ambient Temperature Factor for fluorescent lamps used in abnormally low or high temperatures. Obtain from manufacturers' data for the specific lamp to be used. Omit this entry if the ambient temperature is near 77°F.

20. If the actual voltage supplied to the luminaire varies from the rated voltage of the ballast (for discharge lamps), or the lamp (for incandescent lamps), enter the appropriate light loss factor. Obtain from manufacturers' data.

21. Enter the Ballast Factor for discharge lamps. Omit for incandescent lamps. Obtain from manufacturers' literature or use data provided in Chapter 6.

22. Enter the Luminaire Surface Depreciation Factor for luminaires which are expected to experience a substantial loss of reflectivity over time. This factor is seldom applied since accurate data are not available.

23. Enter the Lamp Lumen Depreciation factor. Obtain from lamp manufacturers' data. If the lighting system is to be group relamped, use the depreciation factor for the total number of accumulated burning hours at the time the lamps are replaced. For example, if a lighting system is to be operated 4000 hours per year, and lamps replaced every 3 years, 12,000 burning hours will elapse between

relamping. Use the depreciation factor at 12,000 hours. If lamps are to be replaced only upon failure, use the depreciation factor at 70% rated lamp life. For example, if the lamp has a rated life of 24,000 hours, use the factor at 70% of 24,000 hours, or 16,800 hours.

24. Enter the Luminaire Dirt Depreciation factor. Obtain from Figure 6-7 or the IES Lighting Handbook.
25. Enter the Room Surface Dirt Depreciation Factor, from Figures 6-12 and 6-13. This factor is normally applied only when the coefficient of utilization is based on the reflectance of freshly painted or new room surfaces. When the CU is based on the reflectance of existing, deteriorated room surfaces, as is frequently done for older existing industrial buildings, the factor is generally omitted since the room surfaces have already deteriorated.
26. In those rare cases where a fixed percentage of lamps will be allowed to fail without replacement, enter the quantity (1–% failed lamps). Note that this is seldom done since it results in the installation of more fixtures, and increases both first cost and operating cost. It is occasionally applied for buildings with extremely difficult and costly relamping.
27. Enter the total light loss factor, which is the product of the individual factors.
28. Enter the design illuminance. Obtain from Entry 2.
29. Enter the area of the room. Obtain from Entry 9k.
30. Enter the total rated lamp lumens per luminaire. Obtain from Entry 8.
31. Enter the coefficient of utilization. Obtain from Entry 15, or 18 if adjustment of the CU for floor cavity reflectance was required.
32. Enter the Light Loss Factor. Obtain from Entry 27.
33. Enter the calculated number of luminaires.

Note: Entries 34 through 39 are made only after the proposed design has been completed, and the average maintained illuminance for the actual number of fixtures to be installed is to be determined.

34. Enter the actual number of luminaires to be installed.
35. Enter the lumens per luminaire. Obtain from Entry 8.
36. Enter the CU for the luminaire. Obtain from Entry 15 or Entry 18.
37. Enter the Light Loss Factor. Obtain from Entry 27.
38. Enter the area of the room. Obtain from Entry 9k.

39. Enter the calculated maintained average illuminance for the design. If the initial illuminance is to be calculated, substitute the Initial Light Loss Factor (the product of Entries 19, 20, and 21) in Entry 37.

IRREGULAR ROOMS

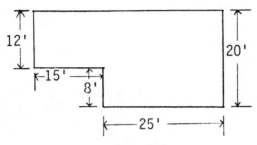

Figure 7-13
One of many irregular room shapes.

Cavity ratios (room, ceiling, and floor), are calculated from:

$$\text{Cavity Ratio} = \frac{5 \, (\text{Cavity Height}) \, (\text{Room Length} + \text{Room Width})}{(\text{Room Length}) \, (\text{Room Width})}$$

which assumes that the room is rectangular. Non-rectangular rooms, such as those in Figure 7-13, can be handled with a simple modification of the equation. The quantity (Room Length + Room Width) is actually one-half of the perimeter, and the quantity (Room Length x Room Width) is the area, so, by substitution:

$$\text{Cavity Ratio} = \frac{5 \, (\text{Cavity Height}) \, (1/2 \, \text{Perimeter})}{\text{AREA}}$$

which simplifies to:

$$\text{Cavity Ratio} = \frac{2.5 \, (\text{Cavity Height}) \, (\text{Perimeter})}{\text{AREA}}$$

For the room in Figure 7-13, assuming a room cavity height of 6.5',
the RCR can be found from:

$$RCR = \frac{(2.5)\,(6.5)\,(40 + 20 + 25 + 8 + 15 + 12)}{(15)\,(12) + (25)\,(20)} = 2.9$$

Chapter 8

CALCULATING ILLUMINANCE AT A POINT

Point calculation methods are used to predict the illuminance at a specific location. This information may be required for a variety of reasons: to assure that sufficient illuminance has been provided at a specific point in the space; to evaluate uniformity of illumination or luminance ratios; or in the illumination of vertical surfaces such as warehouse stacks where average illuminance calculation methods do not apply.

Point calculation methods may be used for both indoor and outdoor lighting. For indoor applications the illuminance at a point consists of two components: light that travels directly from the luminaire to the point, called the direct component, and light that is reflected to the point by room surfaces, called the reflected component. Outdoor lighting is typically concerned only with the direct component. Be aware, however, that cases do exist where a reflected component from an adjacent building or other structure may be significant and will require calculation.

DIRECT COMPONENT

When the illuminance is desired at only a few points, and the number of luminaires which contribute to the illuminance at that point is small, the calculations may be done manually using the Inverse Square Law, which was introduced in Chapter 1. If the illuminance at an array of points is desired or the number of luminaires is large, the calculations are best performed by computer since a large number of redundant calculations is required. For example, the illuminance at one or two points from one or two luminaires can normally be calculated manually in less time than would be required to input the data to a computer program. If an array of 25 points is lighted by 10 fixtures, 275 individual calculations might be required. The calculations would take hours by hand but can be performed in seconds by computer.

A number of simplified methods of performing and applying these calculations manually were developed during the pre-computer era, i.e., the Plan-Scale Method, Angular Coordinate-DIC Method, the IES London Aspect Factor Method, Illumination Charts and Tables, Idealized Source Charts, and Configuration Factors. Of these, only Illumination Charts and Tables are widely used for manual applications.

While the computer is certainly the best method for performing point illuminance calculations involving many points and/or luminaires, there are still many instances where manual calculations can provide the desired information in less time and at lower cost. The computer is simply a very fast calculator, text processor, and library of information. It cannot interpret the results of its calculations, nor evaluate human factors. It is simply a tool to be used by the lighting designer, who must understand the basic principles and methods of calculations to properly interpret the computer's information.

INVERSE SQUARE LAW

The inverse square law is the most widely used method of calculating the illuminance at a point. As discussed in Chapter 1, the horizontal illuminance at a point which is normal (at a right angle) to the source, as in Figure 8-1 can be found from:

$$E = \frac{I}{D^2} \quad \text{or} \quad Fc = \frac{Candlepower}{Distance^2}$$

Figure 8-1
Diagram for inverse square law

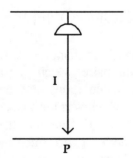

The illuminance must frequently be calculated on a plane which is not normal to the source, as shown in Figure 8-2. In these cases the use of the inverse square law would yield an erroneous illuminance since the area over which the flux is distributed is greater than one square foot, as shown in Figure 8-3, and a correction must be made.

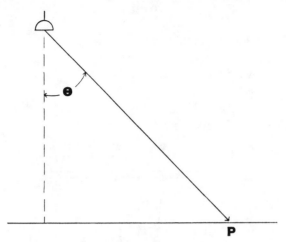

Figure 8-2.
Points at which the illuminance is to be calculated are not always in a plane which is normal to the light source.

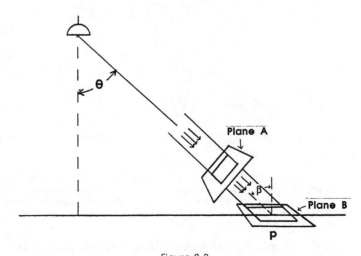

Figure 8-3
When the point at which the illuminance is to be calculated is not in a plane normal to the source the flux spreads out to cover a wider area.

Examination of Figure 8-3 reveals that the area of plane B is trigonometrically related to the area of plane A, which is normal to the source, by:

$$\text{Area of Plane B} = \frac{\text{Area of Plane A}}{\cos \beta}$$

and the illuminance is found from:

$$E = \frac{I \cos \beta}{D^2}$$

This is called the "Cosine Law."

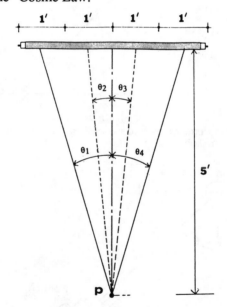

Figure 8-4

If the source has a maximum dimension that is greater than 1/5th the distance from the point at which the illuminance is to be calculated the source is arbitrarily divided into several smaller sources so that each segment is less than 1/5th of the distance.

When the illuminance at a point in a horizontal plane is to be determined, angle β is equal to angle θ, and the equation takes the form:

$$E = \frac{I \cos \theta}{D^2}$$

This form, however, requires the determination of the distance, D, from the source to the point. By the application of basic trigonometry it can be seen that

$$D = \frac{H}{\cos \theta}$$

and by substitution,

$$E = \frac{I \cos \theta \frac{}{H^2}}{(\cos \theta)^2}$$

which, when simplified, yields a form which is much easier to use in practical applications:

$$E = \frac{I \cos^3 \theta}{H^2}$$

Note that the inverse square law assumes a point source: a source so infinitesimally small that it has no dimension. This, of course, does not exist in the real world, but the equation will produce acceptable accuracy as long as the maximum luminous dimension of the source does not exceed 1/5th of the distance from the luminaire to the point at which the illuminance is to be calculated. PAR and "R" lamps, and most incandescent and HID luminaires fall into this category.

For larger sources, such as fluorescent luminaires, it may be necessary to arbitrarily divide the luminaire into sections so that the dimension of each section meets the 1/5th constraint. Using the midpoint of each section as separate luminaire, the angle θ is determined from the section to the point at which the illuminance is to be calculated, as in Figure 8-4. The candlepower of the source, at that angle, is then divided by the number of sections which were arbitrarily created, and that value assigned as the intensity of the section. The contribution of light to the point from each section is the calculated, and the individual values added to determine the total direct illuminance. The candelas are assumed to radiate from the center of each section.

Example: Calculate the illuminance at a point 5 feet below the center of an 8-foot fluorescent luminaire with intensities as given in Figure 8-5.

θ (Degrees)	I	θ (Degrees)	I
0.0	4000	20.0	3679
2.5	3989	22.5	3592
5.0	3975	25.0	3496
7.5	3951	27.5	3398
10.0	3913	30.0	3292
12.5	3863	32.5	3169
15.0	3812	35.0	3045
17.5	3745		

Figure 8-5
Table of intensities, in candelas, used for the example calculation.

Solution:

<u>Step 1.</u> Divide the lamp into 8 sections, each 1 foot long, and calculate 1 thru 4, as shown in Figure 8-6. Note that $\theta_1 = \theta_8$, $= \theta_2 = \theta_7$, $\theta_3 = \theta_6$, and $\theta_4 = \theta_5$, so we need only calculate the contributions from sections 1, 2, 3, and 4, and double these values since the illuminance from the other sections will be identical. If the point is not directly below the center of the fixture or one of the sections, it is necessary to perform the calculations for each section.

$$\theta_1 = \arctan\left(\frac{D_1}{H}\right) = \arctan\left(\frac{3.5'}{5'}\right) = 35.0°$$

$$\theta_2 = \arctan\left(\frac{D_2}{H}\right) = \arctan\left(\frac{2.5'}{5'}\right) = 26.6°$$

$$\theta_3 = \arctan\left(\frac{D_3}{H}\right) = \arctan\left(\frac{1.5'}{5'}\right) = 16.7°$$

$$\theta_4 = \arctan\left(\frac{D_4}{H}\right) = \arctan\left(\frac{0.5'}{5'}\right) = 5.7°$$

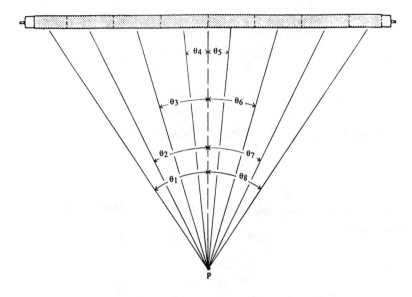

Figure 8-6
Division of the 8-foot lamp into eight segments, each 1-foot long.

<u>Step 2</u>. Determine the intensities at each angle from the photometric data for the fixture. Note that the angles will seldom coincide with the angles at which the luminaire was photometered, and interpolation will be necessary. While interpolated values will not be precise, they are sufficiently accurate for most calculations. Use data from Figure 8-5 for this example.

Section 1: $I_{35°} \doteq 3045$ cd.

Section 2: $I_{26.6°} \doteq 3343$ cd.

Section 3: $I_{16.7°} \doteq 3793$ cd.

Section 4: $I_{5.7°} \doteq 3968$ cd.

<u>Step 3</u>. Divide the intensities in Step 2 by the number of sections into which the luminaire was divided to determine the intensity (I) which will be used in the Inverse Square Law calculation.

Section 1: $I_{35°} \doteq \dfrac{3045 \text{ cd.}}{8} = 381 \text{ cd.}$

Section 2: $I_{26.7°} \doteq \dfrac{3343 \text{ cd.}}{8} = 418 \text{ cd.}$

Section 3: $I_{16.7°} \doteq \dfrac{3793 \text{ cd.}}{8} = 474 \text{ cd.}$

Section 4: $I_{5.7°} \doteq \dfrac{3968 \text{ cd.}}{8} = 496 \text{ cd.}$

Step 4. Calculate the illuminance produced by each section of the luminaire from:

$$E_{TOT} = \dfrac{I_\theta \cos^3 \theta}{H_2}$$

$$E_1 = \dfrac{(381)\,(\cos^3 35°)}{5^2} = 8.4 \text{ fc.}$$

$$E_2 = \dfrac{(418)\,(\cos^3 26.6°)}{5^2} = 12.0 \text{ f}$$

$$E_3 = \dfrac{(474)\,(\cos^3 16.7°)}{5^2} = 16.7 \text{ f}$$

$$E_4 = \dfrac{(496)\,(\cos^3 5.7°)}{5^2} = 19.5 \text{ fc.}$$

Step 5. Add the illuminances produced by each section to obtain the total illuminance. Note that, in the example, only one half of the fixture was used in the calculation since the other half is a mirror image of the first half. Both halves, however, must be added to obtain the total illuminance.

$$E_{total} = E_1 + E_2 + E_3 + \ldots \ldots E_n$$

$$E_{tot} = 8.4 + 12.0 + 16.7 + 19.5 + 19.5 + 16.7 + 12.0 + 8.4 = 113.2 \text{ fc.}$$

For comparison, if the nadir intensity, 4,000 candelas, had been used, and the calculation performed in a single step, the calculated illuminance would be:

$$E = \frac{(4000 \text{ cd.})}{5^2} = 160 \text{ fc.}$$

an overestimate of 41%.

Note that this calculation is for initial illuminance and assumes that all components are operating at rated output. Since we are typically concerned with the actual maintained illuminance, a light loss factor must be applied. For example, if the LLF is 0.68, the maintained illuminance will be 113 fc X 0.68 = 77 fc.

When calculating the illuminance at a point from an array of fixtures it is necessary to calculate for each luminaire which contributes light to the point. This can involve a substantial number of calculations if the building has a large number of fixtures, so some practical cut-off point must be determined. Figure 8-7 shows a plot of $\cos^3\theta$ vs the distance away from a point directly beneath the fixture, in mounting heights.

If $\cos^3\theta$ is considered as an effectiveness factor for horizontal illuminance, it can easily be seen that the effectiveness of the intensity drops off rapidly as the point is moved away from the fixture. A point which is located only one mounting height away has an effectiveness of only about 35%. This means that the intensity at that angle, 45°, must be almost 3 times the intensity at 0° to produce the same illuminance. At 2 mounting heights, about 63°, the effectiveness drops to about 9%. Since few luminaires produce substantial candlepower at angles above 60°, fixtures located more than 2 to 2.5 mounting heights away from the point to be evaluated are seldom included in manual calculations.

The inverse square law may also be used to calculate the illuminance on a vertical surface. This is useful for evaluating the illuminance on walls, stacks in warehouses, and similar applications. For vertical surfaces the equation takes the form:

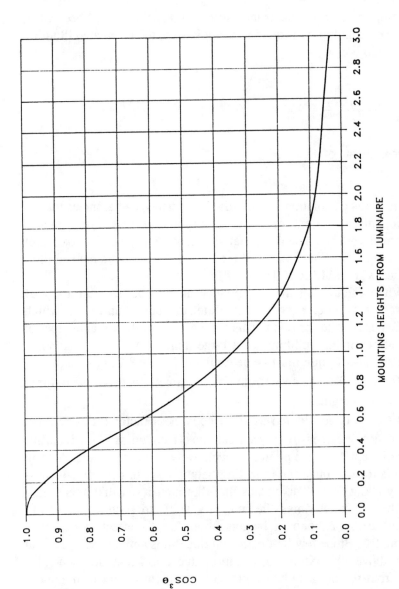

Figure 8-7

Plot of $\cos^3 \theta$ vs mounting heights away from the fixture location.

$$E_v = \frac{I_\theta \, \text{Sin} \, \theta}{D^2}$$

Example: Calculate the vertical surface illuminance at point P for the warehouse rack shown in Figure 8-8. Use the candela distribution for a typical widespread HID luminaire, shown in Figure 8-9.

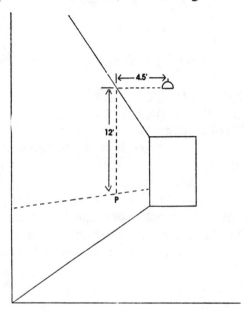

Figure 8-8
Point on warehouse stack used for example calculation of vertical illuminance.

θ	I
0	8564
5	8817
15	10964
25	12534
35	11845
45	9025

Figure 8-9
Candlepower distribution
for fixture used in example
calculation of vertical
illuminance.

The direct line distance from the luminaire to the point is calculated from the Pythagorean Theorem:

$$C^2 = A^2 + B^2 \text{ or } C = \sqrt{A^2 + B^2}$$

$$\text{Distance} = \sqrt{(4.5')^2 + (12')^2} = 12.8'$$

and

$$\theta = \arctan\left(\frac{4.5}{12}\right) = 20.5°$$

From Figure 8-9, the intensity is about 11750 cd, and the illuminance is:

$$E_v = \frac{I_\theta \sin \theta}{D^2} = \frac{(11750)(\sin 20.5°)}{(12.8')^2} = 25 \text{ fc.}$$

Note that the HID luminaire used for this example has a luminous diameter of 18", and the distance from the fixture from the point is 12.8', so the distance exceeds 5 times the maximum luminous dimension of the source and the calculation can be performed in a single step. If the 8-foot fluorescent luminaire used in the earlier example of calculating horizontal point illuminance had been used for this example, the luminaire would be divided into four "2-foot" sections so that each section would be less than 1/5 the distance to the point.

PLAN SCALE METHOD

The plan scale method is simply a method of preparing one set of inverse square law calculations to predict the illuminance at a series of points on a horizontal plane below a fixture, and transferring the results to a scale which can be used to plot the illuminance from each fixture to a given point in the building. These values are then added to determine the total direct illuminance at the point.

The major application of the plan scale method is the use of isofootcandle diagrams for outdoor lighting design, which will be discussed in detail in Chapter 14.

ANGULAR COORDINATE –
DIRECT ILLUMINANCE COMPONENT METHOD

The angular coordinate–DIC method is applicable to continuous rows of fluorescent luminaires. It requires tables of data which are available only for a few generic luminaires, and is seldom used.

Consult the IES Lighting Handbook, Reference Volume, if more detailed information is desired.

ILLUMINANCE CHARTS AND TABLES

Charts and tables are frequently used to predict the illuminance at a point for merchandise, display, and accent lighting equipment. The method is well suited to the rapid evaluation of the suitability of a luminaire for a specific lighting task, and many manufacturers provide charts and tables. Lamp manufacturers also provide tables, graphs, and charts for lamps with integral reflectors such as Par, R, and MR types. Information is presented in different formats so the data should be studied before application to assure that it is applied correctly. Typical manufacturer's data are illustrated in Figure 8-10.

Note that these data are normally for initial illuminance and appropriate light loss factors should be applied.

IDEALIZED SOURCE COMPUTATIONS

Idealized source computations are occasionally used for specific cases: linear, circular or rectangular sources with lambertian distributions. They may be applied to surfaces with non-uniform luminance, but the procedure is too complex for practical applications.

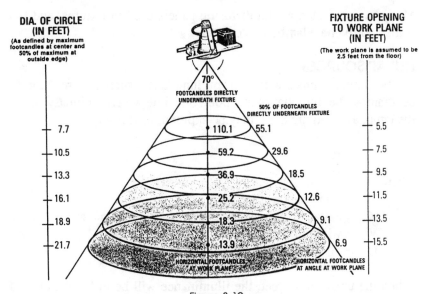

Figure 8-10
Typical pictorial representation of illuminance produced by a luminaire.
(Courtesy Cooper Lighting)

Lambertian surfaces are perfectly diffusing, and have cosine light distributions, as illustrated in Figure 8-11.

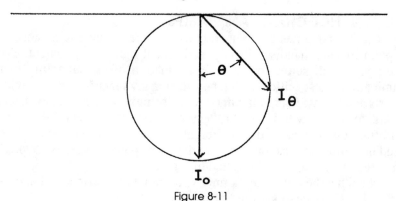

Figure 8-11
Cosine emitter. The intensity at any angle θ is equal to $I_θ$ Cos θ.

This means that the intensity at any angle is equal to the intensity at nadir times the cosine of the angle, or: $I_θ = I_{0°} \, Cos \, θ$
and the total emitted flux will be: $Φ = π \, I_{0°}$ lumens

Many surfaces such as flat diffusing panels used in fixtures, and flat walls approximate lambertian emitters.

LINEAR SOURCES

For linear sources with uniform luminance, such as bare fluorescent lamps, the illuminance at a point on a line which is directly below the lamp, as in Figure 8-12, may be approximated from:

$$E = \frac{MW}{2D}$$

where
   E   = illuminance at the point
   M   = exitance of the source in lm/sq. ft. or sq. meter
   W   = width of the source
   D   = distance from the source to the point

When the units are in feet, the illuminance will be in footcandles. If the meter is used the illuminance is in lux.

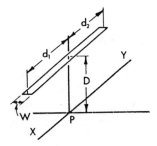

Figure 8-12
Diagram of points used in
calculations with line sources.
*(Courtesy IES Lighting Hand-
book, IESNA)*

This method provides precise answers only if the source is infinitely long; if d_1 and d_2 are both greater than 1.5D the results will be accurate to within 10%. The answer will be accurate to within 5% if both dimensions are greater than 2D.

Example: Calculate the illuminance at a point 10' directly below a continuous row of 6-inch-wide, single-lamp fluorescent fixtures with diffusing panels. Fixture efficiency is 50%, and each 4-foot lamp is rated at 3200 lumens.

Solution: Each lamp lights an area of 0.5' x 4', or 2 square feet. The exitance, M, is:

$$M = \frac{(3200 \text{ lm}) (.50)}{4 \text{ ft.}^2} = 400 \text{ fL}$$

and the illuminance at the point is:

$$E = \frac{(400 \text{ fL}) (.50 \text{ ft.})}{(2) (10 \text{ ft.})} = 10 \text{ fc.}$$

Note that this value is initial illuminance and assumes that all system components are operating as rated. To arrive at a reasonable approximation of the actual illuminance a light loss factor must be applied.

If the point is not directly under the source, or the source is non-lambertian, the calculation becomes complex and is beyond the scope of this text.

CIRCULAR SOURCES

The illuminance produced at a point directly below the center of a lambertian circular source, shown in Figure 8-13, can be found from:

$$E = M \, Sin^2 \, \alpha$$

If the point is not on the axis, the equation becomes:

$$E = \frac{M \, (1 - Cos \, \gamma)}{2}$$

For practical purposes, square and nearly square rectangular sources may be converted to circular sources of the same area, and the above equations will provide sufficient accuracy for most applications.

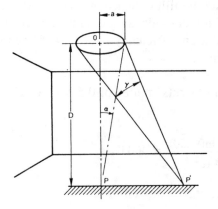

Figure 8-13
Diagram for calculations with circular sources. *(Courtesy Illuminating Engineering Society, IES Lighting Handbook, 1981 Reference Volume)*

REFLECTED COMPONENT

For interior lighting applications the component of illuminance that is reflected from room surfaces must be added to the direct component to determine the total illuminance. The calculation method is similar to the method used to determine average illuminance, except that the reflected radiation coefficient (RRC) is substituted for the coefficient of utilization used in horizontal illuminance calculations, and the wall reflected radiation coefficient (WRRC) is used in place of the coefficient of utilization for vertical surface calculations.

For horizontal surfaces the reflected illuminance is calculated from:

$$E = \frac{(\text{lumens per luminaire}) \, (\text{RRC}) \, (\text{LLF})}{\text{Area per luminaire}}$$

where

RRC	=	WEC + RPM (CCEC − WEC)
WEC	=	Wall exitance coefficient
RPM	=	Room position multiplier
CCEC	=	ceiling cavity exitance coefficient

The calculation of wall exitance coefficients was discussed in Chapter 5, which covered the calculation of coefficients of utilization. Coefficients for typical fixtures are shown in Figure 8-14 and are accessed in the same manner as coefficient of utilization tables.

Room position multipliers are obtained from Figures 8-15 and 8-16.

Figure 8-15 represents the room, and is superimposed with a grid pattern, where each block represents 10% of the room length or width. Letter designations are assigned to each grid line along the length of the room, and numbers are assigned to the lines along the width. Note that the room is effectively broken into quadrants, and the alpha and numeric designators in each quadrant have symmetry. To obtain the coordinates of a specific point in the room, simply locate its position with respect to the grid lines.

To determine the RPM, first locate the section in Figure 8-16 which corresponds to the room cavity ratio. Locate the appropriate letter column and number row, and read the RPM directly from the table.

Example: Determine the reflected component at a point 9 feet from the north wall and 15 feet from the west wall in a room measuring 50' x 30' with an RCR 3. Room surface reflectances are 80% ceiling, 50% wall, and 20% floor. Use fixture # 5 in Figure 8-14, and assume that 24 luminaires will be used.

Figure 8-14

Table of typical wall and ceiling cavity exitance coefficients. (Courtesy Illuminating Engineering Society, IES Lighting Handbook, 1981 Reference Volume)

Wall Exitance Coefficients for 20 Per Cent Effective Floor Cavity Reflectance (ρFC = 20)

ρCC		80			70			50			30			10		
ρW	WDRC	50	30	10	50	30	10	50	30	10	50	30	10	50	30	10
RCR																
0																
1	.208	.168	.096	.030	.164	.093	.030	.156	.089	.029	.148	.085	.027	.141	.082	.026
2	.199	.161	.088	.027	.157	.087	.027	.150	.083	.026	.143	.080	.025	.137	.078	.024
3	.186	.152	.081	.024	.148	.079	.024	.142	.077	.023	.136	.075	.023	.131	.072	.022
4	.172	.142	.074	.022	.139	.073	.022	.134	.071	.021	.128	.069	.021	.124	.067	.020
5	.160	.133	.068	.020	.131	.067	.020	.126	.065	.019	.121	.064	.019	.117	.062	.019
6	.148	.125	.063	.018	.123	.062	.018	.118	.061	.018	.114	.059	.017	.110	.058	.017
7	.138	.117	.058	.016	.115	.057	.016	.111	.056	.016	.108	.055	.016	.104	.054	.016
8	.128	.110	.054	.015	.109	.053	.015	.105	.052	.015	.102	.052	.015	.099	.051	.015
9	.120	.104	.050	.014	.102	.050	.014	.099	.049	.014	.096	.048	.014	.093	.047	.014
10	.113	.098	.047	.013	.097	.047	.013	.094	.046	.013	.091	.045	.013	.089	.045	.013

Ceiling Cavity Exitance Coefficients for 20 Per Cent Floor Cavity Reflectance (ρFC = 20)

ρCC	80			70			50			30			10		
ρW	50	30	10	50	30	10	50	30	10	50	30	10	50	30	10
RCR															
0	.120	.120	.120	.103	.103	.103	.070	.070	.070	.040	.040	.040	.013	.013	.013
1	.112	.099	.087	.096	.085	.075	.065	.058	.052	.038	.034	.030	.013	.012	.011
2	.105	.083	.064	.090	.072	.056	.062	.050	.039	.036	.029	.023	.011	.011	.009
3	.100	.072	.049	.086	.062	.043	.059	.043	.030	.034	.025	.018	.011	.009	.007
4	.095	.063	.039	.082	.055	.034	.056	.038	.024	.032	.022	.014	.010	.008	.006
5	.091	.057	.031	.078	.049	.027	.054	.034	.019	.031	.020	.011	.010	.007	.005
6	.086	.051	.026	.074	.044	.023	.051	.031	.016	.030	.018	.010	.010	.006	.004
7	.082	.047	.022	.071	.041	.019	.049	.028	.014	.028	.017	.008	.009	.005	.004
8	.079	.043	.019	.068	.037	.016	.047	.026	.012	.027	.016	.007	.009	.005	.003
9	.075	.040	.017	.065	.035	.014	.045	.024	.010	.026	.014	.006	.008	.005	.002
10	.072	.037	.015	.062	.032	.013	.043	.023	.009	.025	.014	.005	.008	.004	.002

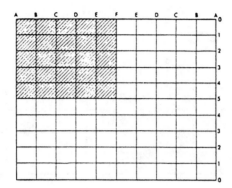

Figure 8-15
Chart for locating points on the workplane in a room. *(Courtesy Illuminating Engineering Society, IES Lighting Handbook, 1981 Reference Volume)*

ROOM POSITION MULTIPLIERS

Room Cavity Ratio = 1

	A	B	C	D	E	F
0	.24	.42	.47	.48	.48	.48
1	.42	.74	.81	.83	.84	.84
2	.47	.81	.90	.92	.93	.93
3	.48	.83	.92	.94	.95	.95
4	.48	.84	.93	.95	.96	.97
5	.48	.84	.93	.95	.97	.97

Room Cavity Ratio = 6

	A	B	C	D	E	F
0	.20	.23	.26	.28	.29	.30
1	.23	.26	.29	.31	.33	.36
2	.26	.29	.35	.37	.38	.40
3	.28	.31	.37	.39	.41	.43
4	.29	.33	.38	.41	.43	.45
5	.30	.36	.40	.43	.45	.47

Room Cavity Ratio = 2

	A	B	C	D	E	F
0	.24	.36	.42	.44	.46	.46
1	.36	.51	.60	.63	.66	.68
2	.42	.60	.68	.72	.78	.83
3	.44	.63	.72	.77	.82	.85
4	.46	.66	.78	.82	.85	.86
5	.46	.68	.83	.85	.86	.87

Room Cavity Ratio = 7

	A	B	C	D	E	F
0	.18	.21	.23	.25	.26	.27
1	.21	.23	.26	.28	.29	.30
2	.23	.26	.30	.32	.33	.34
3	.25	.28	.32	.34	.35	.36
4	.26	.29	.33	.35	.37	.37
5	.27	.30	.34	.36	.37	.38

Room Cavity Ratio = 3

	A	B	C	D	E	F
0	.23	.32	.37	.40	.42	.42
1	.32	.40	.48	.51	.53	.57
2	.37	.48	.58	.61	.64	.67
3	.40	.51	.61	.65	.69	.71
4	.42	.53	.64	.69	.73	.75
5	.42	.57	.67	.71	.75	.77

Room Cavity Ratio = 8

	A	B	C	D	E	F
0	.17	.18	.21	.22	.22	.23
1	.18	.20	.23	.25	.26	.26
2	.21	.23	.26	.27	.28	.29
3	.22	.25	.27	.29	.30	.30
4	.22	.26	.28	.30	.31	.32
5	.23	.26	.29	.30	.31	.32

Room Cavity Ratio = 4

	A	B	C	D	E	F
0	.22	.28	.32	.35	.37	.37
1	.28	.33	.40	.42	.44	.48
2	.32	.40	.48	.50	.52	.57
3	.35	.42	.50	.54	.58	.61
4	.37	.44	.52	.58	.62	.64
5	.37	.48	.57	.61	.64	.66

Room Cavity Ratio = 9

	A	B	C	D	E	F
0	.15	.17	.18	.19	.20	.20
1	.17	.18	.20	.21	.22	.23
2	.18	.20	.23	.24	.25	.25
3	.19	.21	.24	.25	.26	.26
4	.20	.22	.25	.26	.26	.27
5	.20	.23	.25	.26	.27	.27

Room Cavity Ratio = 5

	A	B	C	D	E	F
0	.21	.25	.28	.31	.33	.33
1	.25	.29	.33	.36	.38	.42
2	.28	.33	.40	.42	.44	.48
3	.31	.36	.42	.46	.49	.52
4	.33	.38	.44	.49	.52	.54
5	.33	.42	.48	.52	.54	.56

Room Cavity Ratio = 10

	A	B	C	D	E	F
0	.14	.16	.16	.17	.18	.18
1	.16	.17	.18	.19	.19	.20
2	.16	.18	.19	.21	.22	.22
3	.17	.19	.21	.22	.23	.23
4	.18	.19	.22	.23	.23	.24
5	.18	.20	.22	.23	.24	.25

Figure 8-16
Room position multipliers for calculating reflected component of the illuminance at a point in a room. *(Courtesy Cooper Lighting Co.)*

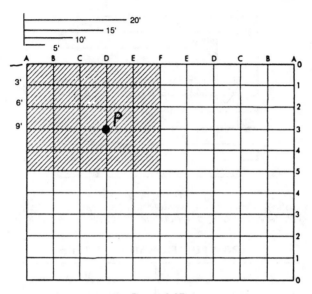

Figure 8-17
Grid location of point in example problem is "D-3."

<u>Step 1</u>. Grid the room, as shown in Figure 8-17, and determine the coordinates of the point at which the illuminance is to be calculated. The coordinates are "D3."

<u>Step 2</u>. Determine the Room Position Multiplier from Figure 8-16, as shown on page 215. The RPM is 0.65 for the example.

<u>Step 3</u>. Determine the WEC and CCEC from the photometric test report for the luminaire, fixture # 5 in Figure 8-14, or Figure 9-62 in the IES Lighting Handbook. Note that the table is accessed in the same manner as a coefficient of utilization table. The WEC is 0.152, and the CCEC is 0.100 for the example.

<u>Step 4</u>. Calculate the RRC.

RRC = WEC + RPM (CCEC − WEC) = (0.152) + 0.65(0.100 − 0.152)
RRC = 0.12

Step 5. Calculate the reflected horizontal illuminance at the point. Assume four F40 lamps rated at 3150 lumens each. The area per luminaire is: 1500 sq. ft./24 luminaires = 62.5 sq. ft. per luminaire.

$$E = \frac{(lumens/luminaire)\ (RRC)}{Area\ per\ luminaire} = \frac{(12600\ lms.)\ (0.12)}{62.5\ ft.^2} = 24\ fc$$

Note that this is the initial illuminance, and assumes that all system components are operating at rated output. The initial value must be multiplied by the light loss factor to predict the actual illuminance. If the initial illuminance is desired, the initial nonrecoverable factors are used. If the maintained illuminance is to be calculated, then the total light loss factor is applied. For example, if a total maintained light loss factor of 0.68 is assumed, the maintained illuminance becomes:

$$E_{maint} = (E_{initial})\ (LLF) = (24\ fc)\ (0.68) = 16\ fc$$

Note also that this value represents only the reflected component of illuminance at the point, and the direct component must be added to determine the actual predicted illuminance.

CALCULATING VERTICAL SURFACE ILLUMINANCE

Calculations to predict the reflected illuminance on a vertical surface are prepared in a similar manner except that the Wall Reflected Radiation Coefficient (WRC) is substituted for the RRC. The equation then becomes:

$$E_v = \frac{(lumens/luminaire)\ (WRRC)}{Area\ per\ luminaire}$$

where

$$WRRC = \frac{WEC}{\rho_w} - WDRC$$

where

ρ_w = wall reflectance

WDRC = Wall Direct Radiation Coefficient

WDRC's for typical luminaires are contained in the IES Handbook, or may be developed for specific luminaires following the procedure outlined in Chapter 5.

Chapter 9

QUANTITY AND
QUALITY OF ILLUMINANCE

In most cases lighting systems are installed for the primary purpose of facilitating the performance of some type of visual task. If insufficient illumination is provided, the performance of the visual task will be impaired and the objective of the lighting system will not be met. Conversely, illumination in excess of the required amount does little to improve visual performance, wastes energy, and needlessly increases the cost of light.

There is a direct relationship between footcandles and cost; a system providing 100 footcandles will cost twice as much to install, operate, and maintain as a system which provides 50 footcandles. Careful analysis of both the task requirements and the worker are required to assure that the quantity of illumination provided is consistent with the lighting need.

In addition to quantitative requirements, the lighting system must also produce illumination of sufficient quality. Light of poor quality may actually impair the seeing process and reduce the quantity and quality of the work performed even though the quantity of light meets or exceeds the recommended level.

SPECIFYING THE
QUANTITY OF ILLUMINANCE

Recommended lighting levels for most tasks are published by the Illuminating Engineering Society, and are generally considered to be the standard to which lighting systems are designed. Current practice, adopted in 1981, differs from prior practice in several ways. The

219

recommended illuminance for indoor lighting is now published as a range of three different illuminances for each listed task, as opposed to the single value which was previously used, and a series of weighting factors has been added as an aid in selecting the proper illuminance from the recommended range. This system recognizes that:

1. The requirement for light increases as we age, so older workers need more light than young workers.
2. The speed and accuracy with which a visual task is performed are related to the amount of light provided, and both speed and accuracy improve as the illuminance increases.
3. Tasks with low contrast require more light than tasks with high contrast.

Current recommended practice for the specification of illuminance also contains a more detailed breakdown of some visual tasks than previous practice.

Note that the tables of recommended illuminances published by the IES are consensus values, and are based on typical tasks performed by typical workers. Unusual conditions, such as partially sighted workers, may merit an increase in lighting levels, and it is the responsibility of the lighting designer to recognize and compensate for these conditions.

Figure 9-1 lists nine generic categories of visual tasks in order of ascending difficulty, assigns an alpha character designation to each description (called the "Illuminance Category"), and lists a range of three illuminances in both footcandles and lux for each Category. In addition to the generic descriptions, detailed listings of Illuminance Categories for nearly 600 individual tasks are contained in the IES Lighting Handbook, 1981 Reference and Applications Volumes, the 1987 Applications Volume, and some other IES publications.

The selection of the recommended illuminance for a specific task is a two-step process; determining the Illuminance Category, and applying weighting factors, as shown in Figure 9-2, to determine which of the three illuminances within the range is applicable. Note that there are two different sets of weighting factors. The first set, which contains factors for occupants' ages and room surface reflectances, applies to Illuminance Categories A, B, and C. These Categories usually represent spaces in which visual tasks are not difficult and can be expected to

I. Illuminance Categories and Illuminance Values for Generic Types of Activities in Interiors

Type of Activity	Illuminance Category	Ranges of Illuminances		Reference Work-Plane
		Lux	Footcandles	
Public spaces with dark surroundings	A	20–30–50	2–3–5	General lighting throughout spaces
Simple orientation for short temporary visits	B	50–75–100	5–7.5–10	
Working spaces where visual tasks are only occasionally performed	C	100–150–200	10–15–20	
Performance of visual tasks of high contrast or large size	D	200–300–500	20–30–50	
Performance of visual tasks of medium contrast of small size	E	500–750–1000	50–75–100	Illuminance on task
Performance of visual tasks of low contrast or very small size	F	1000–1500–2000	100–150–200	
Performance of visual tasks of low contrast and very small size over a prolonged period	G	2000–3000–5000	200–300–500	
Performance of very prolonged and exacting visual tasks	H	5000–7500–10000	500–750–1000	Illuminance on task, obtained by a combination of general and local (supplementary lighting)
Performance of very special visual tasks of extremely low contrast and small size	I	10000–15000–20000	1000–1500–2000	

Figure 9-1

Recommended Illuminance Categories for generic tasks. (Courtesy IES Lighting Handbook, IESNA)

exist throughout the space, such as walking through a corridor or lobby. They may also apply to working spaces, such as CRT areas, where the primary visual tasks are internally illuminated and individual task light is provided at specific locations when needed.

Weighting Factors to be Considered in Selecting Specific Illuminance Within Ranges of Values for Each Category.

a. For Illuminance Categories A through C

Room and Occupant Characteristics	Weighting Factor		
	−1	0	+1
Occupants ages	Under 40	40–55	Over 55
Room surface reflectances*	Greater than 70 per cent	30 to 70 per cent	Less than 30 per cent

b. For Illuminance Categories D through I

Task and Worker Characteristics	Weighting Factor		
	−1	0	+1
Workers ages	Under 40	40–55	Over 55
Speed and/or accuracy**	Not important	Important	Critical
Reflectance of task background***	Greater than 70 per cent	30 to 70	Less than 30 per cent

Figure 9-2

After the proper Illuminance Category has been selected, weighting factors are used to determine which of the three illuminances applies. (*Courtesy IES Lighting Handbook, IESNA*)

Occupants' ages should be based on the highest age group which will use the space, e.g., occupants over 55 years of age can be expected to enter the lobby of most office buildings, so a +1 factor for age is normally applied to lobbies. Factors of 0 (zero) and −1 are applied when it is known that the ages of the occupants will fall within the specified ranges.

The factor for room surface reflectances is based on the weighted average reflectance of the surfaces within the normal field of view. For example, a ceiling with a height of 10' will be within the field of view in all but the smallest lobbies; therefore, the ceiling reflectance would be considered in determining the weighted average room surface reflectance for most lobbies. A ceiling 20' or higher will be outside of the field of view in most lobbies, so in these cases the ceiling reflectance would not be considered.

To apply the factors, simply determine their algebraic sum. If the sum is –2, select the lowest value in the range of recommended illuminances. If the sum is –1, 0, or +1, select the middle value. If the sum is +2, select the highest value. For example, if the factors are +1 for occupants' ages and 0 for room surface reflectances, the sum is +1, and the middle value selected.

Recommendations for Categories A through C represent uniform illuminance since visual tasks exist throughout the space.

The lower set of factors applies to Illuminance Categories D through I. These Categories represent more visually demanding tasks which are usually performed at fixed locations within the space; therefore, they specify illuminance on the task, with lower levels recommended for surrounding, non-task areas.

Categories D, E, and F represent commonly performed visual tasks, in order of increasing difficulty. For example, reading typed originals is relatively easy, and is a Category D task. Copy from a thermal printer is of lower contrast and requires more light, thus it is a Category E task. Drafting with low contrast lead is more difficult, and is classified as an F task.

Category G and higher tasks are highly specialized, are very difficult and visually demanding, and frequently occur over a prolonged time period. Typical examples are very difficult assembly or inspection, usually involving very small parts or details with low contrast.

Three weighting factors are applied to Categories D through I: workers' ages, the importance of speed and/or accuracy in performing the visual task, and the reflectance of the task background.

The factors are added in the same manner as the factors for Categories A through C. If the sum is –3 or –2, the lower illuminance is specified. A rating of –1, 0, or +1 results in selection of the middle value, and a +2 or +3 indicates that the highest illuminance in the range applies.

The workers' age factor should be based on the age group working in the space, if that information is available. Unfortunately, this is frequently not the case, so some intelligent guess work may be required on occasion. Note, however, that there must be at least two negative or two positive factors to indicate selection of the high or low illuminance within the range. Since the reflectance of the task background falls into the 30%-70% range for most tasks in commercial buildings, and for many tasks in industrial environments, the age factor will generally be the deciding factor only when the speed and/or accuracy are unimportant or critical.

Speed may be unimportant, as in the case of an automatic machine where the speed of an operation is governed by the machine, not the worker, or an engineer working on a complex problem involving more thought than reading. Typical office work, such as typing, is an example of a task where speed is important. Speed may be considered critical on a high-volume production line, particularly if a conveyor line is used.

Accuracy is judged by similar metrics. Automatic machines which determine accuracy result in an unimportant rating, while an important rating might be applied to a manual task which requires precision. If an error could result in injury or death, or a substantial economic loss, accuracy would be considered critical. The accuracy factor is related to safety and economics since the specification of higher illuminance will result in higher costs, but the costs may be justified if safety is enhanced or the cost or errors is reduced.

A "+ or −" rating for either speed or accuracy will result in the assignment of a +1 or −1 for the speed/accuracy factor.

The reflectance of the task background refers to the reflectance of the surface immediately behind the visual target. As you read this page, the target is the printed letters, and the background is the paper on which the letters are printed.

The illuminance selection procedure is best explained through examples.

Example 1
Determine the recommended illuminance for the lobby area in a large office building. Employees will enter the lobby and proceed directly to their work locations. Visitors may pause to consult the directory to determine their destinations, so visual tasks will occasionally be performed. A receptionist or guard will also be available to provide directions to visitors, and will occasionally be required to read a directory. The ceiling will be white acoustical tile, and walls will be a medium beige.

Solution:

Step 1. From Figure 9-1, this is a working space where visual tasks are only occasionally performed, and the Illuminance Category is "C." Figure 9-3, a more detailed listing, confirms that lobbies are considered to be Illuminance Category "C." The recommended range of illuminance is 10, 15, or 20 footcandles (100, 150, or 200 lux).

Task	Illuminance Category
Lobbies	C
Reading	
Typed Originals	D
#2 or Softer Pencil	D
#3 Pencil	E
Inspection	
Simple	D
Difficult	F
Very Difficult	G

Figure 9-3
Illuminance Categories for some common tasks. See the IES Lighting Handbook for a complete listing.

Step 2. To determine which of the three illuminances to use, apply weighting factors from Figure 9-2. For Illuminance Categories A thru C, only two factors are considered: occupants' ages, and room surface reflectances. The ages of the occupants will vary, but many will be over 55, so a factor of +1 is applied for "Occupants' Ages."

The room surface reflectances will be in the 30%-70% range, so apply a factor of "0."

Add the factors: +1 +0 = +1.
The sum for the example is +1, so the middle value, 15 fc (150 lux), should be applied.

Example 2

Determine the recommended illuminance for a sales office in the office building used in Example 1. Typical tasks consist of reading handwriting in #2 pencil or ballpoint pen, typed originals, and first- or second-generation xerography. Workers ages range from 20 to 65, speed and accuracy are considered important, and reflectances of the task backgrounds are in the 30%-70% range.

Solution:
Step 1. These tasks are of high contrast, so the Illuminance Category, from Figure 9-1, is "D." This is verified from the more detailed task listing in Figure 9-3. The recommended range for Category "D" is 20-30-50 footcandles (200-300-500 lux).

Step 2. Apply the weighting factors from Figure 9-2. Note that Illuminance Categories "D" thru "I" have three weighting factors, as opposed to the two factors which were applied to the Category "C" task in the previous example.

The factors are:

	Factor
Workers' ages (up to 65)	+1
Speed/Accuracy (important)	0
Reflectance of task background	0
	+1

The sum is +1, so the recommended illuminance is 30 footcandles (300 lux).

Example 3
A worker, age 62, inspects very small electronic circuit boards for defects in etching. The boards are used in the on-board guidance computer in a spacecraft. The board is green, with a reflectance of about 35%.

Solution:
Step 1. From Figure 9-1, the Illuminance Category is "G," so the recommended illuminance range is 200-300-500 footcandles (2000-3000-5000 lux).

Step 2. Apply the weighting factors from Figure 9-2.

	Factor
Workers' age (over 55)	+1
Accuracy (critical)	+1
Reflectance of task background (35%)	0
	+2

The sum of the weighting factors is +2, so the highest illuminance, 500 footcandles (5000 lux), is specified.

Note that this procedure has several constraints. It is used only for interior lighting; recommended illuminances for outdoor lighting are published as a single value. It is not applicable to merchandise and feature display lighting, where luminance ratios and visual attraction to an object are the primary concerns, as will be discussed in Chapter 13. It is also not used for photography, television, theatre, or other similar applications, nor is it applicable to the lighting of artifacts or other objects when deterioration or bleaching are of concern.

FACTORS WHICH AFFECT THE REQUIRED QUANTITY OF ILLUMINANCE

The quantity of light required to perform a specific visual task is dependent on many factors, some physiological (pertaining to the person), and some physical (pertaining to the task). The weighting factor for the workers' age responds to physiological factors, while the factors for reflectance and speed/accuracy address physical factors.

PHYSIOLOGICAL FACTORS

The physiological factors are largely uncontrollable. Other than correct some relatively minor visual problems such as myopia (nearsightedness), hyperopia (farsightedness), presbyopia (loss of focusing ability of the lens), and astigmatism, there is little that can be done to correct the deterioration of the visual system which is symptomatic of the aging process. Over time the eye tends to develop problems such as a yellowing of the lens and clouding of the fluids in the interior of the eye, which reduce our ability to see. As a result, more light is required to facilitate the performance of visual tasks as we age.

The cumulative effects of this deterioration become noticeable in the early to mid-forties and are significant in the early to mid-sixties, as shown in Figure 9-4. Increased illuminance will provide some benefit, as recognized by the weighting factor for age, but no amount of increased illuminance will permit the average 60-year-old eye to see as well as the average 20-year-old eye. Deterioration of the eye varies widely between individuals; however, the need for light can be expected to increase by 2% or more per year past age 20.

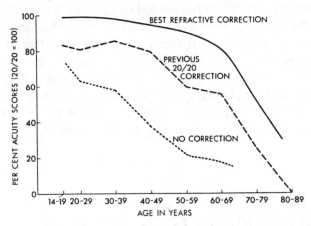

Figure 9-4
Visual acuity declines with age. (*Courtesy IES Lighting Handbook, IESNA*)

PHYSICAL FACTORS

The physical factors of seeing apply to the characteristics of the visual task, and fall into four categories: size, brightness, contrast, and time. These factors will, on occasion, be controllable. If not, they should be recognized and factored into the specification of illuminance.

Size

The size of the visual task refers not to its physical size, but to its visual size. To illustrate this concept, imagine that two identical baseballs are placed side by side at a distance of 5 feet from the eye. Since the balls are identical they both have the same physical size and, since they are the same distance from the eye, have the same visual size. Now imagine that one ball is moved to a distance of 10 feet, as shown in Figure 9-5. Both balls still have the same physical size, but the distant ball subtends a smaller visual angle to the eye, and thus appears to be smaller. The visual size of an object, not its physical size, determines the ease with which the object may be seen, and influences the required illuminance.

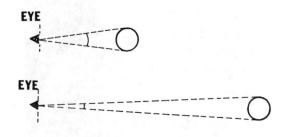

Figure 9-5
Visual size is governed by the angle which an object subtends to the eye. When two identical baseballs are placed at a distance of 5 feet from the eye they appear to be the same size. If one ball is moved to a distance of 10 feet it subtends a smaller angle to the eye than the ball which remains at 5 feet, and thus appears to be smaller.

Brightness

Brightness is a subjective term and refers to the visual sensation produced by an object. As the magnitude of visual sensation increases the object is said to be brighter. The brightness of an object is a function of both the reflectance of the object and the illuminance. For example, white paper has a higher reflectance than gray paper and thus has a greater brightness under the same illuminance. As the brightness of an object increases it becomes easier to see.

Note that the term "Brightness" refers to a visual sensation, and should not be confused with "luminance," which is purely quantitative.

Contrast

Contrast is closely related to brightness, and refers to the difference in luminance between the target object and its background. Black letters printed on a white page have a high contrast against the white background. If gray paper is used the contrast decreases and the page becomes harder to read. Figure 9-6 illustrates this concept.

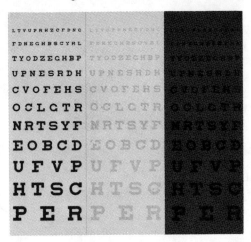

Figure 9-6
Objects of high contrast are easier to see than objects of low contrast.

Contrast is easily calculated from:

$$C = \left| \frac{L_t - L_b}{L_b} \right|$$

where:

L_t is the luminance of the target (i.e., printed letter)
L_b is the luminance of the background (i.e., the paper)

Since the illuminance on the target is invariably the same as the illuminance on the background, the equation may be simplified to:

$$C = \left| \frac{\rho_t - \rho_b}{\rho_b} \right|$$

where:

       ρ_t is the reflectance of the target

       ρ_b is the reflectance of the background

For example, assume that a new printer ribbon prints characters with reflectances of 10%, but the reflectance increases to 40% when the ribbon becomes worn. If the paper has a reflectance of 80%, what is the difference in contrast.

Solution:

$$\text{new ribbon} = \left| \frac{.10 - .80}{.80} \right| = .88$$

$$\text{old ribbon} = \left| \frac{.40 - .80}{.80} \right| = .50$$

The use of the worn ribbon decreases contrast by about 43%.

The use of worn ribbons in typewriters or printers is a common cause of poor contrast which can easily be prevented by proper machine maintenance. Low-contrast tasks require more light than high-contrast tasks, and result in higher installation and operating costs for the lighting system. Task contrast for many reading tasks is frequently controllable by simply replacing machine ribbons when they become worn, and supplying pencils with #2 or softer lead to workers. These practices may result in considerable reductions in lighting costs, and are sound business practice.

Time

    The quantity of illumination required to perform a task is also related to the time available for the task. As the available time decreases the need for light increases. For example, an accountant could read a ledger in a dimly lighted parking lot but the time required to do so would be substantial. The same task, performed in a brightly lighted office, can be accomplished in much less time, and with fewer errors. A bank clerk could sort checks under 10 fc but the speed with which the task could be performed improves dramatically if the illuminance is increased to 50 fc, and we could also anticipate a reduction in the number of errors. Since the cost of the clerk's time is many times the cost of light, the additional expenditure is easily justified.

Recommendations for the quantity of light required to perform various tasks are published by the Illuminating Engineering Society. These values are based on consideration of both the physiological and physical factors just discussed. The recommendations are based on typical conditions and will fit the vast majority of cases. In unusual situations the lighting designer or specifier must rely upon his or her professional judgement and increase or decrease the recommended values accordingly.

EVALUATING
THE QUANTITY OF ILLUMINANCE

The quantity of illumination provided by a lighting system is normally determined through a process of taking meter readings at various locations in the space, and evaluating the metered data. When evaluations of proposed lighting systems are performed the analyses are based on calculated values, and may be verified at the job site after the project has been completed.

PHOTOMETERS FOR FIELD USE

Hand-held photometers, called "Light Meters," are used to measure the amount of light at a specific location. These meters, some of which are shown in Fig. 9-7, vary widely in accuracy, price, and the range of measurements they can perform. Each of these factors must be considered when selecting a light meter to assure the selected instrument is suitable for the type of work to be performed. Compare the following items when considering the purchase of a light meter: accuracy, cosine correction, color correction, suitability for type of measurements to be taken, cost, and availability of repair and calibration service.

ACCURACY

"Accuracy," as applied to light meters, is somewhat ambiguous since there are two different accuracy metrics which are applied.

Absolute accuracy is the ability of the meter to measure a known standard. The best absolute accuracy obtainable with current technology

Figure 9-7
Typical light meters. (Courtesy Minolta, GTE Sylvania, and Lighting Sciences, Inc.)

is about 3%. A typical meter will have an absolute accuracy in the range of 3% to 15%.

Relative accuracy is the ability of the meter to accurately reproduce the same reading when exposed to the same illuminance numerous times. It also refers to the ability to read differences in light levels. Relative accuracy is a function of scale linearity; the ability to read accurately at any point on the scale, and of decade linearity; the ability to produce the same reading on two or more scales if the illuminance falls into a range which can be read on more than one scale. Currently available meters range from about 1/2% to 15% in relative accuracy.

COSINE CORRECTION

When light strikes a surface at any angle other than 90°, the cosine law is applied to calculations to predict the horizontal illuminance, as discussed in detail in Chapter 8. Likewise, light meters must have a built-in correction for light entering the meter at other than a right angle. This is called "Cosine Correction."

Some inexpensive meters have minimal cosine correction, while most high priced meters possess good cosine correction characteristics. The need for cosine correction must be evaluated with respect to the types of measurements to be made. If the meter is to be used to measure illuminance in office spaces with uniform lighting, the need for good cosine correction is not great since fixtures are not widely spaced and several fixtures contribute to the illuminance at a point. Street and parking lot lighting are the converse; luminaires are widely spaced, one or at most two luminaires contribute to the illuminance at a point, and in most cases light strikes the pavement at low angles. In these cases the meter must possess good cosine correction characteristics or the readings will be virtually meaningless.

COLOR CORRECTION

As shown in Figure 1-11 (Chapter 1), the human eye does not respond equally to all colors of light. Light meters must be corrected to properly "weight" various colors so the meter indicates an illuminance which the eye can be expected to perceive. Color correction is accomplished by inserting filters of various colors and densities so that colors have the correct weighting factors, and radiant energy which falls outside the visible spectrum does not produce inaccuracies.

Figure 9-8 shows the photopic eye sensitivity curve and the color correction curve for a color corrected meter.

Note that most light meters are calibrated to an incandescent source, so readings with light sources with spectral energy distributions which are greatly different, such as clear mercury lamps or low-pressure sodium lamps, may have a larger error than indicated by the meters' accuracy rating. The meter manufacturer should be consulted for correction factors for specific lamps.

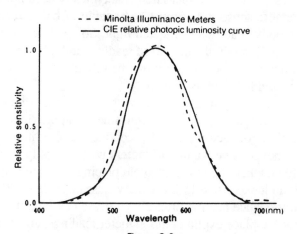

Figure 9-8
Color correction curve for a typical light meter. The error at any specific wavelength is the vertical distance between the two curves. *(Courtesy Minolta)*

SUITABILITY

The selection of a light meter will be influenced by many factors. These may include, but are not limited to: the types of measurements to be made, the desired accuracy, the purpose for making the measurements, portability of the meter, simplicity or complexity of operation, electrical requirements, cost, and the availability of calibration and repair service. The factors may be interactive, and a deficiency in any category may eliminate a meter from consideration even though it is acceptable in all other respects. Most meters have characteristics which make them particularly well suited for some kinds of measurements and less desirable for others.

If a meter is to be used for outdoor area or street lighting the sensor should be equipped with a cable to permit placement of the sensor at the desired location while the user reads the meter at a point far enough away from the sensor so as to eliminate body shadow or light reflected from the user as a source of inaccuracy. The meter should also have scales with sufficient definition to read the low levels of illuminance for these common outdoor lighting applications. A meter with a low range minimum of 5 fc simply cannot be used to measure lower levels of illumination. A meter used for outdoor lighting must also have good cosine correction.

Meters used for interior measurements typically need not be equipped with remote sensors, but must be capable of being read without introducing body shadow or reflected light. Some meters have a hold button which will retain the displayed reading, and this feature may be of help, particularly when determining the reflectance of a wall, which will be discussed later in this chapter.

The degree of required accuracy will depend upon the use of the recorded data. If the purpose is to determine the approximate illuminance in an office or factory, as when conducting a lighting or energy survey, the inexpensive hand-held meters will generally be adequate. Care should be taken, however, to avoid placing absolute values on the readings. If an inexpensive 15% accuracy class meter reads 72 footcandles, the illuminance is more properly expressed as "about 70 fc." It is human nature to place explicit faith in meter readings, and accept them as being exact. This is simply not the case, and a conscious effort must be made to avoid the inference of a higher degree of accuracy than the meter can provide.

If precision measurements are required, as is frequently the case in litigation for personal injury alleging insufficient lighting for safety, a recently calibrated, high-quality meter must be used.

Portability of the meter may also be a consideration. Large, heavy precision photometers may provide precise data, but over the period of a day they may become tiring to manipulate, batteries may run down, and the time required for each set up may increase the time required for the job. If a smaller meter will provide sufficient accuracy for the job requirements it may be a better choice.

Cost is frequently the deciding factor in the selection of a light meter. When the need is for precision measurements and the probe must be remoted from the meter, or the needed scale for very low or very high levels of illuminance are not available on a low-cost meter, there is no other option but to spend the extra money for a meter which will meet the needs. If the characteristics of a meter are simply not acceptable, the money spent will be wasted. Conversely, if the needs can be met by an inexpensive meter there is no justification for purchasing a more expensive one.

Meters do occasionally require calibration or repair. The very inexpensive meters are less costly to replace than repair, so availability of service is of no concern. Higher priced meters, however, should be calibrated at least annually, and may require periodic battery replacement or other repairs. In these cases the availability of service will be of concern, and should be verified prior to purchasing the meter.

PROCEDURES FOR FIELD PHOTOMETRY

Light meters are used in the field to measure illuminance, either horizontal or vertical, and to determine the reflectance or luminance of room surfaces. Illuminance measurements may be used to determine the illuminance at a specific point or work station, or to determine the average illuminance within a space.

Before any measurements are attempted, a new lighting system must be burned in, or seasoned. This burns off impurities in the lamps and allows them to stabilize. Burn-in is accomplished by letting the lamps operate for 100 hours for fluorescent or HID lamps and 20 hours for incandescent lamps. At this time the performance of a lighting system is reasonably predictable, so measured illuminances can be compared to calculated values by applying only the initial, nonrecoverable light loss factors to the calculation of expected illuminance. Note that the comparison, and a determination of the performance that a lighting system can provide, have relevance only if the system is new or is in good condition and has recently been cleaned and relamped. It must also have recently completed the burn-in cycle. If the system has been in operation for a period of months or years the readings tell only the current illuminance, and at best give only an indication of what might be expected with clean fixtures and new lamps. Chapter 18 will expand on this topic and provide guidelines for the intelligent and economic maintenance of a lighting system.

Once burned in, fluorescent or HID lamps should be operated for a minimum period of one hour after they have been turned on to achieve thermal equilibrium. Incandescent lamps will stabilize almost immediately, so no warm-up period is required. The sensor of the meter should also be uncapped and exposed to the light in the space to allow it to stabilize. This normally takes up to 15 minutes. When both the lighting system and meter have stabilized, readings may be taken.

Illuminance measurements are made with the sensor placed at the location where the illuminance is to be determined, and in the plane containing the visual task.

INTERIOR MEASUREMENTS

Ordinary reading and office-type tasks are normally performed in a horizontal plane, so the sensor should be in a horizontal position. When the illuminance is to be measured at a work station the sensor should be at the precise location where the visual task is performed, and the worker should be in the normal work position, as shown in Figure 9-9. This is necessary since body shadow and light reflected from the worker to the task location will be present when the visual task is performed, and must be included if the illuminance at the work location is to be accurately determined.

Figure 9-9
Measuring the illuminance at a work station.

When the average illuminance in the space is to be determined the worker should be absent, and the illuminance measured horizontally in the vertical plane where the task is performed, e.g., the workplane. The room is drawn on a grid pattern, and readings taken at the intersection of each horizontal and vertical grid line. These readings are then averaged to determine the average illuminance in the room. The accuracy will, of course, depend on the size of the grid, and will improve as grid size decreases. Care should be taken to avoid a grid pattern which is harmonic with the fixture spacing to reduce the possibility of obtaining a series of readings at recurring points of high or low illuminance which can be expected to occur at symmetric locations around each fixture, since this can distort the recorded average. For example, if luminaires are located on an 8' x 8' pattern, the grid should be 1' x 1' or 3' x 3', and 2' x 2' or 4' x 4' grids should be avoided.

Measurements taken on a grid pattern can be very time consuming, especially if the room is large, so a simplified method using only a few points in rectangular rooms has been developed by the Illuminating Engineering Society. This method will be accurate to within 10% when compared to a 2' x 2' grid pattern, and is sufficient for almost any job. Note that the method varies slightly for different fixture arrangements.

Room with Two or More Rows of Symmetrically Spaced Luminaires

These rooms are the most common, and fit the general shape and fixture layout of the room in Figure 9-10a. Take readings at all of the indicated points and calculate the average illuminance from:

$$E = \frac{R (N-1) (M-1) + Q (N-1) + T (M-1) + P}{NM}$$

Where

$$R = \frac{(r1 + r2 + r3 + r4 + r5 + r6 + r7 + r8)}{8}$$

N = Number of luminaires per row

M = Number of rows of luminaires $T = \frac{t1 + t2 + t3 + t4}{4}$

$Q = \frac{q1 + q2 + q3 + q4}{4}$ $P = \frac{p1 + p2}{2}$

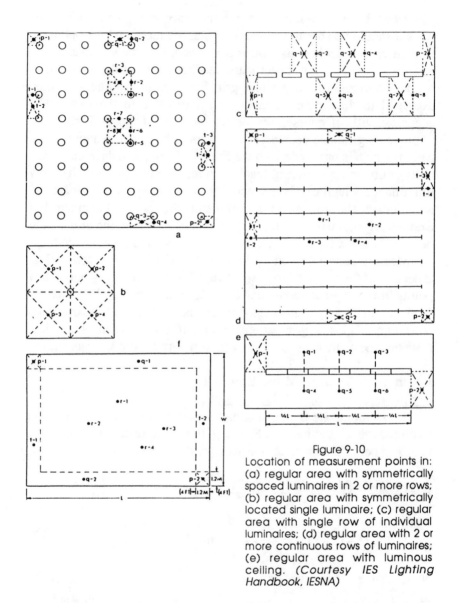

Figure 9-10
Location of measurement points in:
(a) regular area with symmetrically spaced luminaires in 2 or more rows; (b) regular area with symmetrically located single luminaire; (c) regular area with single row of individual luminaires; (d) regular area with 2 or more continuous rows of luminaires; (e) regular area with luminous ceiling. *(Courtesy IES Lighting Handbook, IESNA)*

Example

Determine the average illuminance in the room in Figure 9-10a when the measured illuminance is as shown in Figure 9-11. There are eight rows of luminaires with eight luminaires per row.

				Point					
	1	2	3	4	5	6	7	8	Ave
r	47	53	52	49	48	55	52	52	51
q	43	38	39	36					39
t	39	36	40	37					38
p	36	32							34

Figure 9-11
Illuminance measurements used for example calculation.

$$E = \frac{R(N-1)(M-1) + Q(N-1) + T(M-1) + P}{NM}$$

$$R = \frac{r1 + r2 + r3 + r4 + r5 + r6 + r7 + r8}{8}$$

$$= \frac{47 + 53 + 52 + 49 + 48 + 55 + 52 + 52}{8} = 51$$

N = 8 Luminaires per row

M = 8 Rows of luminaires

$$Q = \frac{q1 + q2 + q3 + q4}{4} = \frac{43 + 38 + 39 + 36}{4} = 39$$

$$T = \frac{t1 + t2 + t3 + t4}{4} = \frac{39 + 36 + 40 + 37}{4} = 38$$

$$P = \frac{P1 + P2}{2} = \frac{36 + 32}{2} = 34$$

So, $E = \dfrac{51(8-1)(8-1) + 39(8-1) + 38(8-1) + 34}{(8)(8)} = 48$ Fc

Rectangular Room with
<u>Single Symmetrically Spaced Luminaire</u>

Rooms lighted with a single luminaire in the center of the room are not common, but they do occur. The average illuminance measurements are quite simple, with only four readings required, as shown in Figure 9-10b. The average illuminance is determined from:

$$E = \frac{p1 + p2 + p3 + p4}{4}$$

Rectangular Room with
<u>Single Row of Individual Luminaires</u>

These rooms, Figure 9-10c, are occasionally encountered. The average illuminance is determined from:

$$E = \frac{Q(N-1) + P}{N}$$

where

$$Q = \frac{q1 + q2 + q3 + q4 + q5 + q6 + q7 + q8}{8}$$

$$P = \frac{p1 + p2}{2}$$ N = Number of luminaires in the installation

Rectangular Room with
Two or More Continuous Rows of Luminaires

These rooms, Figure 9-10d, are fairly common. The lighting systems generally consist of fluorescent luminaires mounted end to end. They are frequently found in industrial buildings and some offices. The average illuminance is found from:

$$E = \frac{RN(M-1) + QN + T(M-1) + P}{M(N+1)}$$

where

$$R = \frac{r1 + r2 + r3 + r4}{4}$$ $$Q = \frac{q1 + q2}{2}$$

N = Number of luminaires per row

M = Number of rows of luminaires $$T = \frac{t1 + t2 + t3 + t4}{4}$$

$$P = \frac{p1 + p2}{2}$$

Rectangular Room with
A Single Continuous Row of Luminaires

These rooms, Figure 9-10e, have only one row of fixtures, typically fluorescent, mounted end to end. The average illuminance is calculated from:

$$E = \frac{QN + P}{N + 1}$$

where

$$Q = \frac{q1 + q2 + q3 + q4 + q5 + q6}{6}$$

N = Number of luminaires in the row

$$P = \frac{p1 + p2}{2}$$

Rectangular Room
With Luminous Ceiling

Luminous ceilings, once popular, are only occasionally installed in new buildings. However, the method of determining the average illuminance for luminous ceilings may also be applied to totally indirect systems when luminaires are installed to provide uniform lighting throughout the room, so the calculation method may be of use. See Figure 9-11f for locations where the measurements are taken. Note that points r1 through r4 are randomly selected in the interior zone, and points q1, q2, t1, and t2 are randomly selected along the walls at locations 2 feet out from the wall. The average illuminance is then determined from:

$$E = \frac{R\,(L-8)\,(W-8) + 8Q\,(L-8) + 8T\,(W-8) + 64P}{WL}$$

where

$$R = \frac{r1 + r2 + r3 + r4}{4}$$

L = Length of the room

W = Width of the room

$$Q = \frac{q1 + q2}{2}$$

$$T = \frac{t1 + t2}{2}$$

$$P = \frac{p1 + p2}{2}$$

VERTICAL SURFACE
ILLUMINANCE MEASUREMENTS

Measurements for vertically oriented visual tasks, such as warehouse racks or vertical merchandise displays, are performed with the sensor placed at the task location, and oriented vertically. Care must be taken to avoid body shadow or reflected light which might distort the accuracy of readings.

REFLECTANCE MEASUREMENTS

The reflectance of room surfaces such as walls may be determined with a light meter by simply measuring the illuminance on the surface, the light reflected by the surface (the luminance), and dividing the reflected reading by the incident reading. The method of taking the readings is shown in Figure 9-12. First, read the vertical illuminance on the wall by placing the meter on the wall with the sensor facing outward. Then read the reflected light by pointing the sensor directly at the wall at a distance of 12 to 18 inches. This is the wall luminance, in footlamberts. You will notice that the sensor can be moved closer to the wall, or farther away over quite a distance before the reading changes.

For example, assume that the illuminance on the wall being measured in the Figure is 50 footcandles, and the reflected reading is 25 footcandles. The reflectance is:

$$\rho = \frac{\text{Reflected reading}}{\text{Incident reading}} = \frac{25 \text{ fc}}{50 \text{ fc}} = .50$$

OUTDOOR MEASUREMENTS

Outdoor lighting measurements differ from indoor measurements in several ways. The levels of illumination are, with the exception of sports lighting and some specialized applications, much lower than indoor levels. This means that the metering equipment must be designed to read low levels, frequently in the range of 0.1 to 10 footcandles. The second major difference is that any given point in most illuminated outdoor areas may receive light from one, or at the most two, luminaires due to wide spacing between poles. Since the light will be entering the meter at low angles the meter must possess good cosine correction.

Meters used for outdoor lighting measurements are frequently placed on rough, irregular, or sloped surfaces such as grass playing fields, parking lots, and roadways. When horizontal illuminance measurements are to be made the sensor must be in a horizontal position, so some means of checking the level must be available. Sensors designed for outdoor lighting measurements may incorporate bubbles, similar to those on a leveling device, or may be mounted in a gimbel ring which provides automatic leveling. The sensors will also be equipped with a cable to permit the operator to move far enough away to prevent body shadow or reflected light from distorting readings.

Figure 9-12
Procedure for measuring wall reflectance. See text for explanation.

Typical sensors designed for outdoor lighting are shown in Figure 9-13.

From the above requirements it can be seen that meters designed with capability for most outdoor lighting measurements will not be of the inexpensive type. If one is not willing to invest in a quality meter with the above features, they would be well advised to refrain from attempting outdoor lighting measurements.

The points at which measurements should be taken for roadway and sports lighting are readily available from the Illuminating Engineering Society, and many outdoor fixture manufacturers. They are straight-forward, and will not be discussed in this text.

DIFFERENCES BETWEEN
MEASURED AND PREDICTED ILLUMINANCE

Actual field measurements of the illuminance at a point and the predicted illuminance at the same point will seldom coincide, and should not necessarily be viewed with alarm.

The differences frequently occur due to manufacturing tolerances in equipment: arc tube skew and base eccentricity on lamps, slight mis-alignment of sockets, and skews in reflectors at a specific location. HID lamps are ANSI rated for a ± 5% variation in light output, and ballasts for these lamps may vary by ± 7.5%, so a lamp/ballast variation may cause variations of up to 12.5% in the worst case.

Small differences in actual versus calculated aiming angles will also affect the difference between calculated and measured illuminances, with the widest variations occurring when fixtures have rapidly changing candela distributions. Site conditions, such as adjacent objects which block or reflect light, are seldom included in calculations since most calculations are based on an assumed clear field around the point where the illuminance is measured.

The bottom line: measurements at a specific point may vary from predicted values by as much as +100% to –50% and still be within the range of expected results. The average for a large number of points should be in the +10% to –30% range, and the average illuminance can be expected to vary up to –20%. When comparisons of actual versus predicted illuminance fall within the above ranges, the system is usually performing as anticipated. If wider variations are encountered there may be a problem.

Figure 9-13
Typical remote sensors for light meters. Top sensor has bubble for levelling, while bottom sensor is gimbel ring mounted for automatic levelling. *(Bottom photo courtesy Lighting Sciences, Inc.)*

QUALITY OF ILLUMINATION

The term "quality of illumination," in its broadest sense, pertains to the distribution of luminance within a visual environment. This encompasses many topics which are far beyond the scope of this text. Our discussion of quality will concentrate on two major components of quality: glare and luminance ratios.

Glare, as defined by the IES, is "the sensation produced by luminance within the visual field that is sufficiently greater than the illuminance to which the eyes are adapted to cause annoyance, discomfort, or loss in visual performance and visibility." A lighting system may provide the specified illuminance but, if excessive glare is present, result in decreased worker productivity due to fatigue, discomfort, or reduced visual effectiveness.

DISABILITY GLARE is glare that interferes with visual performance, such as glare from a glossy magazine page that makes the print unreadable. DISCOMFORT GLARE is glare that produces visual discomfort, such as the sensation produced by bright oncoming headlights while driving at night. While the two types of glare frequently accompany each other they do not always occur together.

Glare may be classified into two categories: direct and reflected. Direct glare is caused by light which enters the eye directly from a bright light source, even if the eye is not looking directly at the source (Figure 9-14).

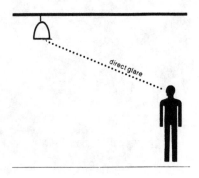

Figure 9-14
Direct glare is caused by bright light source located within the field of view. It is frequently caused by bare lamps, poorly designed fixtures, or high wattage lamps at low mounting heights. *(Courtesy Cooper Lighting)*

This stray light produces a veiling effect on the retinal image of the object to be seen. It is normally caused by a bright light source located within the field of view. The problem can be corrected or minimized by one or more of the following:

1. Remove the fixture from the field of view by moving either the fixture or the worker.
2. Reduce the luminance of the fixture. This can be accomplished by the installation of a lens, louver, or shield on the fixture.
3. Reduce the contrast between the fixture and the background. This may be done by painting the ceiling with a bright, high-reflectance paint which is similar in color to the lamp or illuminated visible surface of the luminaire.

Direct glare is normally obvious and simple to diagnose. Common offenders are bare incandescent or HID lamps, poorly designed fixtures, or improper luminaire placement. Guidelines for minimum mounting heights required to minimize direct glare from HID luminaires in industrial buildings will be discussed in Chapter 11. Fluorescent fixtures, due to their low surface brightness, are seldom significant producers of direct glare.

Reflected glare is the result of a reflection of the light source from a glossy or polished surface, as illustrated in Figure 9-15. It may easily cause visual discomfort and frequently impairs visual performance. Reflected glare is easily diagnosed and the solution is normally obvious. It can be corrected by:

1. Relocating the offending luminaire.
2. Relocating the task (or object reflecting the light).
3. Moving the worker.

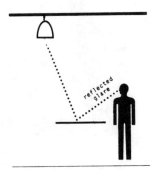

Figure 9-15
Reflected glare is caused by the reflection of light by a shiny surface.
(Courtesy Cooper Lighting)

When designing lighting systems involving specular surfaces, several steps can be taken to minimize reflected glare. Use large-area, low-brightness light sources, matte surfaces on furniture and equipment, and locate luminaires to prevent reflected glare.

Reflected glare is sometimes desirable. It can be used to advantage for some inspection tasks, and enhancing reflections can be used to attract attention and make some merchandise appear more attractive, as discussed in Chapter 13.

VEILING REFLECTIONS are a special form of reflected glare that applies primarily to reading tasks in offices and classrooms. They are so common that most of us experience them to some extent. They are also one of the most insidious of the glare problems since they can produce disability glare so subtly that we are unaware of their existence. Few people recognize them and even fewer do something to correct them even though they substantially detract from the seeing process.

A veiling reflection is the reflection of a large luminous area on the surface being viewed, as in Figure 9-16. Fluorescent luminaires and windows are common offenders. The symptom is an apparent loss of contrast between the object being viewed and the background. A good example is the difficulty we sometimes experience when reading a glossy magazine and have difficulty seeing the print on the page. Tilting or repositioning the magazine usually lessens the problem.

Figure 9-16
Example of a veiling reflection. *(Courtesy Peerless Lighting)*

Veiling reflections are caused by locating a luminaire directly above or slightly in front of the task (Figure 9-17). Corrective measures involve repositioning either the fixtures or the task to remove the light source from the offending zone as shown in the Figure. Fixtures should be positioned to the sides or behind the worker, taking care to avoid shadows. If neither of these options is viable the use of special polarizing lenses may reduce veiling reflections and improve task contrast.

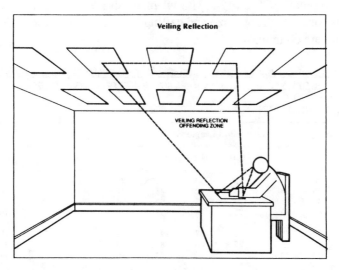

Figure 9-17
Veiling reflections are caused by a large area low brightness source such as a window or fluorescent luminaire, located above and in front of the visual task. (Courtesy GE Lighting)

EVALUATING THE QUALITY OF LIGHT

There are several metrics which are commonly used to evaluate the quality of light provided lighting systems.

The first metric is a prediction of the system's potential for direct glare and is called VISUAL COMFORT PROBABILITY (VCP). VCP is simply a prediction of the percentage of people who can be expected to consider a given lighting system to be visually comfortable when viewed from the least comfortable position in the room. A value of 70% is normally considered acceptable.

The visual comfort of a lighting system depends on many factors: room size and shape; room surface reflectances; illuminance; number and locations of fixtures; luminaire brightness and brightness of the field of view; and location of the observer. Many fixture manufacturers publish VCP tables for their equipment. These tables are not part of normal photometric data sheets but are usually available on request. A VCP table applies only to the fixture for which it was compiled, and for specific room surface reflectances and a specified design illuminance. The table should not be used for other fixtures unless the fixtures are known to have similar luminance and light distribution characteristics.

A typical VCP table is shown in Figure 9-18. To use the table simply locate the room size and ceiling height and read the VCP in the appropriate column.

Example of a Typical Tabulation of Visual Comfort Probability Values*
WALL REFL 50%, EFF CEILING CAV REFL 80%, EFF FLOOR CAV REFL 20%.
LUMINAIRE NO. 000
WORK-PLANE ILLUMINATION 100 FOOTCANDLES

Room		Luminaires Lengthwise				Luminaires Crosswise			
W	L	8.5	10.0	13.0	16.0	8.5	10.0	13.0	16.0
20	20	78	82	90	94	77	81	89	93
20	30	73	76	82	88	72	75	81	86
20	40	71	73	78	82	70	72	76	80
20	60	69	71	74	78	68	70	73	76
30	20	78	82	88	92	77	81	87	92
30	30	73	75	80	85	72	74	79	84
30	40	70	72	75	78	69	71	74	77
30	60	68	69	71	74	67	69	70	73
30	80	67	69	69	72	67	68	68	71
40	20	79	82	87	92	79	82	87	91
40	30	74	76	79	84	73	75	78	83
40	40	71	72	74	77	70	71	73	76
40	60	68	69	70	72	68	69	69	71
40	80	67	68	68	70	67	68	67	69
40	100	67	68	67	69	67	67	66	68
60	40	71	72	74	76	71	72	73	76
60	60	69	69	69	71	68	69	68	70
60	80	68	68	67	69	67	68	66	68
60	100	67	67	66	67	67	67	65	66
100	40	74	75	75	78	74	74	75	77
100	60	71	71	71	72	71	71	70	72
100	80	70	70	68	69	70	69	67	69
100	100	69	68	66	67	69	68	66	67

Figure 9-18
VCP tables predict the percentage of people who will consider a given lighting system to be visually comfortable. They can be a valuable tool when evaluating the glare potential of a lighting system. (Courtesy IES Lighting Handbook, IESNA)

Example

Determine the VCP for a lighting system using the luminaire shown in Figure 9-19 when it is installed in a 30' x 30' room with reflectances of 80% ceiling, 50% walls, and 20% floor. The design illuminance is 100 footcandles, and luminaires are mounted 8.5' above the floor.

Solution

The table in Figure 9-18 is based on the room parameters and illuminance stated in the problem so the table is valid and may be used.

Step 1. Locate the room length and width in the "Room" column on the left side of the table, and go down to the W = 30, L = 30 row.

Step 2. Locate the 8.5 column for "Luminaires Lengthwise," and go down to intersect the row located in Step 1.

Step 3. Read the VCP, 73, directly from the chart. This means that 73% of the population can be expected to find the lighting system comfortable when luminaires are mounted with the long dimension parallel to their line of sight.

Step 4. Locate the 8.5 column for "Luminaires Crosswise," and go down to intersect the row located in Step 1.

Step 5. Read the VCP, 72, directly from the chart. This indicates that 72% of the population can be expected to find the system comfortable when luminaires are mounted with the long axis of the fixture perpendicular to the line of sight.

Since the values exceed the recommended value of 70%, the system meets the VCP criteria.

The second major metric is used to evaluate the effectiveness of a given lighting system in terms of veiling reflections. It is called EQUI-VALENT SPHERE ILLUMINATION (ESI) and is the amount of light from a perfectly uniform lighting system that provides the same visual performance as the system under study. To visualize this concept, imagine two rooms as shown in Figure 9-19. Room A, on the left, is a typical office and will be considered to be the room under study. Room

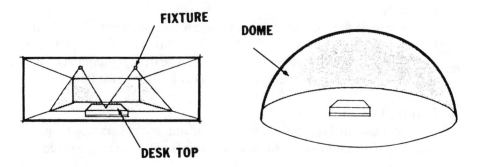

Figure 9-19
The concept of ESI. An actual system is compared to a theoretical system
which is perfectly diffusing, as represented by the hemisphere. The actual
illuminance in the room is compared to the illuminance in the sphere that is
required to provide the same visual performance.

B, on the right, is a large hemispherical dome, and represents the
perfectly uniform system previously described. This room has a
perfectly diffusing white finish and is equipped with a totally indirect
lighting system which lights the inside of the sphere in a perfectly
uniform manner. The lighting system can also be dimmed to any
desired level. Since the room is perfectly diffusing, light will be transmitted to the desk by all parts of the dome's surface, and all parts of the
sphere will have uniform luminance.

Now, imagine that a book is placed on the desk in room A and an
identical copy of the same book is placed on the desk in room B, the
sphere. The dimming control is adjusted in the sphere until the lighting
system produces exactly the same visibility as the system in room A.
Place a light meter on the book in room A and read the illuminance. In a
typical older office with the lighting system as shown, this value might
be 100 footcandles. Now read the illuminance in the sphere, which
might be 16 footcandles for the example comparison. Since 16 footcandles from the perfectly diffusing system produces the same visibility as
the system in room A, the system in room A is said to have an ESI of
16 footcandles. Note that this is true only for the specific location and
direction of view shown in the example. If either of these parameters is
changed the comparison is no longer valid and must be repeated.

The above example is typical of many office lighting systems where
the luminaires are located in front and back of the task location. The low
ESI is caused by veiling reflections from the fixture located in front of

the desk. By simply relocating the fixtures to the sides of the task or rotating the desk 90° the ESI can be improved four-fold as shown in Figure 9-20.

It is important to note that the illumination produced by the lighting system in the sphere is not perfect since there is some luminance in the veiling reflections zone. It is possible, in a limited number of cases, to have ESI values which exceed the raw footcandle values.

While ESI cannot be measured directly with a light meter, it can easily be calculated with one of several available computer programs. The calculation can, of course, be performed manually with a calculator but, due to the number and complexity of the calculations involved, is recommended only for those with highly inquisitive minds or masochistic tendencies.

For office and classroom tasks where ESI is applicable, a reasonable evaluation of lighting quality can be made by simple visual inspection. Any luminaires located in the veiling reflection zone can be expected to degrade the quality of illumination. A simple test can be performed by placing a small mirror at the task location and looking at it from the normal viewing position. Any luminaires or unusually bright surfaces appearing in the mirror can be expected to produce veiling reflections. If a luminaire is observed in the mirror, visibility can be improved by removing some or all of the lamps from that fixture, provided that there are other luminaires in the room which illuminate the task location.

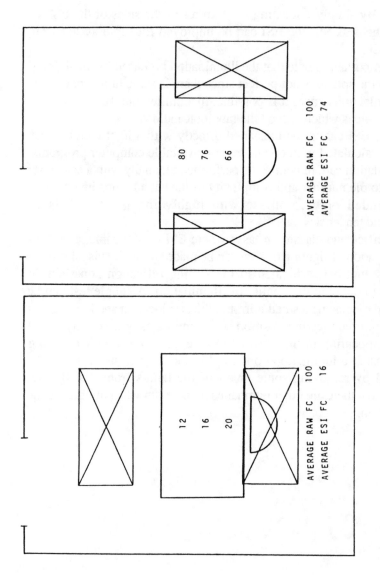

Figure 9-20
Comparison of ESI produced by two luminaire arrangements in the same room. Both systems use 2 four-lamp luminaires to produce 100 footcandles on the desk, but the system on the right produces over 4 times the usable light. The numbers represent ESI footcandles at three points on the desk.

Chapter 10

BASIC
DESIGN METHODS

The final step in the design of a lighting system consists of putting pencil to paper, and creating a drawing which instructs the contractor on how the lighting system is to be installed. Many things must occur, however, before the final drawing is approved, the specifications are written, and the job goes to bid.

The lighting designer is part of a team that conceptualizes, envisions, and engineers the final product—a functional environment. The term "functional environment," of course, has many meanings. As applied to an industrial manufacturing plant it means that the design will facilitate the movement of raw material through the manufacturing process in a safe, orderly, and efficient manner to optimize the fabrication of the finished product. In a high-volume, discount department store "functional" may mean the ability to move a large number of customers through the store in an efficient manner, display merchandise in a way that it can be rapidly evaluated by the potential buyer, permit restocking with a minimal disruption of customer traffic, and points of sale that are readily identified and do not create a bottleneck. To the jeweler operating a low-volume, high-priced establishment which caters to a wealthy clientele, "functional" may mean a relaxing and aesthetically pleasing environment which invites customers to linger and examine the merchandise, encourages them to seek the assistance of sales representatives, and provides security for his substantial investment in wares.

Each of these establishments has different design requirements for the structure and the layout of interior space and equipment. They also have greatly different lighting requirements. The manufacturer wants a system which provides illumination of sufficient quantity and quality for the tasks, at a low owning and operating cost, and typically places a low priority on aesthetics. The high-volume, discount merchandiser typically wants a lighting system which provides a relatively high level of illumination to attract attention to merchandise and facilitate the inspection and acceptance of items without clerk assistance. He may even want a lighting system which appears inexpensive, to infer that overhead costs are not inflated by a lavish building and expensive furnishings. The proprietor of the jewelry store wants a light system that draws customers into the establishment, attracts their attention to feature displays, and creates a comfortable environment which relaxes and pleases them.

The design team must be able to analyze the needs and wants of the client, and produce a design which will meet them. This will be accomplished only when there is good communication and coordination between team members, and a sharing of ideas, problems, and opportunities. Failure to adequately communicate generally results in a project which falls far short of meeting the design objectives, and few members of the design team view the project with the pride of having done a good job.

The size of the design team will vary with the size of the project. Large projects normally have at a minimum an architect, several engineers from various disciplines such as electrical, structural, mechanical, and civil, an interior designer, and the client. The design team for a small project such as the relighting of a building or single room may consist of only the lighting designer and the client.

The lighting designer must work closely with the architect, interior designer, and the electrical engineer, and to a lesser extent with the other members of the design team on specific items such as room surface finishes, ceiling type, or the ability of a structural member to carry the load of a particularly large and heavy chandelier.

In the conceptualization stage the design team determines the form and function of the structure, and if the lighting designer is involved at this stage it is only to develop generalized ideas and to input on any special lighting requirements which may affect the structural design. The real work begins only after the design of the basic structure has been completed and interior spaces have been defined.

Working with the architect, interior designer, and client, the lighting designer must understand the arrangement and function of the spaces to be lighted and the overall objective of the lighting system. At this stage the lighting designer may be able to exert limited influence on the selection of colors and surface textures which favorably affect the visual environment, and a basic lighting strategy is developed: uniform versus task oriented system, light source type (incandescent, fluorescent, or HID), direct or indirect lighting system, and special considerations for electrical controls, flexibility in switching, daylighting, security, and safety. The factors involved in the development of a particular strategy for industrial, office, and merchandise lighting will be discussed in Chapters 11, 12, and 13, respectively.

The actual design of the lighting system begins after the interior design has been completed and plans are available. The balance of this Chapter will be devoted to the mechanics of preparing a lighting layout, and assumes that the design strategy has been defined, that a lamp and luminaire have been selected based on the design criteria, and that the number of luminaires for uniform lighting systems has been determined.

BASIC LIGHTING LAYOUTS

UNIFORM LIGHTING SYSTEMS

The layout of luminaires for uniform lighting systems will depend on the luminaire type, and the type of ceiling used. Due to the wide variety of ceiling configurations only the more common types for which standardized design methods exist will be discussed in detail.

The layout of surface- or pendant-mounted fixtures on a plaster or drywall ceiling may be influenced by ceiling-mounted devices such as HVAC registers, audio systems, sprinklers, and other similar obstructions. Suitable mounting surfaces or structural members capable of supporting the weight of fixtures must also be available. Layouts of recessed fixtures in plaster or drywall ceilings will be subject to the same constraints and will also be greatly influenced by joist patterns. When nonstructural obstructions interfere with the placement luminaires at desired locations, discussions with the responsible members of the design team will frequently result in the relocation of the obstruction.

When recessed luminaires are installed in grid-type ceilings or ceilings with fixed modules, luminaire locations are determined by the ceiling system. Surface-mounted luminaires on the same ceilings may enjoy more flexibility in location; however, it may be necessary to secure them to the grid system for added support.

Industrial luminaires are generally mounted directly on structural members of a combination roof/ceiling, or are suspended on pendants or chains. Luminaire locations may be influenced by structural members such as beams, trusses, and gussets, and by obstructions such as ducts, piping, and other appurtenances. If necessary, special mounting brackets may be fabricated to permit mounting of luminaires in the desired locations if only a small number of locations are affected.

NON-GRID-TYPE CEILINGS
(Flat Plaster or Drywall, Industrial
Roof/Ceiling Combinations, etc.)

SURFACE OR PENDANT MOUNTED FIXTURES

These are normally the simplest layouts since consideration of a fixed ceiling grid is not required and only obstructions and structural members needed for fixture support are of concern. Most industrial layouts and many surface-mounted designs for commercial buildings fall into this category.

Fluorescent fixtures may be mounted individually or end-to-end in continuous rows. High-intensity discharge and incandescent luminaires are normally mounted individually; however, they may be mounted in groups of two or more fixtures at a single location in buildings with very high ceilings such as large aircraft hangers.

INDIVIDUAL FIXTURES

The required number of luminaires is determined using the Zonal Cavity Method, described in Chapter 7. When this has been accomplished the approximate spacing of fixtures is then determined from:

$$\text{Spacing} = \sqrt{\frac{A}{N}}$$

where

    A = area of room
    N = required number of fixtures

The calculated spacing will seldom fit precisely within the room, but it is a starting point, and the number of fixtures will usually require only a minor increase or decrease to provide a reasonable "fit."

Lighting layouts may be described in terms of "rows" and "columns" of fixtures, as shown in Figure 10-1.

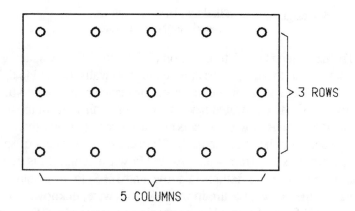

Figure 10-1
Illustration of "rows" and "columns" as used in text.

The next step is to determine the number of luminaires in each row, and the number of columns which will be required. The approximate number of fixtures per row is determined from:

$$N_{row} = \frac{Room\ length}{S_{approx}}$$

and the number of luminaires per column from:

$$N_{col} = \frac{Room\ width}{S_{approx}}$$

The numbers are then rounded up or down to the nearest integer.

The total number of fixtures using this spacing is then determined from:

$$Fixture\ total = (N_{row})\ (N_{col})$$

and compared to the required number of fixtures from the Zonal Cavity calculation. If necessary, the number of fixtures per row or per column is adjusted up or down slightly so the layout uses approximately the required number.

If the proposed layout does not use the required number of fixtures, the percentage of design illuminance may be predicted from:

$$\% \text{ Design fc} = \frac{\text{actual number of fixtures}}{\text{calculated number of fixtures}}$$

The next step is the determination of the actual fixture spacing. It is good practice to locate fixtures close enough to walls to provide light on the walls to brighten them and improve luminance ratios, and to provide light for tasks that are located next to the walls. This is accomplished by locating the fixtures closest to walls at a distance not greater than 1/3 the distance between fixtures. Figure 10-2 illustrates the concept of spacing between fixtures and from walls. An unknown spacing, S_r, is assigned to the spacing between fixtures in a row, and $1/3\, S_r$ to the distance from the wall to the nearest fixture in the row. Likewise, unknown spacings of S_c and $1/3\, S_c$ are assigned to the spacing in columns.

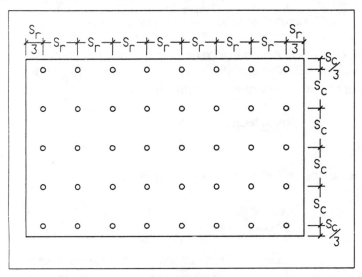

Figure 10-2
Distance between walls and fixtures nearest walls should not exceed 1/3 the distance between rows or columns of fixtures.

From the Figure it can be seen that the total number of S_r distances equals the number of fixtures in the row minus 1/3, and the same holds true for the distances in columns. From this a general-case equation can be developed:

$$S_r = \frac{\text{Room Length}}{\text{Number fixt. per row} - 1/3}$$

and

$$S_c = \frac{\text{Room width}}{\text{Number fixt. per col} - 1/3}$$

The spacings are then adjusted to provide a reasonable fit in the space.

SPACING CRITERIA (Spacing to Mounting Height Ratio)

The final step is to verify that the spacing between fixtures does not exceed the maximum recommended spacing to achieve uniformity of illumination on the workplane. This is done using the luminaire "Spacing Criteria." This specification is a refinement of the "Spacing to Mounting Height Ratio" (S/MH) which was previously used for predicting acceptable uniformity. Either term may be found on photometric reports, depending on the format used by the testing laboratory or the age of the report. For practical applications either value may be used.

Spacing Criteria are determined by the light distribution characteristics of the luminaire. They indicate the maximum recommended spacing between fixtures, as a function of mounting height above the workplane, to achieve uniformity of horizontal illuminance on the workplane. To determine the maximum allowable spacing, simply multiply the Spacing Criteria by the mounting height of the fixture above the workplane. For purposes of applying the SC, the spacing is the distance from the centerline of a fixture to the centerline of the adjacent fixture for incandescent and high-intensity discharge fixtures, and fluorescent fixtures with a maximum length of 4 feet or less. For fluorescent fixtures with longer lengths, the spacing is measured from a point 2 feet in from the end of the fixture. Fig. 10-3 illustrates the dimensions used.

The spacing criteria is not a shortcut design method, and cannot be used to design a system by locating fixtures at their maximum spacing. In fact, a system layed out by this method will probably fall far short of the desired performance. Spacing criteria and spacing-to-mounting-height ratios are simply metrics used to evaluate the probability of achieving satisfactory uniformity of horizontal illuminance of the work-plane. They are not absolute metrics. They consider only the direct component of the illuminance at a point in the room, so the uniformity in rooms with high-reflectance surfaces will be better than uniformity in rooms with surfaces of low reflectance.

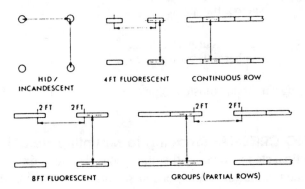

Figure 10-3
Spacing dimensions to be used when applying spacing criteria. Mounting height is the distance from the fixture to the workplane for direct, semi-direct, and general diffuse luminaires, and distance from ceiling to fixtures for semi-indirect and indirect luminaires. (*Courtesy Illuminating Engineering Society, IES Lighting Handbook, 1987 Applications Volume*)

Luminaires such as industrial HID or incandescent downlights with symmetrical light distributions will have only a single published Spacing Criteria. Fluorescent luminaires and other fixtures with asymmetrical light distributions may have two published SC values, such as 1.5/1.3. The first number, 1.5, refers to spacing perpendicular to the long axis of the fixtures, while the second value, 1.3, indicates the end-to-end spacing. This is illustrated in Figure 10-4.

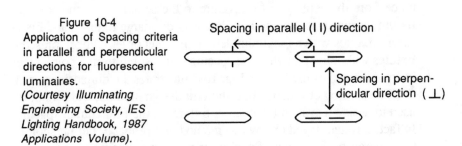

Figure 10-4
Application of Spacing criteria in parallel and perpendicular directions for fluorescent luminaires.
(Courtesy Illuminating Engineering Society, IES Lighting Handbook, 1987 Applications Volume).

Note that the Spacing Criteria is applied only to uniform lighting layouts. It assumes a clear field between luminaires, and obstructions which block light may result in poor uniformity even though the SC indicates that uniformity will be acceptable. The Spacing Criteria system assumes that luminaires with symmetrical distributions will be installed in a square pattern, so the distance between fixtures in rows will be the same as the distance between fixtures in columns. This is not always practical, as in the example problem, but spacing should be kept as symmetrical as possible.

If the proposed layout meets the Spacing Criteria it should then be evaluated for quality using metrics discussed in Chapters 9, 11, 12, and 13. If all aspects are acceptable, the design is committed to paper.

The following example will illustrate the procedure for laying out a surface- or pendant-mounted direct lighting system.

Example 1

A new uniform lighting system is to be installed in an industrial building. The building measures 100' x 150', and has a 21' ceiling. The height of the fixture, including the mounting bracket, is 3' and the workplane height is also 3', so the distance from the bottom of the fixture to the workplane is 15'.

A zonal cavity calculation has indicated that 41.7 luminaires are required. This number is rounded to 42, and the approximate spacing determined:

$$S_{approx} = \sqrt{\frac{A}{N}} \qquad S_{approx} = \sqrt{\frac{(100 \text{ ft}) (150 \text{ ft})}{42}} = 18.9 \text{ ft}$$

The estimated number of fixtures in each row is determined from:

$$N_{row} = \frac{\text{Length of Room}}{\text{Spacing}} = \frac{150 \text{ ft}}{18.9 \text{ ft}} = 7.9 \text{ fixt/row}$$

This is rounded to 8 fixtures per row.
The estimated number of fixtures per column is determined from:

$$N_{col} = \frac{\text{Width of Room}}{\text{Spacing}} = \frac{100 \text{ ft}}{18.9 \text{ ft}} = 5.3 \text{ fixt/col}$$

This is rounded to 5 fixtures per column.

The number of luminaires based on a layout of 5 rows of 8 fixtures per row is 40 fixtures (8 x 5 = 40). Note that the number of rows is the same as the number of fixtures per column.

Since this is not the required number of fixtures based on the zonal cavity calculation, the percentage of designed maintained illuminance is calculated from:

$$\% \text{ design fc} = \frac{\text{actual number fixtures}}{\text{calculated number fixtures}} = \frac{40}{41.7} = 96\%$$

This is acceptable since the design is based on maintained illuminance over a period of many years. The system will be producing a higher illuminance over most of its life since the design illuminance is based on system performance on the day luminaires are washed, and group relamping is performed if lamps are to be group replaced.

The next step is to determine actual fixture spacing.

$$S_{row} = \frac{\text{Length of room}}{(N_r - 1/3)} = \frac{150 \text{ ft}}{8 - 1/3} = 19.6 \text{ ft}$$

and,

$$S_{col} = \frac{\text{Width of room}}{(N_c - 1/3)} = \frac{100 \text{ ft}}{5 - 1/3} = 21.4 \text{ ft}$$

These spacings will be rounded to 20 feet between fixtures in rows and 22 feet between fixtures in columns, and the layout is shown in Figure 10-5.

The luminaire for the example has a Spacing Criteria of 1.5, so the maximum spacing for uniformity of horizontal illuminance is:

$$S_{max} = (1.5) (15 \text{ ft}) = 22.5 \text{ ft}.$$

The maximum spacing for the proposed layout is 22 feet, so the installation can be expected to provide acceptable uniformity.

FIXTURES IN CONTINUOUS ROWS

Continuous-row mounting of fluorescent luminaires in industrial buildings is popular when uniform general lighting systems are in-

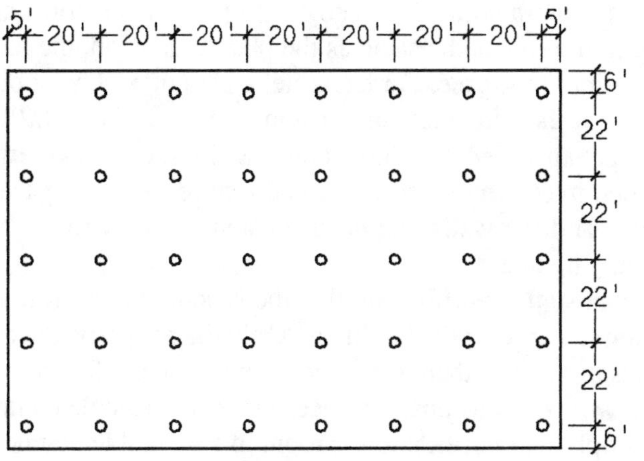

Figure 10-5
Layout for Example 1 using 40 individually mounted fixtures.

stalled. The installation cost of continuous rows is lower than individual fixtures since the fixture channel serves as an electrical raceway and eliminates much of the conduit and labor needed when each fixture must be supplied separately.

First, the required number of luminaires is calculated using the Zonal Cavity Method. The approximate number of luminaires in a row is determined from:

$$N_{fixt/row} = \frac{Room\ length}{Fixture\ length} - 1$$

and the answer rounded upward.

Next, the number of rows is determined from:

$$N_{rows} = \frac{Calculated\ number\ fixtures}{Fixtures\ per\ row}$$

This number is rounded up or down, as required. It may be necessary to slightly adjust the number of fixtures in a row to come close to the calculated number of fixtures. Take care, however, to avoid specifying too many or too few fixtures per row. Actual building dimensions seldom coincide exactly with plan dimensions, and if the total length of a row is the same as the plan dimension, the last fixture may not fit into the space. For example, a plan may show a dimension of 24 feet, but the actual site dimension might be 23' 11-1/2". If three 8' fixtures are specified, the third fixture will not fit. If too few fixtures are specified in continuous rows the end fixtures may be spaced too far from walls, and the walls and tasks located adjacent to them may be insufficiently lighted.

The final step is verification that the spacing between rows meets the Spacing Criteria for the luminaire. Only the SC perpendicular to the long axis is used since there is no end-to-end spacing dimension.

There will be occasional instances where the calculated number of fixtures simply will not fit into the room in a workable manner. Procedures for dealing with this eventuality will be discussed later in this chapter.

Example 2

The room in Example 1 is to be lighted with continuous rows of 8-foot, two-lamp industrial fluorescent fixtures. A Zonal Cavity calculation indicates that 105 fixtures will be required.

First, determine the number of fixtures per row:

$$N_{row} = \frac{\text{Room Length}}{\text{Fixture Length}} - 1 = \frac{150 \text{ ft}}{8 \text{ ft}} - 1 = 17.75 \text{ fixt.}$$

This is rounded to 18 fixtures.

Next, determine the number of rows:

$$N_{rows} = \frac{\text{Calculated number fixtures}}{\text{Fixtures per row}} = \frac{105}{18} = 5.8 \text{ rows}$$

This is rounded to 6 rows, and the proposed installation will consist of 6 rows of 18 fixtures per row, for a total of 108 fixtures. This is only slightly more than the calculated number of fixtures.

The spacing between rows is calculated in the same manner as in Example 1:

$$S = \frac{\text{Room width}}{\text{Number rows} - 1/3} = \frac{100 \text{ ft}}{6 - 1/3} = 17.6 \text{ feet}$$

This spacing is rounded to 18 feet, so the spacing between rows will be 18 feet, and the rows nearest walls will be a nominal 5 feet from the wall, as shown in Figure 10-6.

The final step is verification that the spacing between rows meets the recommended Spacing Criteria. The fixture has an SC of 1.4 in the perpendicular direction, so the maximum spacing should not exceed (1.4) (15 feet), or 21 feet. The layout meets this constraint, and can be expected to provide acceptable uniformity.

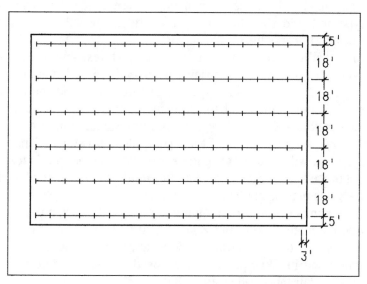

Figure 10-6
Layout for Example 2 using 6 continuous rows of 18 fixtures per row.

RECESSED LUMINAIRES
IN PLASTER OR DRYWALL CEILINGS

While not as common as other systems these layouts are sometimes encountered in commercial buildings, particularly older ones. Layout procedures are similar to the ones used for surface-mounted fixtures, but differ slightly in that spacing in one direction is dictated by the spacing between ceiling joists. The spacing and direction of the joist pattern must be determined before a design is attempted since false assumptions may lead to wasted time if a design must be scrapped because it cannot be installed in the ceiling.

These designs are normally not too difficult since only one spacing dimension is dictated, and luminaires can normally be shifted slightly n the other dimension to accommodate the required number.

GRID-TYPE CEILINGS

Grid ceilings are the most frequently encountered ceiling type in virtually all classifications of commercial space. Grid patterns may vary in type and grid size, but the 2' x 4' grid using either exposed or concealed "T" bar is the most common.

The ceiling grid dictates spacing of fixtures, thus the lighting designer should specify the spacing from walls and orientation of grid members. There are many types of ceiling grids, and the designer should become familiar with the different types to assure that luminaires are compatible with the ceiling system. For instance, a fixture designed for an exposed grid may not fit in a concealed grid system. The 2' x 4' grid system will be used for purposes of explanation in this text, but layout procedures for other systems are similar and the method used for 2' x 4' grids may be applied.

The number of fixtures required to provide the design illuminance can be calculated using the Zonal Cavity Method, and the spacing determined by trial and error to find a pattern which fits the grid. The method used in Example 1 may be used to roughly determine the spacing, and distances adjusted to fit the grid.

When layouts in grid ceilings are regularly done, there are other methods which greatly simplify the job. Regular grid patterns are well suited to standardized fixture layouts, and somewhat simplify the design procedure. Several layouts for recessed fixtures in 2' x 4' T-bar ceilings are shown in Figure 10-7. The number of fixtures in a given room is fixed by the selected layout pattern, with different lighting levels achieved by varying the number and type of lamps, ballast type, and the fixture.

The following method applies to open offices and similar spaces which are not obstructed by partitions or other obstacles. It sounds far

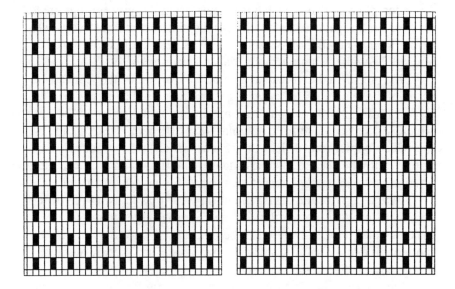

Figure 10-7
Standard layout patterns for typical 2' x 4' "T" bar ceilings. (a) Locates fixtures on 6' x 8' centers to produce average maintained illuminance ranging from 50 fc to 150 fc, (b) uses mounting on 8' x 8' spacing to produce illuminances of 30 fc to 100 fc. Checkerboard layout (c) is similar to (a) in maintained illuminance. Different illuminances within ranges given are obtained by varying number and types of lamps, and ballast type as explained in text.

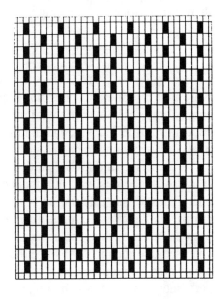

more complicated than it is, but if layouts in similar grid ceilings are to be done on a regular basis it is by far the simplest and fastest way to layout a large number of systems and be as certain of performance as you will be with the more tedious trial and error method.

First prepare a drawing to a scale of 1/8", or some other convenient scale, of the ceiling grid. Draw the grid so that it will be larger than the largest room to be layed out, to a maximum actual size of 8" x 11", which will be adequate for a nominal 60' x 80' room. Make a copy on the office copier, and fill in the spaces which will be occupied by fixtures with the desired spacing arrangement. If more than one spacing arrangement will be used, repeat the process for each arrangement.

Make a copy of the arrangement on clear plastic using the office copier, to produce an overlay. If the room is larger than 60' x 80', make several copies so they can be overlapped on a drawing of the room outline to the same scale as the overlays. Then prepare a drawing of the room outline to the same scale.

Overlay the sheet, or sheets, on the room outline and shift them until the grid fits the room with uniform spacing at opposing walls. This is the proposed fixture layout. Then determine the maintained illuminance from the following:

$$E = \frac{(\text{lamp lumens/fixt}) \ (\text{CU}) \ (\text{LLF})}{\text{Area per Fixture}}$$

Note that this method does not bypass the Zonal Cavity System. It simply presents it in a different format. The effective ceiling reflectance, coefficient of utilization, and light loss factors must still be determined in accordance with the procedures previously presented.

If the first trial does not produce an illuminance which is sufficiently close to the desired illuminance, the number or type of lamp is changed and the calculation repeated. Several trials with different lamps, ballasts, or fixtures may be required to determine a viable option. Tables such as the one shown in Figure 10-8 may be developed and used to dramatically reduce the time required for calculations. The tables may be prepared on a computer with spreadsheet software in a very short time, or may be prepared manually if desired. Note that a table may be used only for a specific luminaire, so a separate table will be required for each different fixture. This is not a difficult or time-consuming task when done by computer.

Illuminance Table

Fixture: ABC Lighting Co. Cat. #. 21x40
Reflectances: 80-50-20
Area per fixture: 64 sq. ft.
Dirt Condition: Clean

FOOTCANDLES

Lamp	F40CW						F40CW/ES					
# Lamps/Fixt.	2		3		4		2		3		4	
Maint. Sch.	1	2	1	2	1	2	1	2	1	2	1	2
RCR												
1	55	53	78	75	100	96	46	44	64	62	82	79
2	51	49	72	69	92	88	42	41	59	57	76	73
3	47	45	67	64	85	82	39	37	55	53	70	67
4	44	42	62	59	79	75	36	35	51	49	65	62
5	41	39	57	55	73	70	33	32	47	45	60	58
6	37	36	53	51	68	65	31	29	44	42	56	54
7	35	33	49	47	63	60	29	27	40	39	52	49
8	32	31	45	43	57	55	26	25	37	36	47	45
9	29	28	42	40	53	51	24	23	35	33	44	42
10	27	26	38	36	49	47	23	22	31	30	41	39

Maint. Schedule:
1 = 12-month fixture wash, spot relamp
2 = 24-month wash, 48-month group relamp

Figure 10-8
Table of average illuminance can be used to quickly estimate the average maintained illuminance produced by 2-, 3-, or 4-lamp fixtures using layout pattern in Figure 10-8 (b), for two different maintenance schedules. Note that the table applies only to specific fixtures. Similar tables may be easily constructed for frequently used fixtures.

INDIRECT LIGHTING SYSTEMS

Indirect systems were popular in the 1950's and 60's for drafting rooms and school classrooms. The first systems used incandescent silver or white bowl lamps, with sizes ranging from 500 to 1000 watts the most popular. These systems, despite their energy inefficiency, can still be found in some older school classrooms. The popularity of fluorescent sources increased during the same period, and they soon replaced incandescent sources for most indirect applications. In recent years both metal halide and high-pressure sodium lamps have been used in indirect systems, both separately and in combination.

Indirect systems are most often used in offices and classrooms. The factors which dictate their use, and considerations for the selection of a particular source and luminaire based on the lighting needs will be discussed in Chapter 12.

Fluorescent luminaires are mounted in continuous rows or as individual units, with continuous rows the most common. HID luminaries are normally installed as individual units. Fixtures may be suspended from the ceiling, attached to office partitions, or mounted in portable kiosks. Ceiling-mounted units are normally used for uniform lighting systems, while partition and kiosk-mounted fixtures are typically used for nonuniform systems.

Procedures for determining the spacing of luminaires are the same as those for surface and pendant mounted units. Refer to Example 1 for individually mounted fixtures and Example 2 for continuous rows.

There is a difference, however, in the application of the Spacing Criteria. With indirect systems the ceiling becomes, in effect, a large "luminaire." If the ceiling has uniform reflectance and is uniformly lighted, the illumination at the workplane will be uniform.

LUMINOUS CEILINGS

Luminous ceilings, once popular for office lighting, have fallen into relative obscurity for commercial use due to their low efficiency. They have gained popularity, however, in residential kitchens so a discussion of design methods is in order. They consist of bare lamp fixtures, typically fluorescent strips, mounted above a ceiling or ceiling section of diffusing or prismatic plastic (Figure 10-9).

Diffusing plastic should have transmittances of about 50% to provide sufficient diffusion to hide the lamps and minimize the appearance of streaks of light on the ceiling. Prismatic plastics are used to provide higher efficiency and better glare control, but many people

Figure 10-9
Typical luminous ceiling.

prefer the visual appearance of the diffusing plastic, particularly in residential applications.

Fixture spacing is also an important factor in controlling the uniformity of ceiling luminance. Fluorescent fixtures should be in continuous rows, and the spacing between rows should be 1.5 to 2.0 times the distance from the lamps to the diffusing media for diffusing ceilings and 1 to 1.5 times the distance for prismatic plastics. The ends of the rows should be as close as possible to the walls of the cavity. Figure 10-10 illustrates this spacing.

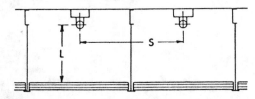

Figure 10-10
Section of a typical luminous ceiling. The spacing between lamps, S, should not exceed 2 L. *(Courtesy Illuminating Engineering Society, IES Lighting Handbook, 1981 Reference Volume)*

The ceiling cavity should be free from obstructions. If obstructions cannot be avoided, luminaires should be located below the level of the obstruction to prevent shadows on the plastic surface. All surfaces within the luminous cavity should be painted with a flat diffusing white paint of at least 80% reflectance.

Coefficients of utilization for generic luminous ceilings are shown in Figure 10-11. Note that these CU's are approximate since there are variables in luminous ceilings, but they are normally of sufficient accuracy for use in calculations. The number of luminaires is calculated using the Zonal Cavity Method.

NON-UNIFORM LIGHTING SYSTEMS

Non-uniform systems are designed in one of two basic formats. The first, and most common, consists of a general, uniform lighting system which provides a fairly low level of ambient light, and an asymmetrical layout of individual luminaires to provide supplemental

Typical Luminaires	$\rho_{CC} \to$	80			70			50			30			10			0
	$\rho_W \to$	50	30	10	50	30	10	50	30	10	50	30	10	50	30	10	0
	RCR ↓	Coefficients of utilization for 20 Per Cent Effective Floor Cavity Reflectance, ρ_{FC}															
ρ_{CC} from below ~65% 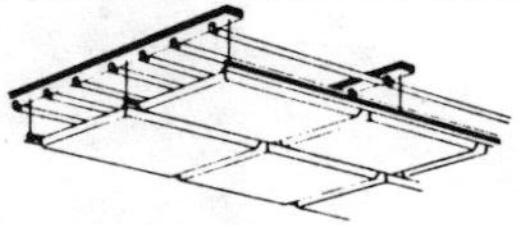	1				.60	.58	.56	.58	.56	.54							
	2				.53	.49	.45	.51	.47	.43							
	3				.47	.42	.37	.45	.41	.36							
	4				.41	.36	.32	.39	.35	.31							
	5				.37	.31	.27	.35	.30	.26							
	6				.33	.27	.23	.31	.26	.23							
	7				.29	.24	.20	.28	.23	.20							
	8				.26	.21	.18	.25	.20	.17							
	9				.23	.19	.15	.23	.18	.15							
	10				.21	.17	.13	.21	.16	.13							
ρ_{CC} from below ~60% 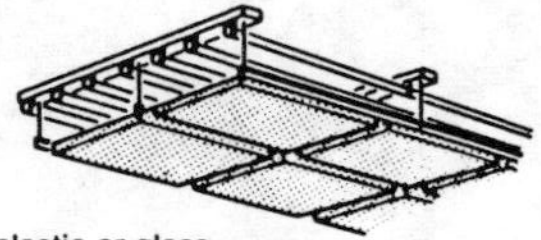	1				.71	.68	.66	.67	.66	.65	.65	.64	.62				
	2				.63	.60	.57	.61	.58	.55	.59	.56	.54				
	3				.57	.53	.49	.55	.52	.48	.54	.50	.47				
	4				.52	.47	.43	.50	.45	.42	.48	.44	.42				
	5				.46	.41	.37	.44	.40	.37	.43	.40	.36				
	6				.42	.37	.33	.41	.36	.32	.40	.35	.32				
	7				.38	.32	.29	.37	.31	.28	.36	.31	.28				
	8				.34	.28	.25	.33	.28	.25	.32	.28	.25				
	9				.30	.25	.22	.30	.25	.21	.29	.25	.21				
	10				.27	.23	.19	.27	.22	.19	.26	.22	.19				

Diffusing plastic or glass

1) Ceiling efficiency ~60%; diffuser transmittance ~50%; diffuser reflectance ~40%. Cavity with minimum obstructions and painted with 80% reflectance paint—use ρ_c = 70.

2) For lower reflectance paint or obstructions—use ρ_c = 50.

Prismatic plastic or glass.

1) Ceiling efficiency ~67%; prismatic transmittance ~72%; prismatic reflectance ~18%. Cavity with minimum obstructions and painted with 80% reflectance paint—use ρ_c = 70.

2) For lower reflectance paint or obstructions—use ρ_c = 50.

Figure 10-11

Coefficients of utilization for typical luminous ceilings. (Courtesy Illuminating Engineering Society, IES Lighting Handbook, 1981 Reference Volume) light at specific locations. Typical examples are: retail stores which use a uniform layout of fluorescent luminaires for general lighting, and incandescent fixtures using a variety of lamps to highlight feature displays; and offices with a ceiling-mounted general lighting system supplemented by furniture-mounted task lights at work stations. Examples of these applications are shown in Figure 10-12.

Figure 10-12
Task lighting is provided by fixtures mounted under bookshelves. Ceiling-mounted general fluorescent provides ambient illumination. *(Courtesy GE Lighting)*

The second method places fixtures at task locations and relies on spill light and reflected light for illumination at non-task locations. It is used primarily where tasks are separated by distances which exceed the Spacing Criteria for the fixture, yet are sufficiently close to achieve adequate illumination in the non-task areas which exist between task locations, and work stations are fixed and seldom moved. This is most often encountered in industrial operations such as machine shops and assembly lines.

TASK LIGHTING IN
CONJUNCTION WITH UNIFORM LIGHTING

The uniform lighting system is designed in accordance with procedures discussed earlier in this chapter.

The task light locations may be fixed, as with furniture-mounted systems in offices, or may be at the discretion of the designer as in feature display lighting in stores. The illuminance at points of interest may be determined using the inverse square law or illuminance charts and tables, as described in Chapter 8. Of these, charts and tables are much easier to use, and are favored by most designers. Typical charts for desk-mounted task lights are shown in Figure 10-13.

These charts indicate only the illuminance provided by the task light so the ambient component must be added to determine the total illuminance on the task.

TASK LIGHTING
WITH AMBIENT SPILL LIGHT

Luminaires are located to provide sufficient illumination on the task, and lesser levels in surrounding locations. The illuminance is calculated by computer to verify that uniformity ratios will be in accordance with recommended practice for the type of environment, i.e., industrial, or office. The use of a computer is recommended due to the multitude of redundant calculations; however, they may be performed manually if necessary.

In buildings with dark room surfaces, such as foundries and many older manufacturing facilities, the reflected component will normally be negligible, and acceptable accuracy may be obtained by calculating only the direct component using the inverse square law for a single luminaire and transferring the illuminance to points of interest on the floor plan.

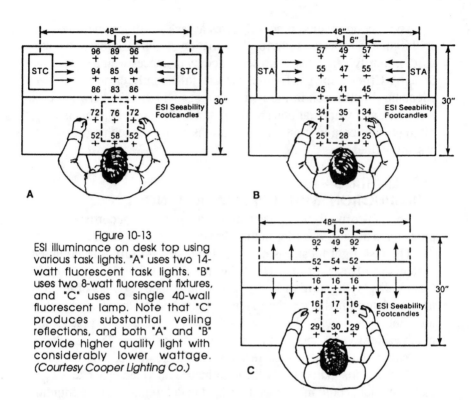

Figure 10-13
ESI illuminance on desk top using various task lights. "A" uses two 14-watt fluorescent task lights. "B" uses two 8-watt fluorescent fixtures, and "C" uses a single 40-wall fluorescent lamp. Note that "C" produces substantial veiling reflections, and both "A" and "B" provide higher quality light with considerably lower wattage. *(Courtesy Cooper Lighting Co.)*

Copies made to scale and reproduced on clear plastic simplify the transfer and permit rapid evaluation. Note that this is the plan-scale method discussed in Chapter 8. As an alternative, isofootcandle diagrams may be prepared and used in the same manner.

WHAT TO DO WHEN NO REASONABLE ARRANGEMENT FITS THE SPACE

When no amount of juggling produces a layout that fits into the room and meets the Spacing Criteria, there is one inescapable conclusion: Try something else!

Sometimes the required change will be relatively minor, such as increasing or decreasing the number of lamps per fixture. On other occasions the luminaire is simply not suitable and a fresh start using another fixture is required. Before abandoning the effort, several things may be tried:

1. If an arrangement using fewer fixtures will fit but falls short of the design illuminance by 25% or less, try the following, either individually or in combination:
    A. More frequent washing intervals.
    B. Group relamping, or more frequent group relamping intervals if group relamping has been specified.
    C. Higher light output lamps of the same type as the specified lamp, i.e., 3450-lumen or 3700-lumen, 4-foot fluorescent lamps or high-output metal halide lamps.
    D. If the design is based on reduced-wattage, energy-saving fluorescent lamps, try standard lamps. Remember to use the ballast factor for standard lamps, not the lower factor for energy-saving lamps.
    E. Higher efficiency lenses.

Note that these alternatives, with the exception of substituting standard fluorescent lamps for reduced wattage lamps, will increase first cost, ongoing maintenance costs, or both. In virtually all cases the reduction in ongoing energy costs will be sufficient to justify the change. In any case, the client should be advised, in writing, of specified maintenance intervals, the lamps to be used as replacements, and any other pertinent information.

Also note that some lamp changes may result in color changes. Make sure that the color characteristics of the new lamp are acceptable.

2. If a layout using fewer fixtures is not viable from a standpoint of uniformity, and the only acceptable layout uses more fixtures, try the following:
    A. Use a reduced-wattage, lower-light-output lamp.
    B. For fluorescent systems, use a reduced-light-output ballast.
    C. Increase the time interval between fixture washing.
    D. Increase relamping intervals.

These options should be considered only after it has been determined that a reduction in the number of fixtures is not viable since more fixtures equals higher first cost and higher ongoing costs.

If none of the above alternatives results in an acceptable layout, the only other choice is to select another luminaire and try again. In fact, if the only suggestions which produce the desired result are increasing maintenance intervals, it is probably in the client's best interests to select a fixture which is better suited to the design requirements. As lighting professionals it is our obligation to do the best job that we can for our client.

RULES OF THUMB, APPROXIMATIONS, AND ESTIMATING TECHNIQUES

Imagine that you are discussing the lighting of a parking lot with a client, and he suddenly says, "The lighting in our factory is pretty old and we should probably replace it. About how much do you think it will cost?"

How do you respond?

One option is to say, "I don't know. Let me go back to the office and work if out. I'll call you tomorrow."

This implies that you are probably going to perform design calculations to arrive at a number of fixtures, and are probably going to spend a substantial amount of time simply to give the client a rough idea of the cost. If he wants to spend no more than $5000, and the preliminary cost estimate is $15,000, you have just wasted your time.

On the other hand, in about 30 seconds, you can say, "Oh, about $20,000."

In half a minute you have given the client a rough idea of the cost, and can now qualify him. If the figure is acceptable, you now have the job. If it isn't, you have saved the time required to perform the necessary calculations.

How did you arrive at those figures? Simple. You know that the factory is a machine shop, about 50,000 square feet, and that the task requires 50 Fc. You also know that a 400-watt HPS lamp is rated at 50,000 lumens, about 50% of the lumens will reach the workplane on a maintained basis from an open industrial type fixture, and that the installed cost of the fixture is about $200.

Rule #1

In open-bottomed fixtures, about 50% of the rated lamp lumens will reach the workplane.

You got the answer by applying the following logic:

The required lumens are:

50,000 lumens x 50 lumens/sq. ft. = 2,500,000 lumens.

In an open industrial type fixture, about 50% of the rated lamp lumens will be delivered to the workplane on a maintained basis, so a 50,000-lumen lamp will provide 25,000 lumens.

2,500,000 required lumens/25,000 delivered lumens
  per fixture = 100 fixtures
100 fixtures at $200 each equals $20,000.

Or try this scenario: You are in an office building, and someone says, "OK, Mr. Lighting Expert, how many footcandles are in this room?" Do you show your ignorance by saying, "Gee, I don't know, I left my light meter in the car"?

No! You casually observe the scene for a few seconds, and respond, "Oh, about 100 footcandles, give or take a few."

Rule #2

In closed-bottomed fixtures, about 40% of the rated lamp lumens will reach the workplane.

You have observed that the fixtures are 4-lamp troffers and are located on 6' by 8' centers. You know that about 40% of the rated lamp lumens will reach the workplane on a maintained basis, and that a standard F40CW is rated at about 3150 lumens. Each fixture covers 48 square feet, and produces 40% of 4 x 3150 lumens. Since you may not want to use a calculator, round off and do it mentally: 4 lamps x 3000 lumens = 12,000 lumens. Forty percent of 12,000 lumens is 4,800 lumens. 4,800 lumens divided by 48 square feet is 100 lumens per square foot, or 100 footcandles.

Or try this one: your boss, the Plant Manager, says, "We have a proposal for skylights in the factory. If we put them in, we can turn off the lights and save money on the power bill. Take a look at this and see if the number of skylights looks reasonable, but don't spend a lot of time on it."

You say, "OK, Chief, have it for you in an hour." Actually, it's going to take less than 5 minutes, but you don't want to let him know that it's that easy. About 45 minutes later you walk into his office and say, "Yeah, the proposal looks OK." This is how you solve the problem:

Rule #3

Footcandles are lumens per square foot, so if you know the illuminance and the area, you know the total required number of lumens.

Some time ago you read an article on skylights that said that the skylights should produce design illuminance within the building when the exterior illuminance reaches 4,000 Fc. Your building is 40,000 square feet, and the design illuminance is 50 Fc, so you need 40,000 lumens x 50 lumens per square foot, or 2,000,000 lumens in the space.

The skylights are 4' x 8', or 32 square feet, and will transmit, according to the proposal, 50% of the light. Since the area of each skylight is 32 square feet, and you can assume that 4,000 lumens per square foot will be available on the exterior surface of the skylight, each skylight will have 32 square feet x 4,000 lumens per square foot, or 128,000 lumens on its exterior surface. Since 50% of the lumens will be transmitted, each skylight will admit 64,000 lumens into the building. This is when the skylight is new, however, and you know that about 30% will be lost due to dirt buildup on the skylight, and estimate that another 5% will be lost due to absorption within the building, so you estimate that each skylight will produce about 41,600 usable lumens.

You need 2,000,000 lumens, each skylight will provide 41,600 lumens, and 2,000,000 lumens/41,600 lumens per skylight = 48 skylights.

The proposal calls for 49 skylights, so the number looks OK.

These are only a few examples of the use of approximating techniques; however, they demonstrate their usefulness in the world of Illumination Engineering.

Over time, you will develop your own rules of thumb based on experience and on the application of sound engineering principles. Many of these rules and estimates will be based on the fundamental units and definitions used in the lighting industry, and a solid under-standing of fundamentals will not only make your estimates more reliable, it will instill a high degree of confidence in your estimates.

Note that approximations are not exact. Final designs should not be based on estimates; however, they will yield answers which are reason-ably accurate, and will provide a basis for answering questions such as the ones at the beginning of this section. They can also be used to quickly verify more detailed calculation methods to assure that a gross mathematical error has not been committed.

Chapter 11

INDUSTRIAL LIGHTING

Industrial lighting systems are installed for the purpose of providing "energy efficient illumination in quality and quantity sufficient for safety and to enhance visibility and productivity within a pleasant environment." (IES Lighting Handbook, 1987 Applications Volume)

It is only within recent years, however, that the term "energy efficient" has played a major role in the selection of lighting equipment and the design of industrial lighting systems. During the 1950's, 60's, and early 70's, electrical energy costs for industrial customers were very low, frequently less than 1 cent per kilowatt hour, and while they were occasionally considered, they were of minor importance. The dramatic escalation of these costs over the past decade has focused attention on operating costs and, coupled with substantial improvements in light source and luminaire efficiencies, has created a myriad of opportunities for the cost effective upgrade of old, inefficient lighting systems in existing buildings, and justification for the installation of higher first cost equipment in new buildings. It is now common practice to replace an old, inefficient lighting system with a new system that provides better lighting, and pay for the new system in one to two years with the savings in energy costs.

BASIC DESIGN CONCEPTS

Industrial facilities encompass a wide variety of working conditions and seeing tasks. These can range from "clean rooms" where the environment (temperature, humidity, and dirt contamination) is highly controlled, and very difficult seeing tasks requiring a high illuminance are performed, to foundries where environmental conditions are largely uncontrollable and visual tasks are generally rather easy and require relatively low levels of illumination.

General lighting systems, Figure 11-1, are used to provide reasonably uniform illumination for general seeing, and task illumination for relatively easy visual tasks with only minimal lighting requirements. They are also used to provide higher levels of task illuminance for more difficult visual tasks when these tasks are located close together, or are subject to frequent relocation within the space.

Task-oriented lighting systems are indicated when task locations are widely spread within the building, when ceiling heights are low, or when some task areas require substantially more illumination than others. Higher levels of task illuminance may be provided by asymmetrical fixture layouts which concentrate luminaires at the task locations, Figure 11-2, or by the use of supplemental luminaires, Figure 11-3. For some applications, such as inspection lighting, specialized fixtures coupled with the specific placement of fixtures relative to the task may be used to create grazing light or reflected glare which enhances the inspection process.

There are no absolute guidelines which may be applied to the selection of a particular lighting strategy for most industrial buildings. It is frequently necessary to employ several different lighting methods to effectively meet the design objectives.

LIGHTING EQUIPMENT SELECTION

LIGHT SOURCES

The selection of a light source is normally the first step in equipment selection. In many cases more than one lamp type will be required; however, it is good practice to keep the number of different lamps to a minimum to simplify relamping.

Lighting systems of the past several decades relied heavily on what were then state-of-the-art sources such as incandescent, mercury vapor, and fluorescent. During the 1960's two new sources were introduced: metal halide, and high-pressure sodium. The early versions of these lamps had characteristics which limited their acceptance, such as short life and poor or inconsistent color, but these drawbacks have been largely overcome and they have gained widespread popularity. Fluores-

Figure 11-1
General lighting system consists of symmetrically located luminaires to provide fairly uniform illumination throughout the space. *(Courtesy Holophane Company, Inc.)*

cent lamps have also seen improvements in life, color, and efficiency, and are well suited for many industrial lighting uses. Incandescent lamps are seldom used for general lighting due to their low energy efficiency and short life, but still have applications for small area task lighting and other specialized use. Mercury vapor lamps are now obsolete for general lighting due to their low efficiency when compared to fluorescent, metal halide, and high-pressure sodium systems.

Figure 11-2
Asymmetrical lighting system concentrates fixtures at task locations and uses wider spacing in non-task areas to provide lower illuminance. *(Courtesy Holophane Company, Inc.)*

Figure 11-3
Supplemental light is provided at the task area by this row of fluorescent luminaires. *(Courtesy GE Lighting Co.)*

The selection of a light source is based on the task requirements, economics, and aesthetic considerations. In many cases one or more factors will be weighted more heavily than others. For example, consider the need to identify slight color differences between color coded wires in a multi-conductor telephone cable. While high-pressure sodium normally has a higher efficiency than either metal halide or fluorescent sources and could be expected to provide a more economical system in terms of operating costs, the ability to discern color would be seriously impaired and it would not be considered a viable source for the task requirements.

HID sources are potential sources of direct glare since they produce large quantities of light from physically small lamps. The probability of excessive direct glare may be minimized by mounting fixtures at sufficient heights, as given in Figure 11-4. Note that the recommendations are for open-bottom, industrial-type fixtures. Luminaires with special lenses may be mounted at lower heights, and manufacturers' literature should be consulted for specific fixtures. Note that the potential for direct glare increases as the spacing criteria or spacing-to-mounting-height ratio increases. This occurs since fixtures with higher spacing criteria emit more light at higher angles. Fluorescent lamps are seldom serious offenders due to their lower luminance, and are generally preferred when fixture mounting heights are below 16 feet.

Lamp	Minimum Mounting Height (feet)
250 W Metal Halide	14
400 W Metal Halide	16
1000 W Metal Halide	20
200 W High Pressure Sodium	15
250 W High Pressure Sodium	16
400 W High Pressure Sodium	18
1000 W High Pressure Sodium	26

Figure 11-4

Minimum recommended mounting heights for typical open bottomed industrial type HID luminaires. Lower mounting heights will normally result in excessive direct glare. Fixtures with refractors or lenses can be used at lower heights; however, the manufacturer should be consulted for recommended minimums.

Some of the other important considerations in lamp selection are:

COLOR

Does the task require color rendering for color matching or identification? If so, does the need exist in the entire facility, or only a part of it? If the need exists in only specific areas, these may be lighted with sources which provide the necessary color rendering, and other areas might utilize higher efficiency sources providing a lesser degree of color rendering.

AMBIENT TEMPERATURE

Fluorescent systems, when used in cold environments, may require special low-temperature ballasts to provide higher starting voltages. An additional light loss factor which compensates for reduced light output may also be needed when calculating the required number of luminaires.

High ambient temperatures will affect the light output of fluorescent systems, and may necessitate the use of high-temperature ballasts. Remote mounting of ballasts may be required when temperatures are extreme.

The effects of high or low ambient temperatures will be discussed in greater detail later in this chapter.

MOUNTING HEIGHT AND DESIGN ILLUMINANCE

Fixture mounting height and the design illuminance are discussed together since they are interactive. HID sources are preferred for high mounting heights since the lamps are physically small packages of light, and reflectors can direct light downward with a high degree of control. Fluorescent lamps are physically large sources, and tend to scatter light; therefore, they are preferred for lower mounting heights.

There are cases, however, where low design levels are desired and ceiling heights range from moderate to high. At first glance, it might appear that an HID source is indicated because of the ceiling height. A more detailed analysis frequently reveals that acceptable uniformity of illuminance cannot be achieved with the calculated fixture spacing due to the high lumen output of HID sources in wattages used for industrial lighting. Lower wattage HID lamps might be used, but this normally has an adverse impact on economics: low wattage HID systems are less efficient than well designed fluorescent systems, and HID fixtures are more costly than most fluorescent fixtures. Conversely, high design illuminances might dictate the use of HID sources at relatively low mounting heights. Figure 11-5 provides a guideline for the selection of HID or fluorescent sources as functions of illuminance and mounting height when economics are a concern. While there are always exceptions, the information contained in the Figure is a reasonable guideline. If a proposal system falls well within the HID or fluorescent ranges, the indications are generally correct. When the systems are close to the line, an individual analysis is indicated.

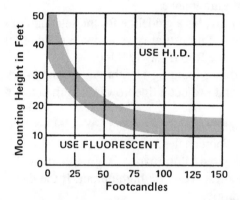

Figure 11-5
Economics play a major role in the selection of an industrial lighting system. This graph aids in the selection of the lowest cost system based on design illuminance and fixture mounting height. Locate the intersection of a vertical line projected upward from the mounting height and a horizontal line projected to the right of the design illuminance. If this point falls well above the curve, an HID system will usually be the most economical. Fluorescent systems will normally be more cost effective if the point is well below the curve. If the point lies close to the line, both sources should be analyzed. (Courtesy Lithonia Lighting)

SLIMLINE vs HIGH-OUTPUT
FLUORESCENT SYSTEMS

A common problem when using fluorescent systems in industrial buildings is the decision on slimline versus high-output lamps. The selection is normally based on both performance and economics. Slimline lamps could be used for almost all industrial applications if performance were the only criterion, but an HO lamp produces about 50%

more light than a slimline. This means that a high-output system would use about 33% fewer fixtures than a slimline system, and reduce first cost by 15% to 20% if an HO system could be used.

Typical industrial-type fluorescent fixtures are designed to be spaced about 1.5 mounting heights apart to provide uniform illumination. This means that fixtures should not be spaced farther apart than 1.5 times their mounting height above the workplane, or dark areas may exist between fixtures. Since high-output luminaires produce more light, fewer fixtures will be required, and fixtures will use a wider spacing. This increases the probability of excessive spacing, particularly for low design illuminances.

Figure 11-6 provides a guideline for the selection of slimline or HO systems using typical industrial-type luminaires. To use the Figure:

1. Calculate the Room Cavity Ratio.

2. Locate the "Mounting Height Above the Workplane" on the horizontal axis, and project a line upwards to intersect the appropriate RCR curve.

3. Project a horizontal line to the left to intersect the vertical axis, labelled "Design Footcandles."

4. If the desired illuminance is greater than the value on the "Design Footcandles" axis, use high output. If the desired illuminance is lower, use slimline.

Example

An RCR 1 building is to be lighted to 30 fc. The fixture mounting height is 15 feet above the workplane. Should slimline or HO lamps be used?

Solution

1. Locate "15" on the horizontal axis, and project a line upwards to the "RCR 1" curve.

2. Project a line from this intersection to the vertical axis, and read the illuminance, 20 footcandles.

3. Select a high-output system since the design illuminance is 30 footcandles, which is greater than 20 footcandles.

If the illuminance on the vertical axis is close to the design illuminance, a site-specific analysis should be prepared.

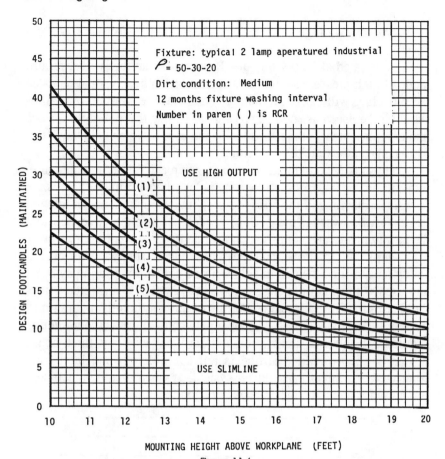

Fixture: typical 2 lamp aperatured industrial
ρ = 50-30-20
Dirt condition: Medium
12 months fixture washing interval
Number in paren ( ) is RCR

USE HIGH OUTPUT

(1)
(2)
(3)
(4)
(5)

USE SLIMLINE

MOUNTING HEIGHT ABOVE WORKPLANE (FEET)

Figure 11-6

If a fluorescent system is to be installed, the designer must usually choose between slimline or high output lamps. This graph will assist in determining the most economical system. See text for instructions.

The following is a recap of some of the important characteristics of lamps used for industrial lighting:

HIGH-PRESSURE SODIUM – The highest efficacy source used for interior lighting. The color is slightly yellowish, but will appear white in the absence of light from other, more balanced sources. While the color is acceptable for most industrial applications, it is not suitable for color matching, and some colors, such as oranges and reds, may appear to be the same.

Common wattages for industrial use are 250W, 400W, and 1000W.

Life is rated at 24,000+ hours. Remember that the typical failure mode of high-pressure sodium lamps is an on/off cycling, and repeated lamp cycling will cause premature igniter failure. Burned-out lamps should be replaced as soon as possible to reduce this possibility, since igniter replacement is time consuming and costly, typically in the $100-$150 range at the time of this writing.

Lamps take approximately 3 minutes to reach full light output after they are turned on. Hot restrike time is less than 1 minute. This is the fastest restrike time of the HID sources, but backup systems or integral quartz restrike systems may still be needed. As an alternative, consider the new twin arc tube lamps which restrike in a few seconds.

Lamps tend to strobe, and may cause visual annoyance or even a safety hazard if connected to single-phase power distribution systems and used with rotating machinery. Lamps should be staggered across the phases of a three-phase power distribution system if rotating machinery is present (Figure 11-7), or supplemental incandescent lighting should be installed in areas where a hazard may exist and three-phase power is not available. Strobe is not a problem when lamps are connected to a three-phase power system, or when rotating machinery is not used. Lumen maintenance is very good.

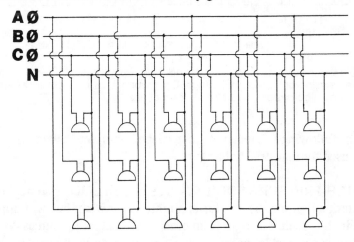

Figure 11-7
Connection of high pressure sodium fixtures to a 3-phase power system to minimize strobe effect.

METAL HALIDE – Has good color-rendering characteristics and high efficiency in the 400 W and 1000 W sizes, which are the most commonly used for industrial lighting. Metal halide should be considered for applications where color requirements preclude the use of high-pressure sodium. Smaller sizes such as 175 W and 250 W are available but seldom used for industrial applications due to short lamp life, relatively low efficacy, and the requirement for enclosed fixtures due to the possibility of violent end of life.

When used in continuous-burn applications, metal halide lamps should be turned off for a period of at least 15 minutes, once per week, to minimize the possibility of violent end-of-life failure. If this cannot be done, an enclosed fixture is recommended as a safety precaution, or use the newer type metal halide lamps employing a separate shield around the arc tube. Lamp manufacturers' recommendations regarding the use and application of their products should always be followed with any lamp, but are particularly important with metal halide lamps.

Rated life is 10,000 to 20,000 hours for lamps used for industrial applications. Note that this rating is based on 10 operating hours per start, and will be affected by the actual operating schedule. Shorter operating hours per start will greatly reduce lamp life, and frequent cycling is not recommended if lamp life is a consideration. For example, life is reduced by about 30% for operation at 5 hours per start.

Lamps take 5-7 minutes to reach full light output from a cold start, and require 15-20 minutes for hot restrike. Back-up systems or integral quartz restrike systems may be required.

Metal halide lamps produce minimal strobe effect and do not normally create strobe problems.

Clear lamps, operated 10 or more hours per start, exhibit fair lumen maintenance characteristics. Phosphor-coated lamps have only fair lumen maintenance at best, and short operating hours reduce their rating to poor. This must be included in the design process.

FLUORESCENT LAMPS – Fluorescent lamps have, for several decades, been an economical means of providing high quality illumination for a wide variety of industrial tasks. Since their introduction in 1938, fluorescent lamps have seen continued improvements in life, efficacy, color, and cost, and are frequently the best choice for many industrial applications. Their principal advantages lie in three areas: 1) good color; 2) high efficiency in a wide variety of sizes; and 3) low equipment and operating costs.

Eight-foot slimline or high-output lamps are the most common, although four-foot lamps are sometimes used. They are physically large packages of light. From this size comes both an advantage and a disadvantage. The physics which govern reflector design dictate that light produced by a physically large source is difficult to control. This means that light from fluorescent fixtures spreads rapidly, and can not easily be concentrated downward. At low mounting heights this is an advantage since it is generally desirable to have a fairly wide light distribution to maintain uniformity. At high mounting heights this scattering frequently becomes a disadvantage since less light is directed downwards to the workplane. For this reason HID sources are frequently preferred at mounting heights of 18 feet of greater unless low levels of uniform illumination are desired.

The 1500-milliampere lamp, commonly called a "very high output" or "VHO," is obsolete for all except a few specialized industrial applications such as freezers which require a special jacketed lamp to maintain adequate bulbwall temperature. This is due to its rapid depreciation in light output which results from high current loading, and results in a lower overall efficiency when compared to slimline or high-output lamps.

Some other important characteristics of fluorescent lamps are:

Rated life ranges from 12,000 to 30,000 hours.

Lamps produce nearly full light almost immediately, and restrike almost instantaneously so back-up systems or restrike devices are not needed. Note that this does not eliminate the need for battery-powered emergency lighting in installations where emergency lighting is required.

Light output is greatly affected by low temperatures, and somewhat affected by high temperatures. Rated lumen output should be adjusted downward when lamps are to be operated in cold storage rooms or other similar low-temperature environments. Special low-temperature ballasts may also be required in cold environments to provide sufficient starting voltage. Figure 11-8 lists the minimum starting temperatures for common fluorescent systems. Note that these ratings apply to standard "full wattage" lamps. Reduced wattage energy saving lamps are not recommended for operation at temperatures below 60°F.

Lumen maintenance is generally good, except for 1500 ma. lamps, which are rated fair to poor.

Lamp	Standard Ballast (°F)	Low-Temp Ballast (°F)
F40	50	0
F96 Slimline	50	0
F96 High Output	50	−20
F96 1500 Ma.	−20	−20

Figure 7-8

Table of minimum reliable starting temperatures for various fluorescent lamp and ballast systems. The use of fluorescent systems in ambient temperatures below those shown may result in failure of the lamp to start, and may shorten lamp life.

LUMINAIRE CONSIDERATIONS

Once a light source has been selected, a suitable luminaire must be specified. There is generally no single "right" choice, but there may be a large number of wrong ones. The selected luminaire must provide light of sufficient quality, and may have to meet other specific requirements such as hazardous environments, damp or wet locations, corrosive environments, and abnormal temperatures. In addition, factors such as the ease of maintenance and installation should be of concern.

HAZARDOUS LOCATIONS

Some environments contain explosive or flammable gases, vapors, dust, or fibers. Examples are paint-spraying booths, gasoline dispensing stations, flour mills, and textile processing areas. These locations require fixtures which are "approved" for use in these environments. "Approved" means that the fixture meets standards which have been established by the local governmental agency that has jurisdiction over building construction. An Underwriters Laboratory (UL) or similar testing agency listing is normally adequate.

The National Electrical Code lists three classifications of hazardous locations: I, II, and III. Each Class is divided into two Divisions, 1 and 2. Class I locations contain flammable gases or vapors in quantities sufficient to produce ignitable or explosive mixtures. In Class I Division 1 locations these vapors are normally or periodically present in normal plant operation. In Division 2 locations they are not normally present, but may exist in the case of accidental equipment failure.

Class II locations contain flammable or explosive mixtures of dust. The dust is normally present in Division 1 locations, and may exist only in sufficient quantity to interfere with heat dissipation or present a fire hazard if ignited by sparks from electrical apparatus in Division 2 locations.

Class III locations contain easily ignitable fibers. Class III Division 1 locations include manufacturing areas; storage and handling areas are considered to be Division 2.

Fixtures used in hazardous locations must be approved for use in the specific type of area, and for the hazard which is present. The National Electrical Code should be consulted for more information when systems are to be designed for hazardous locations. Typical fixtures for use in hazardous locations are shown in Figure 11-9.

Figure 11-9
Typical fixture
approved for use in
hazardous locations.
See text for
explanation.
(Courtesy Holophane
Company, Inc.)

DAMP AND WET LOCATIONS

Some industrial environments, such as process areas where large amounts of water are used, may be subject to moderate degrees of moisture in the air which will condense on equipment surfaces. Other areas may be wet due to splashing water or occasional hosing down of equipment or room surfaces to clean them.

Lighting and other electrical equipment used in these areas must be approved for use in either damp or wet locations, as applicable. Fixtures are sealed against the entrance of moisture to prevent the accumulation of water in the fixture, wireways, or electrical fittings. Fixtures installed in concrete that is in direct contact with the earth must be approved for wet locations. Typical wet and damp location fixtures are shown in Figure 11-10.

Figure 11-10
Fixtures for wet and damp locations must be approved for the intended use. Special sealing and gasketing are provided to prevent the entry of moisture. (HID photos courtesy Holophane Company, Inc. Fluorescent photos courtesy Lithonia Lighting)

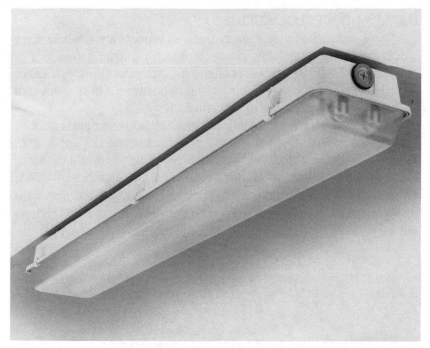

Figure 11-10 (Cont'd)

CORROSIVE ENVIRONMENTS

Plating and anodizing rooms, chemical plants, and other similar spaces may have corrosive fumes or vapors in the atmosphere that can attack and destroy fixtures, so luminaires used in these environments must be capable of withstanding the corrosive action. All exposed surfaces, including screws, bolts, and mounting hardware should be protected by a suitable finish. Special treatment or enclosure of reflecting surfaces may also be required.

ABNORMAL TEMPERATURES

As previously discussed, fluorescent systems used in cold environments may require low-temperature-rated ballasts to provide sufficient starting voltage, and enclosed fixtures may be required to maintain lamp bulbwall temperatures within acceptable limits.

High ambient temperatures may be encountered at ceiling or truss height in foundries, heat treating plants, and other similar heat-producing industries. These conditions may require the use of high-temperature-rated ballasts. In extreme cases ballasts may require remote mounting in cooler locations. Remote mounting is not normally a problem with fluorescent, mercury vapor, or conventional single-ended metal halide systems, but typical igniters used with high-pressure sodium or double-ended metal halide lamps must be located within a specified distance from the lamp to assure reliable starting. Typical maximum recommended distances between lamps and ballasts are shown in Figure 11-11. Manufacturers' data should be consulted for specific applications.

IGNITOR APPLICATIONS
ADVANCE Core & Coil (71A Series) and Potted Core & Coil (73B Series)

Lamp Type	Case Type	Lamp Watts & Type	Type of Ballast Circuit	Standard SHORT RANGE IGNITOR		Standard LONG RANGE IGNITOR	
				Catalog Number	Max. Dist. to Lamp	Catalog Number	Max. Dist. to Lamp
High Pressure Sodium	Round	35W (S76) thru 150W (S55)	CWA Hi-Reactance Reactor	LI551-J4 LI551-H4 LI551-H4	5 Ft. 2 Ft. 2 Ft.	Not Avail. LI551-J4 LI551-J4	— 35 Ft. 15 Ft.
		150W (S56) thru 400W (S51)	CWA, Reg. Lag	LI501-H4	2 Ft.	LI501-J4	50 Ft.
	Oval	250W (M80) 1000W (S52)	HX-HPF CWA	LI520-H5 LI571-H5	5 Ft. 2 Ft.	Not. Avail. LI571-J5	— 50 Ft.
Metal Halide	Oval	70W (M85) 100W (M90) 150W (M81)	HX-HPF HPF-Lag HX-HPF	LI510-H5 LI531-H5 LI520-H5	12 Ft. 20 Ft. 13 Ft.	Not Avail.	—

ADVANCE BALLAST TYPE	LAMP WATTS & TYPE	MAXIMUM BALLAST DISTANCE TO LAMP
F-Can (72 Series)	All HPS 70W H.Q.I. 100W Metal Halide	15 Ft. 5 Ft. 5 Ft.
Post Line (74P Series)	35-70W HPS 100 & 150W HPS	10 Ft. 5 Ft.
Indoor Enclosed (78E Series)	All HPS	50 Ft.
Outdoor Weatherproof (79W Series)	All HPS	50 Ft.

Figure 11-11
Table of typical maximum distances between lamp and igniter for high pressure sodium lamps and metal halide lamps requiring an igniter. These distances should not be exceeded when the igniter is to be mounted remote from the lamp. Verify data with the ballast manufacturer for specific installations. *(Courtesy Advance Transformer Co.)*

Most HID ballasts are rated for operation in ambient temperatures of 55°C; however, higher temperature ratings are available as an option from most fixture manufacturers.

The use of open ventilated fixtures whenever possible will reduce ballast temperature and prolong ballast life.

MOMENTARY POWER FAILURE

HID lamps will not restart immediately if their electrical power supply is interrupted for a fraction of a second or longer. Lamps must cool down before they will restrike. High-pressure sodium lamps will restart in 1 minute or less, while mercury vapor lamps take 3 to 5 minutes to cool sufficiently, and metal halide lamps may require 15 to 20 minutes.

Quartz restrike systems, consisting of a quartz lamp installed in the reflector and an automatic switching device in the ballast, provide some illumination until the HID lamp has restarted. A typical system is shown in Figure 11-12. These systems are available as options in the design process and should be used in sufficient quantity to provide light for safety.

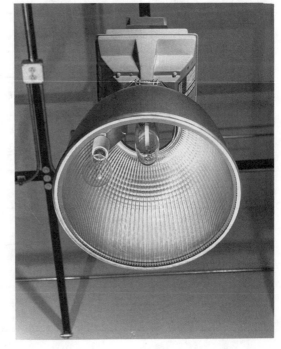

Figure 11-12
Incandescent restrike system used in HID fixtures provides instant light in the event of a momentary interruption of power to the fixture. System shown uses conventional incandescent lamp. Others use smaller quartz lamps. (*Courtesy Holophane Company, Inc.*)

Quartz restrike systems do not replace battery-operated emergency lighting systems which are designed to provide light in the event of a power outage. Quartz restrike systems are not required for fluorescent systems since fluorescent lamps restrike almost instantaneously.

EASE OF MAINTENANCE

Periodic washing of fixtures results in a higher overall lighting system efficiency,which lowers both first cost and ongoing operating and maintenance costs, as will be discussed in detail in Chapter 17.

Open, ventilated fixtures will stay cleaner than open bottom-closed top units, as shown in Figure 11-13, and should be used for most applications in areas of clean to moderate dirt contamination. In areas of heavy or very heavy dirt contamination the use of enclosed, gasketed fixtures which are sealed against the entry of dirt should be considered. An exception might be the use of open, ventilated fixtures in these areas if lamps burn continuously (24 hours per day, 7 days per week), where the continuous flow of air through the fixture keeps them relatively clean. This is particularly true when the dirt is not adhesive.

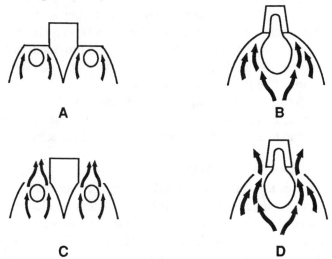

Figure 11-13
Fixtures with closed tops (A & B) trap dirt which collects on reflector and lamp surfaces. Open fixtures (C & D) allow convected air flow which carries dirt up and out of the fixture. The air flow also keeps ballasts cooler and may extend ballast life. These fixtures are preferred in clean to moderately dirty environments.

Enclosed fixtures are normally less efficient than open fixtures, so more fixtures will be required to provide the same illuminance. Poorly designed or ungasketed fixtures will allow dirt to enter the lamp compartment and, since enclosed fixtures have more surfaces on which dirt can deposit, excessive light loss may occur, as shown in Figure 11-14.

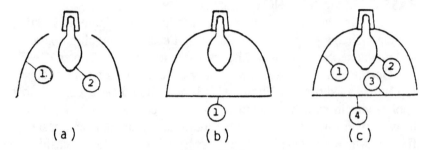

(a)   (b)   (c)

Figure 11-14
Well built enclosed and gasketed fixtures are used under conditions of extreme dirt contamination. Open fixture (a) has two surfaces to collect Dirt, while tightly sealed enclosed fixtures (b) has only one exposed surface. Poorly sealed units (c) collect dirt on four surfaces, and should be avoided.

When specifying enclosed fixtures, make sure than the bottom enclosure can easily be opened for washing and lamp replacement, and that a suitable hinge or retaining mechanism is provided on large or heavy lenses or shields so that maintenance personnel can safely work on the fixture without fear of dropping them.

Open fixtures employing glass reflectors with smooth inner surfaces have been shown to be less susceptible to dirt accumulation than aluminum reflectors or reflectors with painted surfaces.

EASE OF INSTALLATION

Luminaires may be mounted and electrical connections made by a variety of methods, ranging from direct mounting on electrical outlet boxes to special "hook and eye" brackets which are available from most manufacturers as an option. The National Electrical Code permits the attachment of fixtures weighing not more than 50 pounds directly to electrical outlet boxes, provided the box is adequately supported. Fixtures weighing more than 50 pounds must be supported independently of the electrical box.

Fluorescent fixtures are normally mounted directly to the electrical outlet box, or suspended from the ceiling by chain. End-to-end mounting of fixtures in continuous rows (Figure 11-15) is preferred if the required number of fixtures fits a continuous row layout, as discussed in Chapter 10. This mounting arrangement usually has lower installation costs than individual mounting of fixtures since the fixtures serve as electrical raceways, and the number of conduits and connections at fixtures is greatly reduced. Fixtures must be approved for use as raceways, and conductors which pass within 3 inches of the ballast must be rated for 90°C. Type THHN conductors meet this requirement and are normally used.

Figure 11-15
Fluorescent fixtures are mounted in continuous rows when possible to minimize wiring costs. *(Courtesy GE Lighting)*

HID luminaires are normally mounted directly to conduit stems, or to special mounting hooks or brackets available from the fixture manufacturer. Many of these hooks and brackets greatly simplify the installation of fixtures since they can provide support for the fixture while electrical connections are made. Hooks should be equipped with safety screws or other devices which prevent accidental dislodging of fixtures.

Electrical connections are made by running a conduit directly to the fixture, or by means of a flexible cord which is plugged into a receptacle mounted directly above the fixture. "Twist-lock" type cord caps and receptacles are used to prevent accidental disconnection. Cords must contain a grounding conductor, and must be visible for their entire length.

Typical mounting and wiring methods are shown in Figure 11-16.

Figure 11-16 (below)
Typical mounting methods for HID fixtures. (a) uses mounting hook and soft wiring for ease of removal for repair and labor savings during the initial installation; (b) shows conduit stem mounting; (c) shows in and out conduits; (d) shows popular plug and hook arrangement. Note that installation methods must meet local electrical codes. ((Courtesy Holophane Company, Inc.)

(A)

(B)

(C)

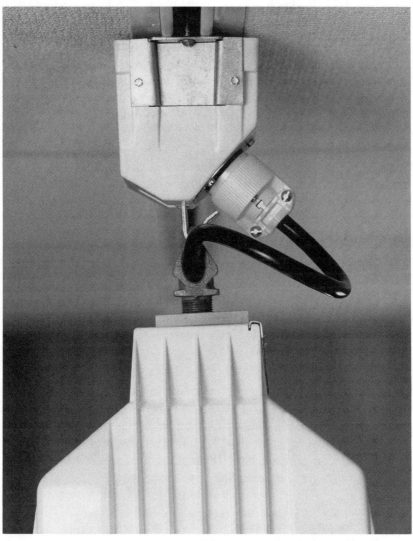

(D)

QUALITY OF CONSTRUCTION

Examine the fixture for potential installation and maintenance problems such as sharp edges or projections which might cause injury to personnel. Workers are seldom enthusiastic about on-the-job injuries, and the cost of sending electricians or maintenance personnel for emergency treatment, plus the cost of filling out accident reports and filing insurance claims may be considerably higher than the cost of purchasing better quality fixtures.

Make sure grounding screws are accessible—some fixtures require removal of the ballast or other components to gain access to tighten the screw. Make sure that holes for grounding screws are tapped. Captive screws on access plates to wiring compartments are a great help when installing or working on fixtures from a ladder. Flimsy latches or hinges may require frequent repair. Knock-outs, the holes provided for the connection of a conduit, are sometimes improperly stamped, and the opening must be drilled with a hole saw at greatly increased labor costs.

Unfortunately the problems outlined are not confined to so-called "cheap" fixtures, and a few minutes spent evaluating a sample fixture prior to its specification can result in substantial savings in installation and maintenance costs.

LIGHT DISTRIBUTION

Industrial lighting systems generally use direct distribution fixture types, which direct 90% to 100% of the light downward. These fixtures are available in beam spreads ranging from concentrating to wide-spread, as shown in Figure 11-17. Fluorescent fixtures typically have fairly wide beam spreads, but HID units are available in beam spreads ranging from highly concentrating to wide-spread.

Luminaire Classification	S/MH Ratio
Highly Concentrating	Up to 0.5
Concentrating	0.5 to 0.7
Medium Spread	0.7 to 1.0
Spread	1.0 to 1.5
Wide-Spread	Over 1.5

Figure 11-17
Fixtures are classified by their spacing criteria. See text for use of this information.

Note that the spacing criteria, discussed in Chapter 10, is related to the beam spread. Concentrating units require close spacing, i.e., 0.9:1, 1.0:1, while widespread units have a spacing criteria of 1.8:1 to 2.0:1.

The selection of a beam spread is influenced by the ceiling height, spacing between fixtures, and plane (horizontal, vertical, or tilted) in which the seeing task is oriented. All three of these factors must be considered interactively.

Concentrating fixtures are preferred for horizontal or near horizontal seeing tasks when ceiling heights are equal to or greater than the room width. The degree of concentration of the beam spread should increase as the ceiling height increases. This increases the quantity of light which travels directly to the workplane, and reduces light loss due to absorption by wall surfaces. At low design illuminance, greater spacing between fixtures will be required, thus wider beam spreads are indicated. As the orientation of the visual task changes from horizontal to vertical, increasingly wider beam spreads are desired to increase the vertical surface illuminance.

Locating fixtures at their maximum spacing, as determined by the spacing criteria, should be approached with caution, particularly when using HID luminaires. A layout may look fine of paper and meet the spacing criteria, but shadows may exist when workers are stationed between the luminaire and the task. HID lamps in sizes typically used for industrial applications produce large quantities of light, which means that fixtures are normally spaced widely apart and only one or two fixtures may make substantial contributions to the illuminance at some points in the building. If a worker is stationed between a task location and the luminaire which is the primary light source for that location, the worker may block light, and excessive body shadow may occur as shown in Figure 11-18.

When luminaires are to be suspended below the ceiling it is important to select fixtures with some uplight component to illuminate the ceiling. Failure to do so may create a feeling of working in a cave, and may increase the visual perception of direct glare since the bright fixture will be contrasted against a dark background.

Figure 11-18
Care must be taken with HID layouts to avoid body shadow (shown), or shadowing created by machines or other obstructions in the space. This condition is most likely to occur when mounting heights are low and fixtures are widely spaced.

DESIGN PROCEDURES

GENERAL AND TASK-ORIENTED SYSTEMS

Designs for general lighting systems which provide uniform horizontal illumination throughout the space, and task-oriented systems consisting of asymmetrical layouts which provide horizontal task illuminance and lower ambient levels for non-critical seeing in non-task areas are performed in accordance with the procedures outlined in Chapter 10.

WAREHOUSE STACK LIGHTING

The lighting of vertical warehouse stacks requires special treatment since the classical average illuminance calculation methods provide only information on the horizontal illuminance on the floor surfaces between racks, and the areas of primary interest are the vertical surfaces of the racks.

Point calculation methods, discussed in Chapter 8, are used to predict the illuminance on a grid pattern, as shown in Figure 11-19. The calculations should be for the plane in which the visual task is located. For example, reading labels on the sides of boxes or identifying material which is vertically oriented requires the evaluation of illuminance in a vertical plane. If the task involves looking down into a box after it has been removed from the rack, the horizontal illuminance should be evaluated. Many warehousing operations contain visual tasks in several planes, so multiple calculations may be required. Due to the large number of calculations required, the job is best done on a computer. Many fixture manufacturers can provide these analyses.

Fluorescent fixtures are generally used for racks with heights of 15 feet or less, with luminaires centered in the aisles at heights slightly greater than the stack height. A typical system is shown in Figure 11-20. Industrial type fixtures with apertured reflectors (Figure 11-21) are preferred over strip fixtures since the reflectors direct more light downward to illuminate the lower portions of the stacks.

HID luminaires are used at higher mounting heights since they direct more light downward to illuminate the lower part of the racks. A typical system is shown in Figure 11-22. Fixtures with asymmetrical distributions which direct light up and down aisles allow wider spacing between fixtures. As with fluorescent systems, luminaires should be mounted slightly above the tops of the racks, about 1/2 the width of the aisle.

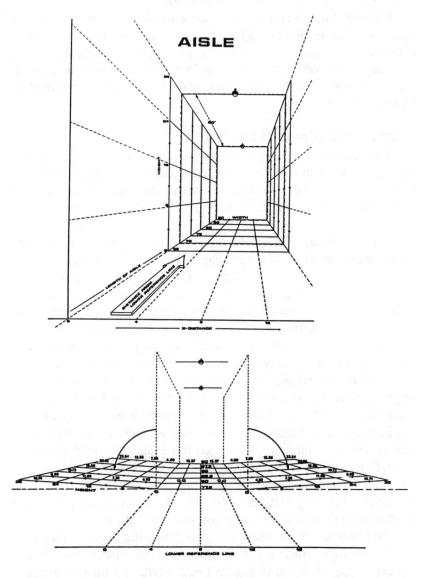

Figure 11-19
Illuminance for warehouse stacks is calculated on a grid pattern as shown. These calculations are typically done by computer since many calculations are required. *(Courtesy Illumination Computing Service)*

Figure 11-20
Fluorescent system used for warehouse lighting. *(Courtesy GE Lighting Co.)*

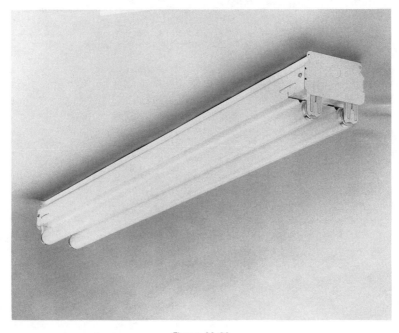

Figure 11-21
Fluorescent fixtures for industrial lighting may be strips (above), unaperatured reflectors (top of following page), or aperatured reflectors (bottom of following page). Aperatured reflectors are generally preferred. *(Courtesy Lithonia Lighting)*

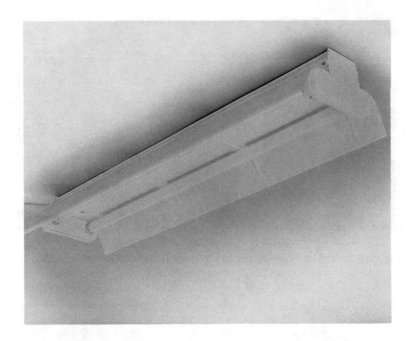

Figure 11-22
Typical HID warehouse lighting system. *(Courtesy Holophane Company, Inc.)*

SUPPLEMENTAL LIGHTING

Supplemental lighting is required when specific tasks require higher illuminance than that provided by the general lighting system. One or more luminaires are located to provide the desired illuminance on the task. Systems may be as simple as a single incandescent fixture mounted on a machine (Figure 11-23), or may consist of one or more ceiling-mounted or suspended fluorescent or HID luminaires as in Figure 11-24. When incandescent lamps are mounted on machines and subjected to vibration, vibration service lamps should be used, since vibration will cause conventional lamps to fail rapidly.

Figure 11-23
Incandescent lamps are frequently used in machine-mounted fixtures to concentrate light where it is needed. Lamps for these applications must be vibration service rated.

Luminaires should be located to avoid direct or reflected glare, which generally means they should be located to the sides of the task rather than above or in front of the task.

Point calculation methods are used to evaluate the anticipated illuminance. Remember that lighting levels are additive: the illuminance produced by the general lighting system must be added to the illuminance from the supplemental system to determine the total illuminance. In most cases the reflectances of surfaces in industrial buildings are low, and sufficient accuracy may be obtained by calculating only the direct component of illuminance from the general and supplemental luminaires.

Figure 11-24
Fluorescent fixtures are frequently used for supplemental light at task locations.
(Courtesy GE Lighting)

Chapter 12

OFFICE LIGHTING

In past decades office lighting generally consisted of blanketing the ceiling with 2' x 4', four-lamp troffers to provide uniform general illumination of 100-200 footcandles. Many designers gave little thought to lighting quality, and brute force was used to compensate for a lack of good design practice. The available lighting equipment was limited to a few lamps and one basic ballast type, and four-lamp fixture types were dominant. Many so-called lighting designs consisted of simply installing these fixtures on 6' x 8' or 8' x 8' spacing since experience had shown that these "standard" layouts would produce a system that people would accept.

The energy crisis of the mid 1970's gave rise to three occurrences that have had a profound effect on office lighting practice. First, manufacturers responded by introducing a plentitude of new energy-efficient products. Then legislation placed wattage and equipment efficiency constraints on lighting manufacturers and designers. And finally, the recommended lighting levels published by the Illuminating Engineering Society were revised to be more responsive to lighting needs as functions of workers' ages, the importance of speed and accuracy in performing visual tasks, and the light reflecting characteristics of the task and the space.

These factors, coupled with the widespread use of computers, with their unique lighting requirements, have necessitated changes in the way office lighting systems are designed, and have mandated a higher level of expertise on the part of the lighting designer.

Office work encompasses a wide variety of seeing tasks, ranging from reading and writing hard copy to simple conversations or business machine operation. In addition, offices range from large bull-pen arrangements to private executive offices, and each has its own set of lighting requirements. In classical textbook problems the determination of a design strategy is normally quite simple since task locations are fixed and all of the required information is readily available. Unfortunately, this is seldom the case in real world design problems since task locations are seldom known and it is a given fact that they will probably be moved frequently even if they have been determined prior to the original design.

UNIFORM vs TASK-ORIENTED SYSTEMS
UNIFORM SYSTEMS

In open office plans a uniform system is generally employed to allow flexibility in locating work stations and to accommodate the relocation of task areas as is common in many offices. While this means using standard fixture spacing, the designer does have considerable latitude in equipment selection to provide appropriately sized packages of light, and the designer's decisions will greatly impact the ongoing operating and maintenance costs of the system. Figure 12-1 shows how equipment selection can affect the visual appearance of the space.

Figure 12-1 (a)

Equipment selection greatly influences the visual perception of the space. This figure shows the same space lighted with (a) standard 12 pattern lenses, (b) open parabolic, and (c) small cell parabolic wedge louvers. The shielding media is the only change, yet the space is rendered differently by each system. *(Courtesy Peerless Lighting)*

Figure 12-1 (b)

Figure 12-1 (c)

EQUIPMENT SELECTION
FOR UNIFORM SYSTEMS

The trend towards lower lighting levels has essentially obsoleted the four-lamp, 2' x 4' troffer for office lighting. Three-lamp fixtures are now preferred for office areas with demanding visual tasks such as drafting, and two-lamp fixtures are preferred for lighting offices with typical reading and writing tasks. When three-lamp fixtures are specified they should employ only two-lamp ballasts in the Slave/Master configuration, as in Figure 12-2.

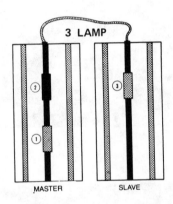

Figure 12-2
Three-lamp fixtures using three two-lamp ballasts. Factory supplied whip connects slave fixture to master and contains both power supply to third ballast and secondary ballast conductors for middle lamp. (*Courtesy Lithonia Lighting*)

The driving force in determining the lamp lumens required to produce some desired lighting level is the maximum spacing between fixtures that produces the desired uniformity of illumination. For example, four-lamp troffers using standard 40-watt, cool white lamps spaced on 8-foot centers in a typical large office can be expected to produce an average maintained illuminance of about 100 footcandles with good uniformity. Three-lamp fixtures in a similar arrangement provide over 75 footcandles, and dropping to two lamps per fixture results in about 55 footcandles average illuminance. A four-lamp system producing about 55 footcandles would require locating fixtures on 10' x 12' centers, which would exceed the recommended spacing criteria for most fixtures on ceiling heights of less than 11 feet. Note that these examples are based on typical fixtures and room characteristics and, like published automobile mileage ratings, may vary.

During the 1950's, 60's, and most of the 70's the lighting designer was hampered by a lack of available equipment, and the typical design used standard F40 lamps, standard magnetic ballasts, and troffers with #12 pattern lenses or white egg crate louvers. Today's menu includes energy efficient magnetic ballasts, electronic ballasts, T-8 and T-10 lamps employing exotic rare earth phosphors which combine good color rendering with high lumen output, and improved lens, louver, and fixture designs. The recent introduction of single ended compact fluorescent lamps in sizes up to 40 watts (Figure 12-3) permits the design of high efficiency fixtures in 2' x 2' sizes. By careful selection and application of this equipment a skilled designer can create a highly efficient system which produces illumination of the highest quality and enhances the visual environment.

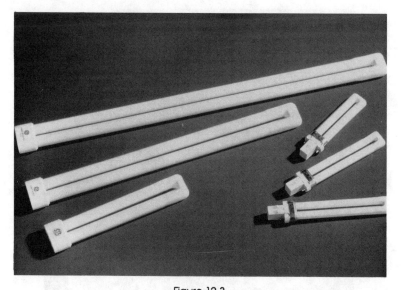

Figure 12-3
Compact single ended fluorescent lamps are available in sizes up to 40 watts. These highly efficient lamps are rapidly becoming popular for office lighting. (Courtesy GE Lighting)

This vast array of equipment can, however, be very confusing to the new or part-time lighting specifier. When faced with selecting or specifying new and unknown equipment it is human nature to stay with the standard systems and layouts that have been proven over years of use, and thus many opportunities for improved efficiency and lower life cycle cost systems are lost.

TASK-ORIENTED SYSTEMS

Task-oriented systems in offices usually consist of a general, uniform system designed to produce a low level of ambient light for visual comfort and general seeing in non-task areas, and a higher level of task illumination at the work stations. This is accomplished by designing a uniform system to the desired circulation level, and installing supplemental task lights at work stations.

Task-oriented systems are particularly applicable to modular office furniture arrangements, executive offices, and other spaces where desk and other task locations are fixed and not subject to frequent relocation.

Figure 12-4
Ceiling-mounted 2' x 4' troffers provide general illumination, with under-counter task lights providing supplemental light as needed. *(Courtesy GE Lighting)*

In private offices, fixtures can be concentrated around the task areas and spill light, wider fixture spacing, or lower wattage fixtures used to provide lower lighting levels in non-task or circulation areas. This also enhances the designer's opportunities to highlight artwork and other display items on walls or in display cases, and to create visual contrast within the space.

When modular furniture is used, common practice is to design a general lighting system to provide a lower level of circulation area illuminance, and install furniture-mounted task lighting, Figure 12-5, as described in detail in Chapter 10.

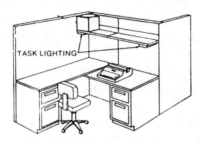

Figure 12-5
Application of supplemental task light. *(Courtesy Cooper Lighting)*

Care should be taken to assure that walls are adequately lighted to create a feeling of spaciousness since dark walls tend to visually "shrink" the space and create a closed-in feeling. Task-oriented systems are indicated when task areas are widely spread throughout the space, or when modular office furniture is used.

LAMPS FOR OFFICE LIGHTING

Fluorescent lamps, supplemented by incandescent lamps for specific highlighting or display applications, are the preferred source for most office lighting. Offices have been successfully lighted using metal halide and high-pressure sodium sources, but these are the exception, not the rule.

Metal halide lamps still suffer from non-uniform color, and while the color rendering characteristics of high-pressure sodium are acceptable for most industrial applications, they generally meet with less than enthusiastic acceptance in offices. In addition, the ceiling heights and design illuminances in offices generally dictate the use of fairly low lumen packages to achieve uniformity, and the efficacies of lower wattage metal halide and high-pressure sodium systems may be less than a comparable wattage fluorescent system. These problems may be ameliorated in the future but the lamps are currently considered to be rather poor choices for the average office.

Lamp selection is based on the desired color rendering, the lumen rating of the lamp, available fixture types, and economics. For example, cool white is the most common color, and thus has the lowest cost. The color rendering of cool white lamps, however, is considered to be rather poor, and the slightly higher investment in improved color rendering lamps may pay big dividends in worker satisfaction. The new breed of high color rendering "compact" lamps in wattages ranging up to 40 watts permit the use of 1' x 1' or 2' x 2' fixtures, as previously discussed. When possible, the number of different lamp types and colors should be held to a minimum to reduce stocking of replacement lamps and the probability of the incorrect lamp being used as a replacement.

White it is common to light new offices with reduced wattage, "energy saving," 34-watt lamps, this practice is frequently less energy efficient than using standard 40-watt lamps. This is due to the lower ballast factor with krypton-filled fluorescent lamps, which is discussed in detail in Chapter 16.

FIXTURE SELECTION

Fixture selection is based on availability, cost, aesthetics, light distribution characteristics, glare control, and the type of ceiling to be installed. While manufacturers list many fixtures in their catalogs, they normally carry large inventories of only their fast-moving fixtures. Lead times of weeks or even several months are normal for the less common fixtures. If the job is on a fast track it may be necessary to substitute a less desirable fixture, so a few minutes checking availability may be time well spent. It is far better to base a design on the fixture that will actually be installed than to specify one fixture and later settle for a substitute that may have inferior performance characteristics.

The issue of cost is frequently the driving force in the selection of a fixture, particularly in speculative or tenant-occupied buildings. It is an unfortunate fact of life that lighting is frequently the area in which costs are cut in order to bring the project on line within budget. It is important to ascertain, in advance, the amount that is available for lighting. In working with the client it should be stressed that fixture price is only one element in the cost of a lighting system, and savings in installation, operating, and maintenance costs must be weighed against the fixture price. Many of the low-priced "economy" fixtures are in reality high-priced if the added installation and maintenance costs are considered.

Be aware that the movement to use more efficient compact fluorescent lamps in applications previously employing incandescent sources has prompted the "conversion" of some incandescent fixtures to use the fluorescent lamps by simply installing a ballast and socket, with no modification to the reflector. In many cases both aesthetics and performance suffer. Figure 12-6 shows two downlights using fluorescent lamps. The fixture on the left is an incandescent "conversion," while the fixture on the right was actually designed for the compact fluorescent. Note the high glare and scattering of light by the "conversion" as compared to the low brightness and good control of the fixture designed for the lamp.

Figure 12-6
Comparison of incandescent downlight converted to use a compact fluorescent lamp (left), versus a downlight specifically designed to utilize the fluorescent lamp. See text for explanation.

The light distribution characteristics of the fixture will determine the maximum permissible spacing between fixtures to achieve uniformity of illumination, and the potential for causing direct glare. Uniformity may be evaluated with respect to the spacing criteria, discussed in Chapter 10, and glare is discussed in Chapter 9. These chapters should be reviewed as necessary.

The ceiling system will also be a determining factor in the ultimate selection of a fixture. Grid ceilings are the most common, yet some plastered ceilings are still encountered. Fixtures designed for plastered ceilings have flanges that are not compatible with grid ceiling systems, and different grid systems may require special fixtures. Make sure that the specified fixture is compatible with the ceiling system.

Fixture layouts and calculations for spacing for uniform systems are also covered in Chapter 10.

DIRECT vs INDIRECT LIGHTING SYSTEMS

Most offices are lighted with direct lighting systems using recessed troffers or fixtures that are mounted directly on the ceiling surface. These fixtures project all (or nearly all) of their emitted light downward to the work surface, and are referred to as "Direct" lighting systems. An alternate system uses fixtures that are suspended some distance below the ceiling (Figure 12-7) and direct most or all of the light upward to the ceiling. The ceiling then reflects the light downward to the workplane. Each system has its strengths and weaknesses.

Direct systems are more energy efficient in producing raw footcandles since a higher percentage of the lamp lumens reach the workplane. They tend to produce higher veiling reflections, however, so the quality of light may be reduced.

Indirect systems are less efficient than direct systems since some light is lost through absorption by the ceiling, but they produce fewer veiling reflections. This means that Equivalent Sphere Illuminance is improved, and the system need not provide the same raw illuminance to provide equal visual performance. Uniform indirect systems produce a very uniform illuminance throughout the work space, and this may create a very bland visual perception of the environment.

The primary application of uniform indirect lighting systems is the electronic office with a large number of CRT's. The uniform ceiling luminance reduces the perception of glare in the screen and makes it

Figure 12-7
Indirect system uses light reflected by ceiling to provide high quality, low glare illumination. These systems are particularly well suited to offices with high concentrations of CRT's. *(Courtesy Peerless Lighting)*

easier to see. Care must be taken, however, to assure that the fixture body is light colored so it will blend into the ceiling. Dark fixtures will appear as dark lines reflected in the screen and most workers will find the condition annoying.

Indirect fixtures for lighting the electronic office must have light distributions that are wide enough to create a relatively uniform ceiling luminance. Uniformity ratios of 4:1 or less are recommended. A narrow distribution will concentrate light on the ceiling directly above the fixture, and create alternating areas of light and dark ceiling that will reflect from CRT's and be objectionable. The intent of an indirect system is to cast a uniform ceiling reflection on the CRT screen, thus minimizing the visual perception of reflected glare.

Offices employing CRT's may also be lighted successfully with deep-cell, parabolic-type luminaires with good cut-off characteristics, as shown in Figure 12-8. These fixtures provide good horizontal illumin-ance for reading hard copy, yet the sharp cut-off minimizes reflected glare from the computer screen.

Figure 12-8

Fixture on right is a common 3-lamp parabolic unit used for many office lighting applications. Fixture on left is also a parabolic, but employs a special low brightness louver design. The low brightness unit is well suited for offices with CRT's, and other applications where a low brightness ceiling is desired.

Chapter 13

MERCHANDISE LIGHTING

Retail stores encompass a variety of lighting tasks ranging from the display of merchandise to warehousing and office functions. This chapter will concentrate on the lighting of the retail sales area, and other chapters should be consulted for lighting other areas.

Retail stores have a single objective: to generate a profit for the owner or owners by selling merchandise. The methods by which this is accomplished will vary from store to store, but in nearly all cases the performance of the lighting system will influence sales, and thus influence profit. A poorly designed system which does not provide sufficient illumination for the identification or evaluation of merchandise will result in lost sales, while a system that provides adequate lighting will enhance sales and improve the potential for profit.

Lighting can represent 50% or more of the electrical energy costs in a retail store. Since operating costs can cut deeply into profit, it is in the best economic interest of the owner to obtain the desired lighting effects with energy efficient equipment and design techniques which keep the cost of light at a minimum.

BASIC DESIGN CONCEPTS

For purposes of developing a basic design concept, retail sales areas may be divided into three broad categories based on the customer traffic volume: high activity, medium activity, and low activity. Each category has characteristics which distinguish it from the other categories, and each calls for a different lighting strategy to meet the design objective of the owner.

The activity level and ambience of the space determine the design illuminance, as shown in Figure 13-1. Note that the recommended

Areas or Tasks	Description	Type of Activity Area*	Illuminance** Lux	Illuminance** Foot-candles
Circulation	Area not used for display or appraisal of merchandise or for sales transactions	High activity Medium activity Low activity	300 200 100	30 20 10
Merchandise† (including showcases & wall displays)	That plane area, horizontal to vertical, where merchandise is displayed and readily accessible for customer examination	High activity Medium activity Low activity	1000 750 300	100 75 30
Feature displays†	Single item or items requiring special highlighting to visually attract and set apart from the surround	High activity Medium activity Low activity	5000 3000 1500	500 300 150
Show windows 　Daytime lighting 　　General 　　Feature 　Nighttime lighting 　Main business districts (highly competitive) 　　General 　　Feature 　Secondary business districts or small towns 　　General 　　Feature			 2000 10000 2000 10000 1000 5000	 200 1000 200 1000 100 500
Sales Transactions	Areas used for employee price verification and for recording transactions	Reading of copied, written, printed or electronic data processing information	See Fig 2-1	
Support Services	Store spaces where merchandising is not the prime consideration	Alteration rooms, fitting rooms, stock rooms, locker areas, wrapping and packaging	See Fig. 2-1	

* One store may encompass all three types within the building, i.e.:

High activity area — Where merchandise displayed has readily recognizable usage. Evaluation and viewing time is rapid, and merchandise is shown to attract and stimulate the impulse buying decision.

Medium activity — Where merchandise is familiar in type or usage, but the customer may require time and/or help in evaluation of quality, usage or for the decision to buy.

Low activity — Where merchandise is displayed that is purchased less frequently by the customer, who may be unfamiliar with the inherent quality, design, value or usage. Where assistance and time is necessary to reach a buying decision.

** Maintained on the task or in the area at any time.

† Lighting levels to be measured in the plane of the merchandise.

Figure 13-1

Recommended illuminances for merchandising areas. (Courtesy Illuminating Engineering Society, IES Lighting Handbook) levels for feature display lighting are five times the recommended levels for general lighting. These levels are required to make the display stand out and draw the customers' attention.

HIGH ACTIVITY

High-activity areas are primarily self-service (as opposed to clerk-assisted) and merchandise is readily accessible to customers to permit both visual and tactile evaluation. Few feature displays are used, and those that are used are seldom highlighted. The emphasis is on volume sales through low prices, and frills are generally kept to a minimum. Examples of high-activity stores are supermarkets, discount department stores, drug/department stores, and discount specialty stores. Points of sale (cashiers) may be available in specific departments, but they are generally clustered at one location.

Lighting systems are generally fluorescent, with surface-mounted or suspended strip fixtures the most common. They are popular for several reasons: low first cost, ease of maintenance, high efficiency, and a psychological connotation of low cost which subconsciously tells the customer that money is not wasted on costly buildings so prices are lower. They also create a high-brightness ceiling which, when viewed from outside the store, tells customers that the store is open. Eight-foot high-output or slimline lamps are generally used.

Fluorescent strip systems are followed in popularity by recessed troffers, particularly in stores with lower ceiling heights. Troffers have a lower surface brightness than bare lamp strip fixtures, and thus have a lower potential for causing direct glare.

A new breed of high-volume discount store, patterned after industrial-type warehouses, has recently emerged. These facilities use warehouse lighting designs and frequently employ industrial type HID luminaires with metal halide lamps.

High-activity stores rely on impulse purchases for a significant percentage of their sales. This means that the probability of a sale is related to the amount of merchandise the customer sees, and the more merchandise viewed, the higher the probability of a sale. Consider the layout of a supermarket: meat and dairy products are almost universally located at the rear of the store. This is not by chance. Almost everyone buys meat and dairy products, and must pass many other items to reach these staples. How many times have you gone to a market to buy a quart of milk, yet returned with several other items? By locating the milk at the rear of the store you were forced to pass many other items, and probably bought one or more on an impulse.

Brightly lighted rear walls act as a beacon and help attract customers, so particular attention should be given to lighting these walls if they are visible from other parts of the store and the arrangement of merchandise permits it. Figure 13-2 shows a supermarket using a combination of fluorescent and metal halide equipment.

Figure 13-2
Supermarket lighted with a combination of metal halide and fluorescent equipment. Metal halide is used in the produce department to impart sparkle and glitter to fruit and vegetables, with fluorescent used to provide general lighting throughout the balance of the store. Note the high luminance of the rear wall. *(Courtesy Lithonia Lighting)*

MEDIUM ACTIVITY

Medium-activity stores are usually a combination of self-service and clerk-assisted. Most merchandise is readily accessible to the customer for evaluation, but salespeople are available to assist in the location and evaluation of specific items. Numerous feature displays may be used, and are generally highlighted to draw the customer's attention.

While the primary emphasis is on sales, prices are generally slightly higher than high-volume stores to justify the added time spent with customers. More emphasis is placed on decor to create a more pleasing and comfortable environment to encourage customers to linger, and thus increase the possibility of a sale. In large department stores the points of sale are located in each department, as opposed to a central location. Examples are neighborhood hardware stores, most sporting goods stores, and department stores catering primarily to the mid- to upper-middle-income consumer.

Lighting systems generally employ a combination of fluorescent equipment for general lighting, and incandescent equipment for display lighting. Color rendering is usually important, so high color rendering fluorescent lamps are desirable. Conventional "Deluxe" lamps are seldom used due to their low efficacy (a deluxe lamp produces about 25% less light than a standard color), but the new generation of high color rendering lamps employing exotic phosphors provides excellent color rendering and high efficiency. While they are slightly higher priced than the conventional colors, benefits in improved color more than justify added cost. Warm colors are generally preferred over cool colors as they provide a better color match with incandescent display lighting.

Fluorescent fixtures are generally recessed troffers, either 2' x 4' or 2' x 2'. The 2' x 4' configuration is preferred if the size and shape are compatible with the desired visual appearance of the ceiling since lamps are less costly and available in more colors. This trend may change, however, due to the introduction of single ended compact fluorescent lamps in higher wattages which offer high efficacy and good color in lamps that are well suited to 2' x 2' fixture types. Parabolic louvers provide a lower surface brightness than most plastic lenses, and may be desirable in areas where a low-brightness ceiling is needed. The use of small (2" x 2" or less) louvers creates a very dark ceiling surface; however, fixture efficiencies are considerably lower than large cell parabolic fixtures or plastic lenses. For this reason they should be applied only when the visual appearance of a very low-brightness ceiling is required.

Feature display lighting consists of highlighting specific displays of merchandise to levels of five times the ambient level (Figure 13-3). It is normally done with reflectorized incandescent lamps with light control characteristics that place most of the light on the display and limit

Figure 13-3
Display (top) is lighted with a combination of line and low voltage incandescent lamps to create a high level on the merchandise and lower surrounding ambient level. Bottom photo shows the track system used for the display. Two 60PAR/HIR floods and three MR-16 lamps provide good performance at low energy cost.

spill light in adjacent areas. Lamps may be either line or low voltage, with low voltage lamps preferred for most applications due to their superior light control and higher efficacy.

LOW ACTIVITY

Low-activity stores are generally clerk-assisted, and little or no merchandise is readily accessible to the customer. They are generally specialty shops or separate departments in medium-activity department stores. Merchandise is generally high-quality and carries commensurate price tags, which are frequently not visible. Low-volume stores frequently cater to a specific customer class, and place emphasis on creating an environment in which the customer feels comfortable. The point of sale is generally the sales person, and the cash register may not be visible to the customer. Typical examples are fine jewelry or furniture stores.

Low-activity stores are generally lighted to fairly low ambient levels, 30 footcandles or less for general illumination. This creates a restful environment which encourages customers to linger. Incandescent equipment is usually used for both general and display lighting. On those occasions when fluorescent equipment is employed the appearance of the fixture must fit in with the rest of the store, so low-brightness louvers are preferred over lenses.

A single store may contain more than one category of merchandising space, such as a medium-activity department store with a fine jewelry department (low activity) and a "bargain basement" (high activity). Some caution must be applied when categorizing retail sales operations, since some stores will appear to fall into a specific category, yet an analysis of the lighting needs may dictate a design which is contrary to the general design guidelines for that category. For example, a dealer in heavy construction equipment may have only one or two prospective customers a week, and sell only one or two pieces of equipment each month. By definition, this is a low-volume store, but the required lighting design strategy may well fit the general recommendations for a high-activity store.

Discussions with the store owner are mandatory if the lighting designer is to do the job properly. If there is a communications gap and the designer fails to understand or properly evaluate the lighting needs, the lighting system will probably fail to meet the owner's needs and expectations. This point was illustrated recently when I took a group of lighting students to a shopping mall to evaluate lighting systems in stores. After stopping outside a clothing store I asked the class for their impression of the store. The response was unanimous: high class, high quality, and expensive. Their impressions were based on a visual evaluation of relatively low general lighting levels, impressive feature display lighting, an elegant and obviously expensive decor, and no customers in the store. Upon entering the store we discovered that the merchandise was of moderate quality and price. The store manager complained of a lack of customers, indicated that many of those who entered left without making a purchase after indicating that they were looking for better quality clothing, and sales were barely adequate to justify continuing operation. The problem was quite simple: the decor, enhanced by the lighting system, communicated the wrong message to potential customers.

DESIGN METHODS

The design of a system for retail store lighting has several facets. The general lighting system is designed, and feature display lighting, wall washing, and point of sale identification are then integrated into the basic design.

GENERAL LIGHTING

General lighting systems provide relatively uniform illumination throughout the space to permit selection and evaluation of merchandise. In small stores, or those with a high density of merchandise and limited circulation area, a uniform arrangement of luminaires is typically used. Layouts in stores with wide aisles for circulation, the locations of which are fixed and seldom change, may utilize asymmetrical layouts which provide higher levels in merchandising areas and lower levels in the aisles. General lighting is usually used in high- and medium-activity stores.

The number of luminaires is calculated using procedures in Chapter 7, and layouts determined in accordance with the methods described in Chapter 10.

FEATURE DISPLAY LIGHTING

Feature display lighting highlights specific displays to draw attention to the merchandise, Figure 13-4. Feature displays take many forms, such as mannequins, clothing attached to a wall, or an arrangement of merchandise which is set off from surrounding merchandise. The objective of feature display lighting is to illuminate the merchandise to levels five times the ambient level, and avoid excessive spill light in the surrounding area which would diminish the visual impact of the highlighted display.

Incandescent lamps are typically used for their color rendering and good light control characteristics (Figure 13-5), but recent advances in metal halide and color-improved high-pressure sodium lamps have given the lighting designer high-efficiency HID sources which are acceptable for many display lighting applications.

Incandescent lamps may be line voltage (120v) or low voltage, typically 5.5v or 12v. (See Chapter 2 for the differences between line and low voltage lamps.) PAR lamps, either standard or halogen cycle, are preferred over "R" lamps when line voltage lamps are used due to their superior light control characteristics. When low-voltage lamps are used the MR-16, MR-11, and PAR 36 12v lamps are preferred for moderate sized displays. When small objects are displayed individually, the very tight beam control provided by 5.5v PAR 46 lamps may be preferred.

The illuminance on the display can be predicted using the point calculation methods explained in Chapter 8. The most common method, however, is the use of charts and tables which are provided by most lamp and fixture manufacturers. These provide a quick, simple means of estimating the illuminance which a specific lamp or fixture will produce on the display.

WALL WASHING

Walls are lighted in most stores for two basic reasons: merchandise is frequently located on walls for purposes of display, and well illuminated walls tend to create a visual impression of spaciousness, thus making the space appear to be larger.

Wall washing consists of mounting fixtures on or near walls, and directing light downward to more or less graze the walls. The closer the fixtures are to the walls, the more pronounced the grazing effect will be,

Figure 13-4
Unlighted floral display (top) appears bland and lacks appeal. Same display
stands out and attracts attention when lighted with a single 42-watt MR-16 lamp
(bottom).

Figure 13-5
Low voltage MR-16 lamps provide good control to highlight specific items in this display case. Note that not all merchandise is highlighted. *(Courtesy GE Lighting)*

and the greater the shadowing on uneven surfaces. If rough textures are used and shadowing is desired, fixtures should be located as close to the walls as possible. If the wall has unintentional rough areas and a smooth appearance is desired, the fixtures should be ceiling-mounted some distance away from the wall to minimize shadowing.

Fluorescent sources are generally used, both for their energy efficiency and uniformity of illumination. Incandescent sources may be used, but energy costs will be higher than a fluorescent system, and the illuminance will not be as uniform. A typical system uses a single continuous row of F40 lamps, with the ends overlapped, as shown in Figure 13-6. The lamp ends are overlapped to prevent shadows where the lamps butt together.

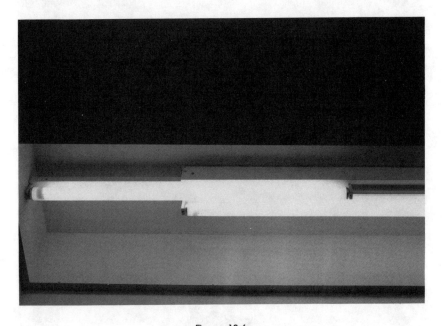

Figure 13-6
Ends of lamps should be overlapped to achieve uniformity when using fluorescent sources for wall washing. The overlap in this photo is excessive but necessary due to the length of the cavity. An overlap of about one inch is adequate.

Chapter 14

FLOODLIGHTING — PARKING LOT — STREETLIGHTING

Outdoor lighting calculations may be performed using a variety of methods which vary widely in both complexity and accuracy. The methods typically employed, in order of ascending sophistication, are:

1. The "Watts per square foot method," where the area to be lighted, in square feet, is multiplied by an estimated "watts per square foot" which are required to produce some specified illuminance. The method is based on average efficacies for various light sources, and average values for the percentage of lamp lumens delivered to the area to be lighted. This is generally the simplest and least accurate method used for outdoor lighting and is frequently used as an estimating tool or to determine a starting point for more complex design methods.

2. Isofootcandle diagrams are pictorial representations of the distribution of light on a horizontal plane below the luminaire. They are generally prepared using the inverse square law to determine the illuminance produced by a specific luminaire at various points, and the lines of equal footcandles are traced on the diagram. These "pictures" are drawn to a convenient scale and published by many luminaire manufacturers as an aid in the design process. In practice, the diagrams are reproduced on clear plastic or traced on velum, and the reproductions are used as overlays on the area to be lighted to evaluate the uniformity and distribution of light. By shifting the transparencies on a

scale drawing of the area to be lighted, the pole locations and uniformity of illuminance can be estimated.

3. The Beam Lumen Method is similar to the lumen method used for interior lighting design. A "coefficient of beam utilization" is substituted for the familiar coefficient of utilization, and the average lumens per square foot are calculated within the confines of the area to be lighted. Since the method provides only the average illuminance, the lighting levels at several points should be determined using isofootcandle diagrams or point calculations to evaluate the uniformity of illumination.

4. Point calculations are prepared using the inverse square law for a representative sample of points within the area to be lighted.

Outdoor lighting design frequently involves much trial and error, and the less complex calculation methods may prove to look fine on paper but leave much to be desired in actual application. There is no "one best" method for calculations, and a combination of different methods may be employed to arrive at a final design. For example, the beam lumen method requires that the locations and aiming points of the luminaires be known in order to determine the percentage of beam lumens that reach the area to be lighted. It would appear obvious that if this information were available, the system would have already been designed, and the calculations unnecessary.

A reasonable approach to the initial design might be to approximate the number of fixtures needed using either the "watts per square foot" method, or by estimating the beam coefficient of utilization, laying out the system using normal design practices, and then verifying the average illuminance using isofootcandle diagrams. As a final check, point calculations might be utilized as a validation of the proposed design.

Floodlighting calculations are usually performed using the beam lumen method or the inverse square law. Parking lot lighting may be designed using either of those methods, by the use of isofootcandle diagrams, or estimated on a "watts per square foot" basis. Street lighting design typically employs the inverse square law, isofootcandle diagrams, or "Standardized designs" using only a predetermined luminaire type, mounting height, and spacing between poles.

GENERAL DESIGN CONSIDERATIONS

FLOODLIGHTING

Floodlights differ from typical parking lot and street lighting luminaires in that they are mounted on adjustable brackets or other devices and are capable of being aimed at some specific point as determined by the lighting designer. Parking lot and street lighting fixtures typically employ a fixed mounting arrangement and are aimed at a predetermined point which is specified by the luminaire manufacturer. Note that while some minor adjustments may be made in the field, these types of luminaires offer only limited adjustment when compared to the wide range of adjustments possible with floodlights.

Floodlights are classified by the National Electrical Manufacturers Association (NEMA) on the basis of beam spread, as shown in Figure 14-1. The beam angle is defined as the angle at which the intensity of the beam will have dropped off to a value of 10% of the maximum candlepower contained within the beam (Figure 14-2). For example, assume that a floodlight had a maximum intensity of 100,000 candelas. The limits of the beam, for classification purposes, are the angles at which the intensity is 10,000 candelas. Note that some light falls outside the angle that defines 10% of the maximum intensity; however, this is normally a small fraction of the total flux projected by the luminaire—the maximum candlepower for floodlights is normally found at or near the center of the beam. Note that in all other countries, the beam angle is defined as the angle at which the candlepower drops to 50% of maximum. For theatre lighting the 50% value is used to define beam angle, and the term "field angle" is applied to the 10% value.

Luminaires which have a symmetrical distribution (Figure 14-3) are described by a single number, e.g., NEMA type 2 for a floodlight that projects a circular pattern which is contained inside an angle of not less than 18° but not more than 29°. Symmetrical distribution luminaires are generally used in narrow distribution types for sports lighting and other similar applications involving long throw distances.

Asymmetrical distribution type luminaires (Figure 14-4) have beam spreads which differ between horizontal and vertical angles, as illustrated in Figure 14-5. A typical asymmetrical distribution luminaire, designated "NEMA 7x6" would have a horizontal beam spread of 130° or greater, and a vertical spread of 100°-130°. The first number denotes the horizontal spread and second number refers to the vertical spread. Asymmetrical distribution luminaire types are generally used for area and building floodlighting.

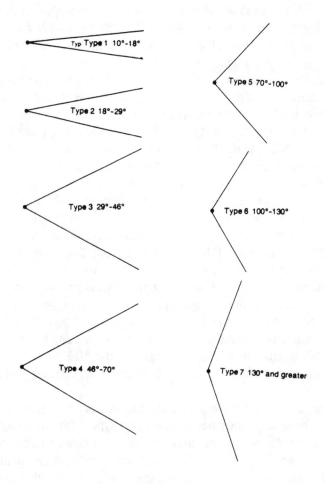

Figure 14-1
Angles used for NEMA beam spread classifications.

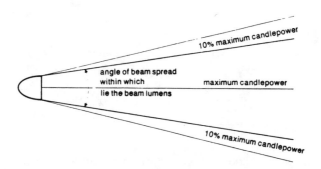

Figure 14-2
Beam lumens are the lumens contained within the angle at which the intensity
has dropped to 10% of the maximum intensity. (*Courtesy Cooper Lighting*)

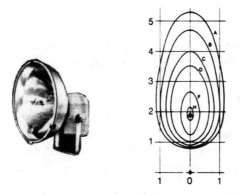

Figure 14-3
Typical NEMA 2 floodlight, and the projection of light on the ground. Note that,
for this illustration, the fixture is aimed at a point 2 mounting heights in front of the
fixture location, causing the pattern to be oblong. (*Courtesy GE Lighting*)

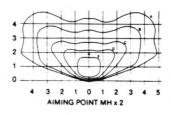

Figure 14-4
Asymmetric distribution luminaire projects an irregular beam pattern which can
be useful for lighting many outdoor areas. (*Courtesy GE Lighting*)

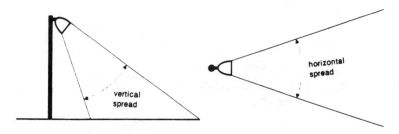

Figure 14-5
Vertical spread describes the spread of light from front to rear of pole.
Horizontal spread describes the light distribution to the sides. (*Courtesy Cooper Lighting*)

The following general guidelines apply to the selection of most floodlighting equipment:

1. Use narrow beam spreads for long throw distances and wide beam spreads for short throw distances.

2. When using wide angle,non-aimable luminaires (e.g., parking lot types), limit the spacing between poles to 4 times the mounting height of the luminaire and locate the first pole in an arrangement not more than 2 times the mounting height from the edge of the area to be lighted. (See Figure 14-6.)

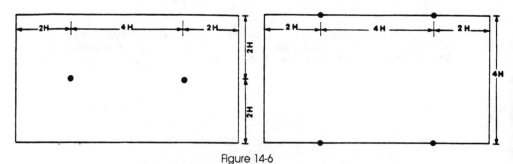

Figure 14-6
Poles for floodlighting are typically located not more than 2 mounting heights away from the edges of the area to be lighted, and not more than 4 mounting heights apart. This is done to provide uniform illumination.

3. Installations consisting of a few large poles and high-wattage lamps will generally be more economical than installations of many short poles and low-wattage lamps. Use the highest poles

and largest lamps consistent with the job requirements and the available maintenance equipment. Note, however, that most lighting maintenance companies can work heights of 40-45 feet with existing equipment. Poles of greater height will normally increase maintenance costs dramatically unless the poles are equipped with costly lowering devices for luminaires. High-mast lighting, involving pole heights of 80 feet or greater and 1000-watt lamps can provide substantial economies, but heights between 45 feet and 80 feet require careful analysis. Recommended minimum mounting heights necessary to limit glare when using unshielded type luminaires are shown in Figure 14-7. Fixtures with lenses or sharp cut-offs may frequently be mounted at lower heights, and manufacturers' literature should be consulted.

	Mercury		Metal Halide		High Pressure Sodium	
	Wattage	Mtg. Ht	Wattage	Mtg. Ht	Wattage	Mtg. Ht
Minimum	–	–	–	–	35	6
Recommended	–	–	–	–	50	7
Mounting	–	–	–	–	70	8
Heights*	100	8	–	–	100	12
Guidelines	175	10	175	16	150	18
	250	15	250	20	250	25
	400	23	400	25	400	30
	1000	30	1000	35	1000	38

Figure 14-7
Recommended minimum mounting heights for typical non-shielded flood and street light luminaires to limit direct glare. Shielded and cut-off types may be used at lower mounting heights. (*Courtesy Cooper Lighting*)

4. Select beam patterns to provide sufficient overlap to achieve uniformity. As a general rule, a 50% beam overlap, as shown in Figure 14-8, is desirable.

5. When the size and geometry of the area to be lighted are such that poles could be located either inside the area or around its perimeter, locate poles within the space if people and the visual task are both located within the area; locate poles around the perimeter if the primary use involves viewing within the area by viewers located outside the area, or if the primary purpose of the lighting is security.

6. To avoid shadows, the area should be lighted from more than
    one direction.

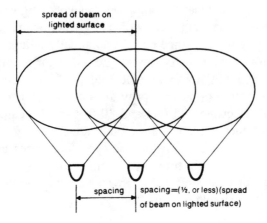

Figure 14-8
Beam patterns should overlap by 1/2 for uniformity. (*Courtesy Cooper Lighting*)

PARKING LOT LIGHTING

Parking lots may be lighted using floodlighting equipment, street
lighting equipment, or fixture types which are specifically designed for
parking lot applications. A typical fixture designed for lighting parking
lots and similar areas is shown in Figure 14-9.

These fixtures typically throw a somewhat rectangular beam pattern
to meet the requirements of lighting rectangular areas, are usually
mounted in a horizontal or near horizontal position, and are generally
used at mounting heights below 45 feet.

Large parking lots or similar areas may also be lighted with "high
mast" equipment using pole heights of 80 to 150 feet and arrays
consisting of 10 to 20 luminaires using 1000-watt high-pressure
sodium or metal halide lamps, as illustrated in Figure 14-10. These
systems employ special lowering equipment which permits lowering of
the fixture array to ground level for maintenance. Due to the wide
distribution of light over an area, these systems are used only for large
areas where the light can be confined to the area to be lighted. The "per
pole" cost of high-mast lighting is high, typically in the range of
$15,000-$25,000 per pole, including luminaires and installation, but
the number of poles is dramatically reduced from the number required

for 45-foot or lower poles, and the total cost of a high-mast system can be much lower.

Figure 14-9
Typical rectilinear fixture used for parking lot and area lighting. *(Courtesy Holophane Co.)*

Figure 14-10
Typical high mast lighting system.

Note that in parking lot lighting special care should be taken to provide adequate illumination at entrances and exits. Care must also be taken to minimize direct glare to avoid safety hazards. This is normally achieved by using cut-off type luminaires which reduce or eliminate high-angle brightness, or mounting luminaires at heights which are sufficient to remove the luminaire from the field of view. Recommended minimum mounting heights for unshielded luminaires are shown in Figure 14-7.

The general guidelines listed for floodlighting also apply to parking lot and similar area lighting.

Figure 14-10 (Cont'd)

STREET LIGHTING

Street lighting normally employs equipment that is specifically designed for the task–throwing a long, narrow beam of light on the roadway. Since street widths are normally somewhat standardized, the

design of street lighting system is usually accomplished using standardized designs provided by fixture manufacturers or utilities.

Street lights are classified into 5 different types, based on their light
distribution, as illustrated in Figure 14-11. Type I luminaires are
typically mounted in the center of the roadway and throw light up and
down the street. Types II, III, and IV are located near the curb line and
have successively narrower patterns. Type V has a symmetrical
distribution and is designed for use at the center of an intersection. Note
that Types I and II are also available in 4-way distributions for use at
intersections.

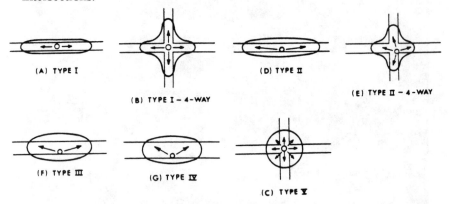

Figure 14-11
Street light luminaires are classified by the way light is distributed on the
roadway. *(Courtesy IES Lighting Handbook, IESNA)*

Street lights are also classified by their projection or "throw" of
light along the roadway: short, medium, and long. Short distribution
luminaires project their maximum candlepower into a zone which lies
from 1 to less than 2.25 mounting heights from the luminaire. Medium
distribution luminaires project maximum candlepower from 2.25 to less
than 3.75 mounting heights, and long distributions project from 3.75 to
less than 6 mounting heights. This is illustrated in Figure 14-12. In
general, the maximum spacing between luminaires is 4.5 mounting
heights for short distribution fixtures, 7.5 mounting heights for
medium, and 12 mounting heights for long distribution luminaires.

Roadway lighting has a language of its own, as can be seen in
Figure 14-13. Street lights see the street as having a fixed width, called
the "transverse distance," and an infinite length, called the "longitudinal

distance." These distances are expressed in terms of the mounting height of the fixture, e.g., a 50'-wide street, lighted by a luminaire mounted at 25', would have a transverse distance of 2 mounting heights (2 MH). Most street lights overhang the roadway slightly and have asymmetric distributions designed to project light along and across the roadway and, to a lesser extent, rearward to illuminate the narrow strip of road which lies behind the fixture. The direction in front of the luminaire is called the "street side," and the direction behind the fixture is called the"house side."

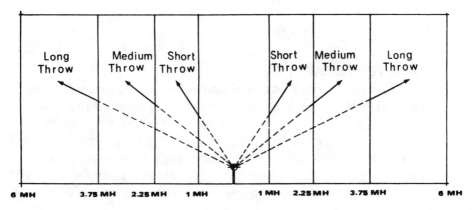

Figure 14-12
Street lighting luminaires are also classified as long, medium, or short throw, depending on the distance they project light along the highway.

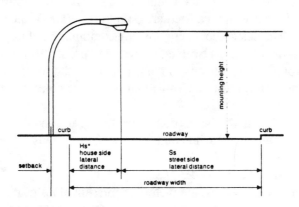

Figure 14-13
Terminology used in street lighting. *(Courtesy Cooper Lighting)*

LIGHT LOSS FACTORS

Outdoor lighting installations are subject to the light loss factors discussed in Chapter 6. These factors, with the exception of luminaire dirt depreciation and room surface depreciation, are of the same magnitude in both indoor and outdoor installations, and may be applied as previously discussed.

Luminaire dirt depreciation factors for outdoor lighting may be found in Figure 14-14 for typical fixtures.

CALCULATION METHODS

WATTS PER SQUARE FOOT

This method is based on typical lamp efficacies and fixture efficiencies, and assumes that the light distribution characteristics are compatible with the geometry of the space to be lighted. It frequently serves as a starting point for more rigorous design procedures.

The process consists of simply determining the total square footage of the area to be lighted, and multiplying the area by a factor found from Figure 14-15 to determine the total lamp watts required to light the space.

For example, assume that a 30,000-square-foot area is to be lighted to a level of 4 footcandles, maintained, using a high-pressure sodium source. From Figure 14-15 the watts per square foot required are 0.08, and the total lamp watts required are 0.08 times 30,000 square feet, or 2,400 watts. The design could employ any combination of lamps which total 2,400 watts, such as twelve 200-watt lamps, six 400-watt lamps, or twenty-four 100-watt lamps. If the geometry of the space called for 5 poles, an installation of ten 250-watt lamps (2 per pole) might also be considered.

Note that this method is approximate at best, and should be verified by the use of isofootcandle diagrams or point calculations.

ISOFOOTCANDLE DIAGRAMS

Isofootcandle diagrams are pictorial representations of the illuminance at various points on the ground beneath the luminaire. A typical isofootcandle diagram is shown in Figure 14-16.

Roadway Lighting

Luminaire Type	Cleaning Every Year Environment:			Cleaning Every Three Years Environment:			Cleaning Every Six Years Environment:		
	Clean	Medium	Dirty	Clean	Medium	Dirty	Clean	Medium	Dirty
Roadway	0.95	0.92	0.87	0.86	0.77	0.67	0.77	0.62	0.49
Area	0.95	0.92	0.87	0.86	0.77	0.67	0.77	0.62	0.49
Architectural	0.95	0.92	0.87	0.86	0.77	0.67	0.77	0.62	0.49

Floodlighting

Luminaire Type	Cleaning Every Year Environment:			Cleaning Every Three Years Environment:			Cleaning Every Six Years Environment:		
	Clean	Medium	Dirty	Clean	Medium	Dirty	Clean	Medium	Dirty
Flood	0.95	0.92	0.87	0.86	0.77	0.67	0.77	0.62	0.49
Flood, Hermetically Sealed	0.98	0.98	0.98	0.98	0.96	0.84	0.96	0.78	0.61

Clean: Clean pavement, grass, no open loose ground, little or no adhesive qualities in atmosphere. Average car and truck traffic. Downtown open areas, intermediate roads and freeways in open areas.

Medium: Same as *clean* except slightly more exposure to adhesive qualities in atmosphere. Residential, intermediate roads and local minor roads. More than average truck traffic.

Dirty: Confined. Greater than *medium*. Cars and trucks on expressways and freeways. Downtown. Major roads. Adhesive dirt.

Figure 14-14

Typical luminaire dirt depreciation factors for outdoor luminaires. *(Courtesy Cooper Lighting)*

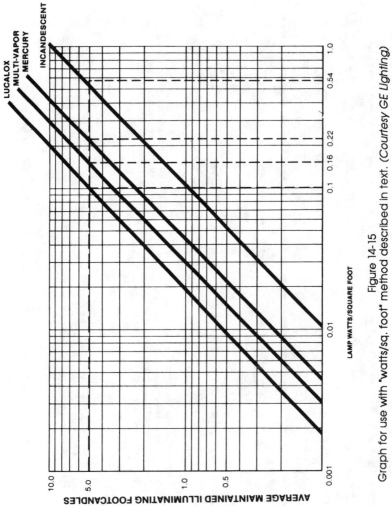

Figure 14-15

Graph for use with "watts/sq. foot" method described in text. *(Courtesy GE Lighting)*

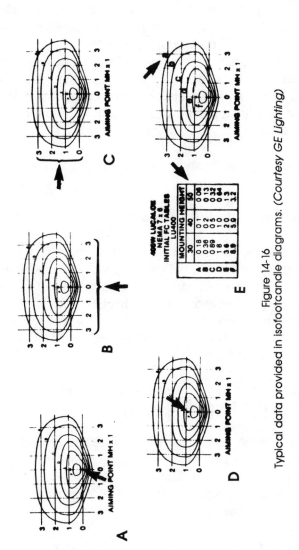

Figure 14-16
Typical data provided in isofootcandle diagrams. (*Courtesy GE Lighting*)

The Figure contains a great deal of useful information about the illuminance produced by the fixture, and can be used in the design process to locate luminaires and assess the predicted uniformity of illumination. Note, however, that the diagrams are based on the assumption that the fixture is performing precisely as designed, which is seldom the case. In practice, variations in illuminance will occur due to manufacturing tolerances in the alignment of arc tubes in lamps and of sockets and reflectors in fixtures, and numerous other factors. The diagrams provide an approximation of the actual illuminance which might be expected.

The following explanation will be helpful in interpreting the diagram.

A. The origin, point 0, 0, represents the location of the pole.

B. The numbers 0, 1, 2, etc. on the horizontal (X) axis represent mounting heights away from the pole base. Zero indicates the base of the pole; 1 represents a point which is located one mounting height to the left or right of the pole, and 2 represents 2 mounting heights away from the pole. For example, if the pole height is 30 feet, 1 mounting height is 30 feet to the left or right of the pole.

C. The numbers on the vertical (Y) axis represent mounting heights in front or in back of the pole.

D. The triangle represents the point on the ground where the fixture is aimed. Changing the aiming point can drastically alter the light distribution pattern, so some manufacturers produce isofootcandle diagrams for several different aiming points with the same fixture.

E. The solid line, called the isofootcandle line, represents lines of equal footcandles at various points on the ground. The lines are referenced to a table which lists the illuminance as a function of the mounting height of the fixture. For the fixture shown, all points on line "a" will have an illuminance of 0.06 fc if the mounting height is 50', 0.1 fc if the height is 40', and 0.18 fc for a mounting height of 30'.

The use of diagrams is outlined in the following steps:
1. Prepare a drawing of the area to be lighted to some convenient scale; e.g., 1 inch = 40 feet.

2. Select a luminaire which has a light distribution pattern that appears to provide the desired coverage.

3. Draw a copy of the isofootcandle diagram to the same scale as the drawing of the area to be lighted. Note that some fixture manufacturers can provide scaled isofootcandle diagrams for common engineering scales such as 1:40.

4. Make several transparencies of the scaled isofootcandle diagram. This can be done on a copying machine using clear plastic sheets.

5. Overlay the transparencies on the site drawing and shift them around until the desired coverage and uniformity are achieved. Several different distribution types of fixtures and aiming points may be required to achieve the desired results. Be sure to add the contribution of light from all applicable fixtures when determining the illuminance on the ground. When the overlays have been positioned to provide the desired lighting effect, the pole locations are transferred to the site plan.

Isofootcandle diagrams are frequently used in conjunction with the"Watts per Square Foot Method" and the "Beam Lumen Method" to evaluate the uniformity of illumination.

BEAM LUMEN METHOD
The beam lumen method is similar to the lumen method used for interior lighting calculations except that the percentage of beam lumens striking the area to be lighted is substituted for the coefficient of utilization. This percentage, called the COEFFICIENT OF BEAM UTILIZATION (CBU), is determined from:

$$CBU = \frac{\text{Initial Lumens reaching the area directly from fixt.}}{\text{Total Beam Lumens}}$$

and the average maintained illuminance is:

$$E = \frac{\text{(No. Fixtures) (Beam Lumens) (CBU) (Light Loss Factor)}}{\text{Area}}$$

The number of floodlights required to produce a specific design illuminance can also be found be rearranging the above equation:

$$\text{No. Fixtures} = \frac{\text{(Footcandles) (Area)}}{\text{(Beam Lumens) (CBU) (LLF)}}$$

Beam lumens are defined as the lumens contained within a specified angle, as projected from the fixture. This information is obtained from the photometric data sheet for the luminaire.

COEFFICIENT OF BEAM UTILIZATION

The coefficient of beam utilization for a floodlight depends upon the location of the light, its aiming point, and the distribution of light within the beam. For example, lights located some distance from the boundary of the area will throw a larger percentage of light outside of the area, and will thus have a lower CBU than lights located within the boundary. The CBU can be determined only after the number of floodlights to be used, their locations, and their aiming points are known. This paradox–the number of lights needs to be known to determine the CBU, but the CBU must be known in order to determine the number of lights required–can be circumvented by first estimating a CBU, and then estimating the number of lights based on the estimated CBU. Most floodlight applications will have CBU's ranging from 0.60 to 0.90. Values lower than 60% generally indicate that the floodlight has too wide a spread for the application, and will throw excessive light outside the boundaries of the area to be lighted. Values higher than 90% usually indicate that the beam spread is too narrow, and the illumination will be spotty, with excessive luminance ratios.

As an alternative, the number of floodlights required can easily be estimated using the "watts/square foot" method to determine the total lamp wattage to be utilized. The number of poles required and their locations can then be estimated, and aiming points established in order to determine the CBU.

When the locations and aiming points of the luminaires have been established, a CBU can be calculated by plotting the outline of the area to be lighted on the photometric data sheet for the luminaire, summing the lumens which fall within the outline, and dividing this value by the total beam lumens. The procedure is best explained with an example.

Assume that a field, as illustrated in Figure 14-17, is to be lighted using four 250-watt high-pressure sodium luminaires as shown. The photometric data for the luminaire is given in Figure 14-18.

Note that the photometric data sheet is presented in the form of lumens contained within zones formed by subtending various angles from the fixture. The aiming point is represented by the grid coordinates 0, 0. The numbers 0, 10, 20, etc. on the horizontal axis and the vertical axis represent angles, in degrees, between the grid point, the floodlight, and the aiming point.

To determine the CBU:

1. For each floodlight, determine the angles which define points L, A, G, B, E, C, H, and D on the boundaries of the area to be lighted, with respect to the aiming point and the floodlight. See Figure 14-19.

Point   Angle

L $LFO = Arctan\left(\dfrac{LO}{LF}\right) = Arctan\left(\dfrac{50}{30}\right) = 59°$

A $AFL = Arctan\left(\dfrac{LA}{LF}\right) = Arctan\left(\dfrac{50}{30}\right) = 59°$

G $GFO = Arctan\left(\dfrac{OG}{OF}\right) = Arctan\left(\dfrac{50}{50+30}\right) = 40.6°$

B $BFE = Arctan\left(\dfrac{EB}{EF}\right) = Arctan\left(\dfrac{50}{100+30}\right) = 25.5°$

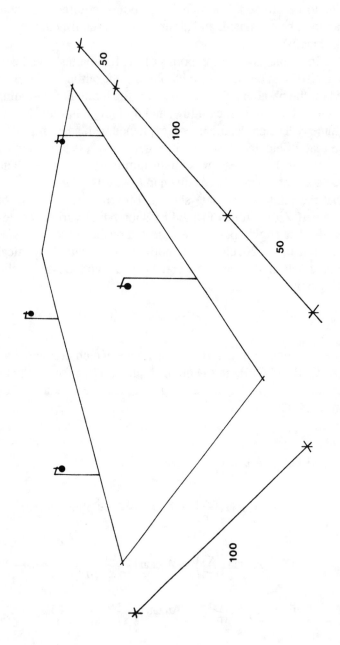

Figure 14-17
Field used in explanation of the "Beam Lumen Method" of outdoor lighting calculations.

HUBBELL LIGHTING PHOTOMETRIC REPORT

9865-XXXHS

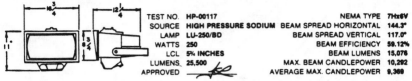

TEST NO.	**HP-00117**	NEMA TYPE **7Hx6V**
SOURCE	**HIGH PRESSURE SODIUM**	BEAM SPREAD HORIZONTAL **144.3°**
LAMP	**LU-250/BD**	BEAM SPREAD VERTICAL **117.0°**
WATTS	**250**	BEAM EFFICIENCY **59.12%**
LCL	**5¾ INCHES**	BEAM LUMENS **15,076**
LUMENS	**25,500**	MAX. BEAM CANDLEPOWER **10,292**
APPROVED		AVERAGE MAX. CANDLEPOWER **9,369**

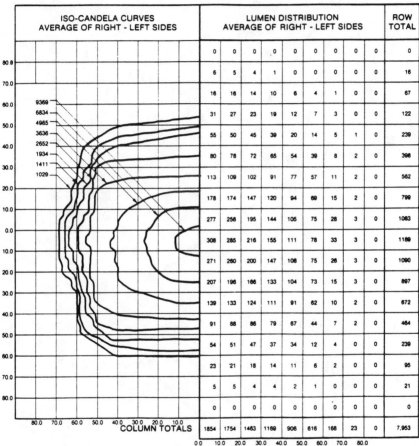

ISO-CANDELA CURVES AVERAGE OF RIGHT - LEFT SIDES	LUMEN DISTRIBUTION AVERAGE OF RIGHT - LEFT SIDES									ROW TOTAL
	0	0	0	.0	0	0	0	0	0	0
	6	5	4	1	0	0	0	0	0	16
	16	16	14	10	6	4	1	0	0	67
	31	27	23	19	12	7	3	0	0	122
	55	50	45	39	20	14	5	1	0	239
	80	78	72	65	54	39	8	2	0	398
	113	109	102	91	77	57	11	2	0	562
	178	174	147	120	94	69	15	2	0	799
	277	256	195	144	105	75	28	3	0	1083
	308	285	216	155	111	78	33	3	0	1189
	271	260	200	147	108	75	26	3	0	1090
	207	196	166	133	104	73	15	3	0	897
	139	133	124	111	91	62	10	2	0	672
	91	88	86	79	67	44	7	2	0	464
	54	51	47	37	34	12	4	0	0	239
	23	21	18	14	11	6	2	0	0	95
	5	5	4	4	2	1	0	0	0	21
	0	0	0	0	0	0	0	0	0	0
COLUMN TOTALS	1854	1754	1463	1169	906	616	168	23	0	7,953

TESTED TO CURRENT IES AND NEMA STANDARDS UNDER STABILIZED LABORATORY CONDITIONS. VARIOUS OPERATING FACTORS CAN CAUSE DIFFERENCES BETWEEN LAB DATA AND ACTUAL FIELD MEASUREMENTS.

HUBBELL lighting division

Lighting Division HARVEY HUBBELL INCORPORATED   Electric Way, Christiansburg, Virginia 24073   •   (703) 382-6111

OCT. 1976   ® 1976 HARVEY HUBBELL, INC.   Printed in U.S.A.

Figure 14-18
Typical Photometric Data Sheet for flood lights. *(Courtesy Hubbell Lighting)*

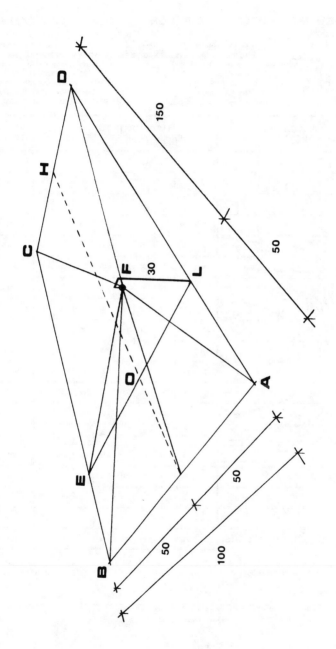

Figure 14-19
Angles used to calculate the coefficient of beam utilization.

E $EFO = \angle EFL - \angle LFO = Arctan\left(\dfrac{EL}{FL}\right) - 59° =$

$$Arctan\left(\dfrac{100}{30}\right) - 59° = 14.3°$$

C $CFE = Arctan\left(\dfrac{EC}{EF}\right) = Arctan\left(\dfrac{150}{100 + 30}\right) = 62.4°$

H $HFO = Arctan\left(\dfrac{OH}{OF}\right) = Arctan\left(\dfrac{150}{50 + 30}\right) = 68.8°$

D $DFL = Arctan\left(\dfrac{LD}{LF}\right) = Arctan\left(\dfrac{150}{30}\right) = 78.7°$

2. Transfer the points from Step 1 to the photometric data sheet and connect the points to create an outline of the area to be lighted. Note that the horizontal lines are straight, but the vertical lines are curved. See Figure 14-20.

3. Sum the lumens which fall inside the area. Note that several zones are cut so that only a portion of the lumens within the zone strike the lighted area. The lumens from these zones are estimated, based on the percentage of the zone which is included in the area. The total lumens which fall on the lighted area are about 10,978 and the total beam lumens are 15,076, so the floodlight has a CBU of 10,978/15,076 or 0.73. Since the fixtures are spaced symmetrically around the field, each light will have the same CBU, and the average CBU for the installation is also 73%. If additional poles were set in between the poles shown in Figure 14-17, Step 3 would be repeated for each location and the weighted average CBU determined.

4. Determine the average initial illuminance for the installation from:

$$E = \dfrac{(\# \text{ Fixtures}) \text{ (Beam Lumens) (CBU)}}{AREA} =$$

$$\dfrac{(4) \ (15076) \ (0.73)}{3000} = 14.7 \text{ fc}$$

HUBBELL LIGHTING PHOTOMETRIC REPORT

9865-XXXHS

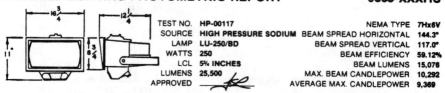

TEST NO.	HP-00117	NEMA TYPE	7Hx6V
SOURCE	HIGH PRESSURE SODIUM	BEAM SPREAD HORIZONTAL	144.3°
LAMP	LU-250/BD	BEAM SPREAD VERTICAL	117.0°
WATTS	250	BEAM EFFICIENCY	59.12%
LCL	5¾ INCHES	BEAM LUMENS	15,078
LUMENS	25,500	MAX. BEAM CANDLEPOWER	10,292
APPROVED		AVERAGE MAX. CANDLEPOWER	9,369

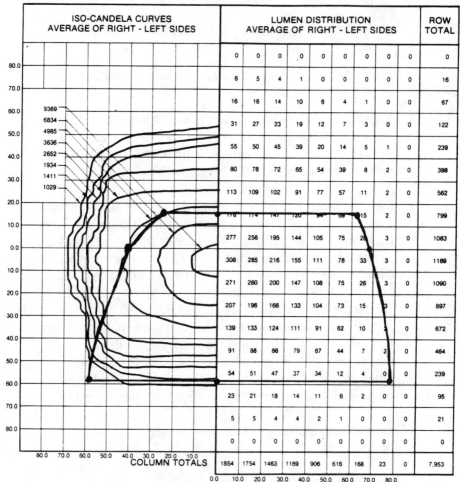

LUMEN DISTRIBUTION AVERAGE OF RIGHT - LEFT SIDES									ROW TOTAL
0	0	0	0	0	0	0	0	0	0
6	5	4	1	0	0	0	0	0	16
16	16	14	10	6	4	1	0	0	67
31	27	23	19	12	7	3	0	0	122
55	50	45	39	20	14	5	1	0	239
80	78	72	65	54	39	8	2	0	398
113	109	102	91	77	57	11	2	0	562
178	174	147	120	94	69	15	2	0	799
277	256	195	144	105	75	28	3	0	1083
308	285	216	155	111	78	33	3	0	1189
271	260	200	147	108	75	26	3	0	1090
207	196	166	133	104	73	15	3	0	897
139	133	124	111	91	62	10	2	0	672
91	88	86	79	67	44	7	2	0	464
54	51	47	37	34	12	4	0	0	239
23	21	18	14	11	6	2	0	0	95
5	5	4	4	2	1	0	0	0	21
0	0	0	0	0	0	0	0	0	0
COLUMN TOTALS									
1854	1754	1463	1169	906	616	168	23	0	7,953

ISO-CANDELA CURVES AVERAGE OF RIGHT - LEFT SIDES

9369, 6834, 4985, 3636, 2652, 1934, 1411, 1029

TESTED TO CURRENT IES AND NEMA STANDARDS UNDER STABILIZED LABORATORY CONDITIONS. VARIOUS OPERATING FACTORS CAN CAUSE DIFFERENCES BETWEEN LAB DATA AND ACTUAL FIELD MEASUREMENTS.

Figure 14-20

Outline of area to be lighted is drawn on the photometric data sheet to determine the total lumens which strike the area to be lighted. *(Sheet courtesy Hubbell Lighting)*

Note that this is the initial illuminance, assuming that all system components are operating as rated. Since this is seldom the case, an appropriate light loss factor should be applied.

INVERSE SQUARE LAW

The mechanics of the inverse square law were discussed in detail in Chapter 8 and will not be repeated here. It is significant to note, however, that many luminaires used for outdoor lighting have asymmetrical distributions which vary widely over only a few degrees. Care must be taken to assure that the proper angles are used when determining the candela intensity which is used for the calculation.

The inverse square law is generally used in manual calculations to check only a few points of interest in the area to be lighted. Its main use is in computer programs which calculate the illuminance at an array of points.

MISCELLANEOUS METHODS

A greatly simplified version of the Beam Lumen method employing coefficient of utilization curves can be used for floodlighting and street-lighting calculations. Some manufacturers publish these curves, shown in Figure 14-21, along with photometric data for floodlights and street-lights. Note that the formats vary slightly but contain essentially the same information.

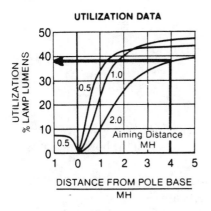
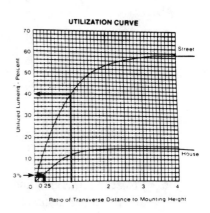

Figure 14-21
Utilization curves are published by some manufacturers as a design aide.
Figure on left is for flood lights; right figure is for street lights. *(Courtesy GE Lighting)*

These curves, in conjunction with information on beam spreads or luminaire distribution types (e.g., II-M, etc. for streetlights), can be useful in both luminaire selection and the initial design.

To use the floodlight data, first divide the width of the area to be lighted by the mounting height of the fixture. Note that the width is defined as the distance from the pole base to the opposite side of the field. Locate this value on the horizontal axis of the chart. Then locate the curve which represents the aiming point of the luminaire as a function of its mounting height, and project a vertical line from the point on the horizontal axis to the curve. Project a horizontal line from the intersection of the vertical line and the curve to the vertical axis and read the CU. The initial illuminance can be found from:

$$E = \frac{(\text{\# fixtures}) \, (\text{Lumens per lamp}) \, (\text{CU})}{\text{Area}}$$

For example, assume that an area 100' wide x 300' long is to be lighted using three 250-watt, high-pressure sodium luminaires mounted at 25' and aimed at the centerline of the field (50'). The utilization curve is given in Figure 6-21.

1. Determine the width/MH ratio

   Ratio $= \dfrac{100'}{25'} = 4$

2. Locate "4" on the horizontal axis.
3. Determine the aiming distance/MH ratio

   Ratio $= \dfrac{50'}{25'} = 2$

4. Locate the curve for aiming distance/MH ratio = 2.
5. Project a vertical line upwards from "4" on the horizontal axis to the curve.
6. Project a horizontal line from the intersection of the vertical line and the curve to intersect the vertical axis.
7. Read the CU as 38%.
8. Calculate the initial illuminance from:

$$E = \frac{(\text{\# Fixt.}) \ (\text{lumens/fixt}) \ (CU)}{\text{Area}} =$$

$$\frac{(3) \ (27500) \ (0.38)}{3000} = 10.5 \ \text{fc}$$

Note that this is the initial illuminance and assumes that all lighting system components are performing as rated. A more realistic estimate of the actual illuminance would include the light loss factor.

Street lighting calculations may be approximated in a similar manner using utilization data for the appropriate luminaire. Note that these data are presented in slightly different format and contain 2 curves, one for the street side of the luminaire, and one for the house side. Using a similar procedure to the one just explained, determine the coefficients of utilization for both the street side and the house side, and add them to obtain the total CU. Note that the distance used to determine the width/MH ratio is the street side lateral distance for the street side CU and the house side lateral distance for the house side CU. All distances are measured from a point directly below the fixture.

The coefficient of utilization method provides only an estimate of the average illuminance, and uniformity should be verified through the use of isofootcandle diagrams or point calculations.

Chapter 15

LIGHTING CONTROLS

Lighting controls are an integral part of a lighting system. Without controls, nothing can happen. Lights cannot be turned on or off, and moods which require changing of lighting levels cannot be created. The installation of a lighting control system in a building does not, however, guarantee that the needs or desires of the occupants or owner will be met. The system must be responsive to the functional and aesthetic requirements placed upon it, and should perform these duties in an energy efficient manner. This chapter will discuss the hardware that is used to control lighting systems, and the development of an overall control strategy.

Lighting control equipment is divided into two broad categories, on/off and dimming. Each category is then subdivided into two additional categories, manual and automatic. On/off controls simply turn lights on or off, while dimming controls permit the adjustment of lighting levels over a range. Manual systems, such as the single pole switch or residential type dimmer for incandescent lights, must be activated by a person, while automatic controls, such as timeclocks or photocells, can be preset to activate at a certain time or in response to a variable condition such as darkness. Personnel sensors can detect the presence of people and turn the lights "on" only when the room is occupied. Computerized control systems can be programmed to operate lighting and other building systems in accordance with a predetermined schedule, and can also utilize inputs from sensors such as photocells and thermostats.

Some devices, such as single pole switches, simple incandescent dimmers which replace a wall switch, and inexpensive photocells, are self contained units which operate independently and cannot be integrated into remotely controlled systems. Other devices, such as relays and contactors, may be remotely controlled manually, or connected to a sensing or timing device for automatic operation. When remote operation is desired, some form of communication between the controller and the actual switching device must be provided. This may be accomplished using line or low-voltage control circuits, power line carrier systems, telephone systems, or radio frequency transmitters and receivers. Figure 15-1 illustrates the basic concept of the components and their relationships.

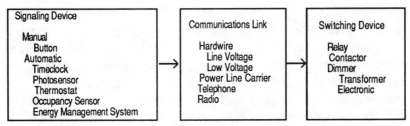

Figure 15-1
Relationships between the three component groups of a remote switching system. A command is initiated by the signaling device and transmitted, via the communications link, to the actual switching or dimming device.

CONTROL EQUIPMENT

ON/OFF CONTROLS

The simplest type of on/off control device is the single-pole switch (Figure 15-2) commonly found in residences and small commercial buildings. Single-pole switches used for commercial or industrial service are normally rated for higher current and are of higher quality than residential grade switches. These switches are called "Specification Grade." Switches can be used to provide multiple points of control for a single lighting circuit if 3-way or 4-way switches are specified. Manual switches have one major drawback–a person must activate them. If the occupant leaves a room and neglects to flip the switch, the lights will remain on.

SIMPLE AUTOMATIC SWITCHES AND TIMERS

A more practical approach to controlling lights is the use of automatic switches. The common "twist timer" is well suited for use in spaces which are only occasionally occupied, such as inactive storage

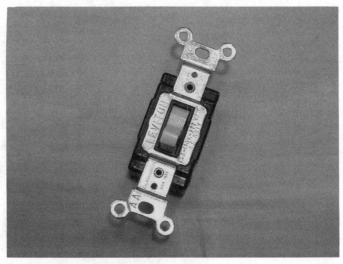

Figure 15-2
A single pole switch—the original energy conservation device.

rooms, some mechanical equipment rooms, and janitors' closets. These timers are adjustable for any desired "on" interval within the range of the timer, typically from 5 minutes to 12 hours. As an alternative, jamb switches, which turn lights on only when the door is open, may be used for small storage rooms and janitors' closets if the door is left open whenever the room is in use.

The state-of-the-art counterpart of the twist timer looks like a common single-pole switch but contains an electronic timing device and automatic trip to turn lights off automatically after a preset elapsed time. Some are adjustable in discrete steps; e.g., 5, 15, or 30 minutes, while others have only one built-in "on" time interval, ranging from 15 minutes to 4 hours. Installation is simple, and consists only of removing the manual switch and replacing it with the timer. Electronic timers are generally rated for a maximum of 500 or 600 watts, while the mechanical types are typically rated for 1800 watts. Common automatic switches are shown in Figure 15-3.

Figure 15-3
Simple automated switches such as twist timers are preferred over single pole switches in locations such as janitors' closets and infrequently used storage areas. Jamb switches or electronic wall timers may also be used. *(Courtesy Intermatic)*

PHOTOCELLS

Photocells are used to turn lights on or off in response to available daylight. They may also be used to dim lights in conjunction with daylighting in buildings. Photocells fall into two categories: simple, inexpensive cells used for outdoor lighting, and more costly precision types which are used inside buildings in conjunction with daylighting.

Most inexpensive photocells use cadmium sulfide as the active element for light detection. Cadmium sulfide changes its electrical resistance as a function of photon bombardment, and this change in resistance determines the time at which switching occurs. Over time, usually several years, the sensitivity of the cadmium sulfide changes, and the photocell turns lights on earlier in the evening and off later in the morning. This drift causes lights to operate for longer periods. Electronic photocells use a light-sensitive silicon diode to detect daylight, and provide more precise control. They are also more stable and minimize unnecessary operation of outdoor lights during daylight hours. Electronic photocells (Figure 15-4) are more costly than cadmium sulfide cells but savings in energy costs generally justify their installation.

Figure 15-4
Electronic photocells are used to obtain precision switching in daylighting applications or to maintain precise control of outdoor lighting. *(Courtesy Energy Conservation Engineering)*

Precision photoelectric control systems typically use a light-sensitive cell to activate an electronic controller which in turn activates a relay, contactor, or dimming device. They are generally used for control of lights in daylit buildings where a precision control is needed. The controller has adjustments for sensitivity so the device may be set to turn lights off in the building when sufficient daylight exists, and back on as daylighting levels drop. They should be equipped with a time delay feature to prevent frequent switching on partly cloudy days. A typical control system for a building with skylights is shown in Figure 15-5.

When on/off photocell controls are used in daylighted buildings the system should also be equipped with a timeclock or programmable lighting controller to turn lighting circuits off at quitting time. This is a necessary precaution since the daylight controller may have turned the lights off prior to quitting time, and failure to manually turn the circuit off may result in inadvertent operation of the electric lights at night when the photocontrol senses darkness and calls for light.

TIMECLOCKS

Timeclocks, either mechanical or electronic, can be used to control lights when their operation is in accordance with a fixed operating

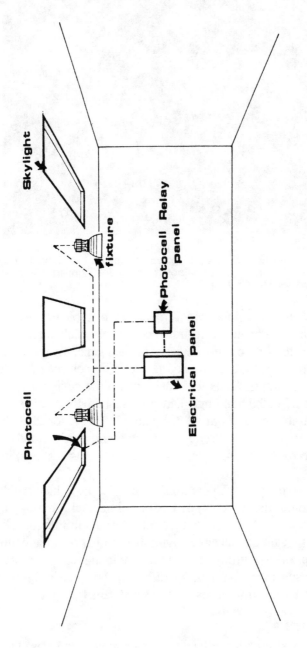

Figure 15-5

Typical physical layout for the components of a precision photocell switching system in a daylit building.

schedule. Timeclocks range from simple single-pole, single-throw (one circuit on/off control), to complex electronic systems which can control 16 or more individual circuits, and can be programmed to activate two or more times per day. The mechanical systems are typically capable of switching line voltages (120v or 277v), or can be used to activate low-voltage relays. Relays will be discussed later in this chapter. Electronic timeclocks are usually rated for low current, up to 5 amps, and are used to activate relays.

Timeclocks used to control outdoor lights should be of the "Astrological" type, which have built-in compensation for seasonal changes in daylight hours. This eliminates the necessity for monthly changes in the "ON" and "OFF" times as daylight hours increase in spring and summer and decrease in fall and winter. Timeclocks with built-in interfaces for photocells are available to fine tune sunset "ON" times or sunrise "OFF" time settings.

The least expensive timeclocks simply repeat a preset on/off cycle on a daily basis. For example, parking lot lights which are to be turned on at dusk and off at 10 p.m. (or 11 p.m. or midnight), 7 days a week. If one or more days are to be omitted from the schedule a "Skip-a-Day" or "7-Day" timeclock should be used. These timers can be set to omit the on/off cycle on specific days. Some sophisticated timeclocks may be programmed for an entire year, and can provide different schedules for holidays and weekends. Typical timeclocks are shown in Figure 15-6.

LIGHTING CONTACTORS

Lighting contactors (Figure 15-7) are simply electrically activated remote control switches. A typical contactor consists of two or three switching contacts, called "poles," for the lighting loads, and a separate magnetic switching device which opens or closes the contacts. The magnetic device is activated by a remote wall switch, timeclock, photocell, computer, or other signaling device.

Unlike the switches previously discussed, contactors can handle both large and small electrical loads. Current ratings typically range from 20 amps to 300 amps, and they may have up to 12 poles. Voltage ratings are 120v, 208v, 240v, 277v, and 480v. Large contactors may

(a)

Figure 15-6
Timeclocks range from the (a) simple single pole-single throw mechanical unit and (b) astrological clocks which compensate for seasonal changes in the length of daylight hours, to (c) programmable electronic micro-computers. *(Courtesy Intermatic)*

(b)

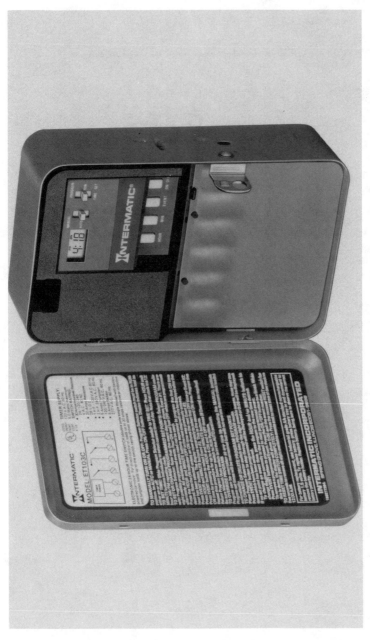

be used to control entire lighting distribution panels, while small contactors control individual circuits.

Contactors are generally located at or near the electrical distribution equipment for economy in wiring costs.

Figure 15-7
Typical lighting contactor.
(Courtesy GE Wiring Devices, Warwick, RI)

LOW-VOLTAGE RELAYS

Like contactors, low-voltage relays are remotely controlled switches. Unlike contactors, relays used for lighting control are generally rated for 20 amperes or less. Each relay can control one lighting circuit. While the relay is actually switching line voltage, the control portion of the system is operating at low voltage, typically 12v or 24v. The low voltage is obtained from the transformer as shown in Figure 15-8. In operation, a momentary contact switch is activated to send a low-voltage pulse to the relay. This pulse generates an electromagnetic field in the low-voltage winding, which causes the rod inside the relay to move, thus opening or closing the switch contacts, as illustrated in Figure 15-9.

Relays may also be activated by signals from timeclocks, photocells, thermostats, or computerized building operating systems.

Relays are enclosed in special relay panels that have internal barriers to isolate the low-voltage wiring from higher line voltages, as shown in Figure 15-10. These panels vary in size and complexity

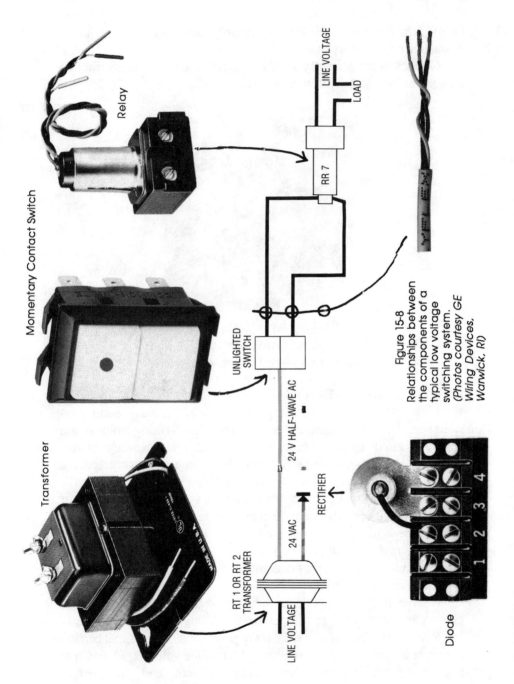

Figure 15-8
Relationships between
the components of a
typical low voltage
switching system.
*(Photos courtesy GE
Wiring Devices.
Warwick, RI)*

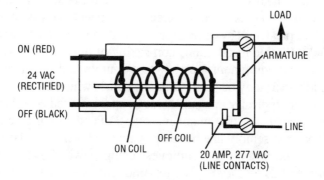

Figure 15-9
Diagram of the interior of a typical low voltage relay. *(Courtesy GE Wiring Devices, Warwick, RI)*

Figure 15-10
Interior of a typical low voltage relay cabinet. Note barrier between line and low voltage sides of the cabinet, as required by the National Electrical Code. This panel also contains a microprocessor to provide automated control of the lighting system. *(Courtesy GE Wiring Devices, Warwick, RI)*

according to the number of relays and additional components which are required. These components may include transformers, rectifiers, time-clocks, and interfaces for thermostats and photocells. Automatic sequencing of a number of relays with the push of a single button is also feasible.

Individual relays can be mounted in standard electrical switch boxes when only one relay is to be installed. Relay panels may be purchased as complete pre-wired systems or as components which are assembled in the field. Simple installations con-sisting of only one or two relays and no special equipment are easy to assemble and either prefab or field-build panels may be used. Complex panels which are assembled in the factory are typically less expensive in total installed cost than field-assembled panels for two major reasons: 1) factory-fabricated panels are built on an assembly line basis by workers who are familiar with the equipment and assembly techniques, using power-driven tools for assembly, and 2) factory production-line labor costs are substantially lower than those for a journeyman electrician.

The major benefits to low-voltage control systems are ease of installation, and flexibility in lighting control strategies. These benefits become even more pronounced in existing buildings where rewiring costs can be considerable. Since the systems operate at voltages of 24v or less, conduit is not required in most buildings, and a cable approximately 1/2 inch in diameter and containing up to 26 individual 18 ga. or 24 ga. wires can be used to control up to 12 individual relays. This means that both installation labor and material costs are reduced since conduit and large wires are not required. An additional benefit lies in the small physical size of low-voltage switches. Some manufacturers can provide up to eight individual switches in a single switch "Block" for use in an enclosure which can contain only one or two line voltage switches.

Switches operate in parallel so extra switches can be easily added for multiple point control of a single relay.

Some switches incorporate pilot lights which indicate if a relay is on or off. This permits easy identification of energized circuits, particularly when the lighted area within the building is not visible from the control location.

Note that while most relays will operate on alternating current, operation on direct current is more reliable, and is preferred.

OCCUPANCY SENSORS

Occupancy sensors are produced in three basic types: ultrasonic, infrared, and sonic. Ultrasonic sensors emit high-frequency sound waves. Motion within a space distorts the waves, and the sensor detects this distortion. When no motion has been detected over a preset time interval the sensor sends a signal to a remote relay and turns lights off. When motion is detected, a signal is sent to the relay activating it to turn lights on. Some units have an automatic "off," manual "on" feature.

Infrared detectors operate in a similar manner to ultrasonic units except that they sense changes in background infrared radiation. Infrared detectors also use relays as a switching device. Unlike ultrasonic units, which should be used only indoors, infrared detectors may be used outside.

Occupancy sensors are available with several different beam spreads and patterns for use in different types of areas. Sensors used in offices should have rectangular, circular, or tear-drop shaped patterns that cover the entire room. Sensors for use in corridors or warehouse aisles should have long, narrow patterns to provide maximum penetration into the aisle. Typical patterns and areas of coverage are shown in Figure 15-11.

Sensors are available for either ceiling or wall mounting. Wall-mounted units can replace existing wall switches and require no special wiring. Typical units are shown in Figure 15-12.

Sonic detectors are activated by audible sound. They are available in either modular units which plug into a wall receptacle or in hardwired units which replace standard wall switches. The units are adjustable for both the intensity of sound required to activate the unit, and for time delay in turning lights off. The time delay prevents deactivation of the circuit for short time intervals between noises. The plug-in types are typically rated for a maximum of 300 watts and are designed to control table lamps and similar loads. The wall-mounted versions are available in 120v or 277v and can control an entire 20-amp lighting circuit.

Ultrasonic and infrared occupancy sensors are well suited to private offices, school classrooms, conference rooms, warehouses with rack storage, and other spaces with similar occupancies. They are generally less effective when used in large, open offices or other areas in which one or more occupants are present through the majority of the work day.

Most occupancy sensors have adjustments for sensitivity and time delay. The sensitivity feature permits tuning the unit to the space so that the movement of a person walking past a door will not activate the unit,

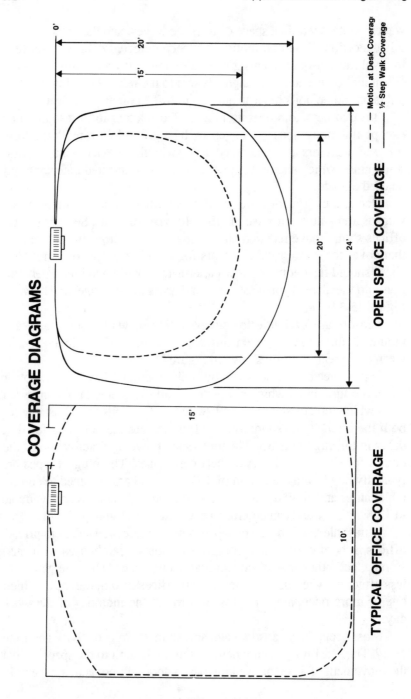

COVERAGE DIAGRAMS

OPEN SPACE COVERAGE

– – – Motion at Desk Coverage
——— ½ Step Walk Coverage

TYPICAL OFFICE COVERAGE

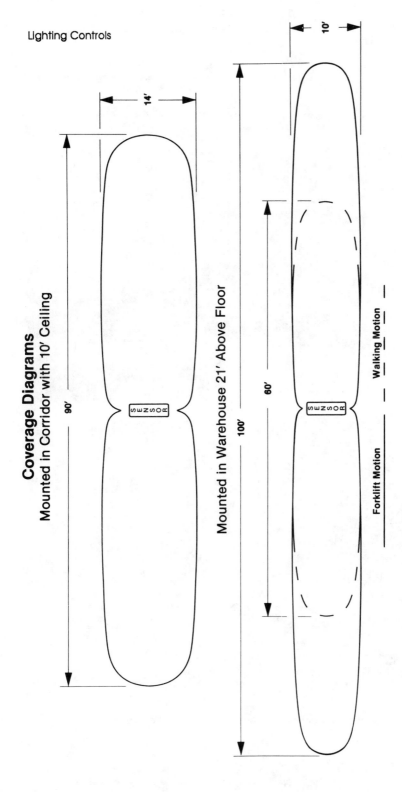

Coverage Diagrams
Mounted in Corridor with 10' Ceiling

14'

90'

SENSOR

Mounted in Warehouse 21' Above Floor

10'

100'

60'

SENSOR

Forklift Motion —

Walking Motion — —

Figure 15-11
Coverage patterns for typical occupancy sensors. Patterns vary considerably between types, so manufacturer's literature should be consulted to assure that the selected unit provides adequate coverage. *(Courtesy Novitas, Inc.)*

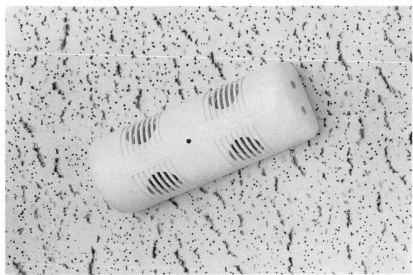

Figure 15-12
Typical occupancy sensors. *(Courtesy Novitas, Inc.)*

but the entry or movement within the space by the occupant will be sensed to prevent turning lights off when the room is occupied. A time delay is provided to prevent unwanted switching if an occupant is motionless or the room is unoccupied for short time periods. A time delay of 5 to 10 minutes will prove acceptable in most installations. Care must also be taken to assure that switching of the lights does not disturb other occupants in nearby areas.

BI-LEVEL SWITCHES

The switching of fixtures in many buildings is accomplished with wall switches which control all of the lights in a large area with a single switch, making it impossible to reduce lighting levels for janitorial service or other functions which do not require full light output from the system. The installation of a special bi-level switch in fixtures with two incandescent lamps or two fluorescent ballasts may solve the problem

The switch (Figure 15-13) turns all lights on the first time the wall switch is turned on. If the wall switch is turned off and then on again, only one half of the lights are energized. Turning the wall switch off and back on restores power to all lamps. The device should be applied in situations where frequent switching during normal operating hours is not performed, but it is desired to provide reduced lighting levels during specific times periods or non-business hours. It may also be of use in meeting the switching requirements of some State energy codes which require switching capability to reduce lighting levels by 50% when the full design lighting level is not needed.

DIMMING CONTROLS

Dimming of electric lights can be accomplished with a variety of controls. The type of control is determined by the light source to be dimmed, and the desired result of the dimming. Commonly used equipment falls into one of two general categories: electronic dimmers, and dimming transformers. In the past, some dimming was accomplished with variable resistors called rheostats. This practice has become obsolete due to the fact that no energy was saved, the equipment was bulky and expensive, and considerable heat was dissipated by the dimmer.

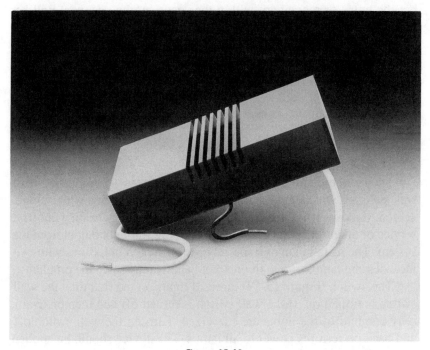

Figure 15-13
Switching device mounts inside fixture and permits switching of two lamps in a four-lamp fixture simply by activating the wall switch. *(Courtesy Scientific Component Systems)*

INCANDESCENT DIMMERS

Incandescent lamps operated at line voltage (120v or higher) are typically dimmed with electronic dimmers made specifically for incandescent products (Figure 15-14). These dimmers can not be used with low-voltage incandescent, fluorescent, or HID lamps since they will not function properly and may cause premature transformer or ballast failure. Electronic dimmers for incandescent lamps contain a small semi-conductor switching device, called a "triac." They work by actually turning the power to the lamp off for a portion of each cycle. They are typically rated at either 600 watts or 1800 to 2000 watts. Most electronic dimmers for incandescent lamps are wall-mounted and are used in lieu of standard wall switches.

When electronic dimmers are used with incandescent lamps the rapid on/off switching may cause filament vibrations which produce audible noise in the form of a high-pitched hum. They may also generate a high-pitched noise in the switching device. This condition may be corrected by installing a filter across the switch leads. Filters are available from most major lighting distributors.

Figure 15-14
Inexpensive residential type dimmer for incandescent lights. These dimmers cannot be used for fluorescent or HID systems, or for low voltage incandescent systems.

Variable transformers simply reduce the magnitude of the voltage supplied to the lamp. Transformers are quite large when compared to electronic dimmers, and can be costly. Most transformers are designed for dimming of one or more circuits. Some transformers are equipped with a motor to permit remote operation, while others require manual operation at the unit.

Low-voltage incandescent systems require special dimmers which will not damage the low-voltage transformers. Electronic dimmers are made for this purpose, or variable transformers may be used.

Incandescent lamps are generally dimmed for aesthetic reasons rather than energy savings, since large reductions in light output are accompanied by only small reductions in power usage, as was discussed in Chapter 2.

FLUORESCENT DIMMING

Fluorescent lamps may be dimmed with special dimming ballasts and electronic controls, dimming transformers, or electronic devices. Magnetic dimming ballasts are used when a wide dimming range is desired since lamps can be reduced to 1%-2% light output. This is the preferred method when dimming is to be done for aesthetic reasons.

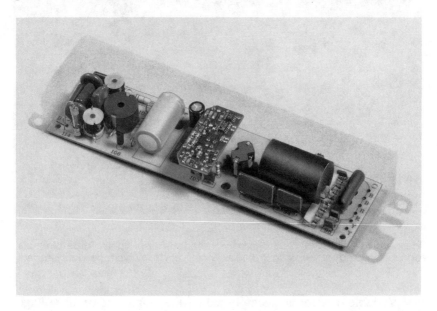

Figure 15-15
New electronic dimming ballasts provide energy efficient operation at all light output settings. *(Courtesy Advance Transformer)*

Some electronic ballasts are also capable of providing dimming, Figure 15-15, either in response to a photocell input to provide variable dimming, or external adjustments which provide a fixed level of dimming. Dimmable electronic ballasts can provide substantial savings in equipment costs when only a few fixtures are to be dimmed, as in the case of many small offices with daylighting potential.

Transformers reduce the voltage supplied to the lamp and can provide only 30%-40% dimming since lamps will extinguish due to low voltage below this level. For this reason, transformer type dimmers are used on fluorescent systems primarily to reduce energy cost since the available range of dimming is insufficient for most aesthetic applications. Excessive voltage reductions may also result in insufficient cathode heating, and a reduction in lamp life. Reduction in lamp life is an economic issue, and in some cases savings in energy costs may more than offset the increased relamping costs, particularly if lamps are group replaced before the majority of lamps in an installation have failed.

Dimming transformers are available in two types: variable and fixed reduction. Variable transformers are generally used in conjunction with daylighting, and are automatically activated by photo controls to vary the light output of the electric lighting system as necessary. They may also be used to dim incandescent lamps for aesthetic reasons. Fixed reduction transformers provide a fixed percentage of voltage reduction, typically 15% to 20%, so lights operate at either full voltage or a fixed 15% to 20% reduction. A fixed reduction transformer is shown in Figure 15-16. The control circuit for either type of transformer should provide full voltage to discharge lamps during the starting cycle and until lamps have sufficiently warmed up before voltage is reduced. Incandescent lamps may be turned on at any voltage setting.

Dimming fluorescent lamps with transformers can be accomplished on large systems with reasonable payback periods. Systems involving loads of less than 10 kW are seldom cost effective due to the high cost per kva for small transformers. Transformers are fairly large, and may be noisy, so adequate provisions must be made for their installation.

Special electronic dimmers for fluorescent lamps generally control one or more circuits per dimmer, although several manufacturers offer smaller dimmers which can control individual fixtures. These devices work by turning the power to the system off for a part of each cycle, and are different than the electronic dimmers for incandescent lamps. A typical electronic dimmer is shown in Figure 15-17.

Caution should be exercised in the selection of electronic dimmers since some types cause distortion of the input voltage and current sine waves, and result in a reduced power factor. They may also produce considerable audible noise, and may require mounting in a remote location to avoid complaints.

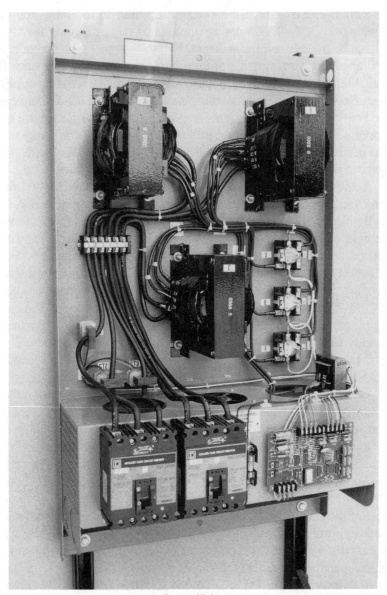

Figure 15-16
Fluorescent systems can be dimmed as an energy conservation measure by a
fixed percentage, normally 10% to 20%, with transformers. *(Courtesy Energy
Conservation Development Corp.)*

   Dimming of fluorescent lamps as an energy conservation measure
can be cost effective since the reductions in both light output and power
consumption are of about the same magnitude over normal dimming
ranges, as discussed in Chapter 3.

   Variable transformers and electronic dimmers can provide additional savings in energy costs over those associated with daylighting
control when a routine maintenance program, which will be discussed
in Chapter 17, is implemented. Lighting systems are designed to
produce the desired illuminance at some future point in time after lamps
have depreciated and dirt has built up on fixtures. They are producing
more light than is necessary over most of their life, and are also
consuming more power. A dimming system will reduce the light output
and power input to a system to maintain the illuminance for which the
control is calibrated. This can provide energy savings if lamps are
replaced and fixtures washed periodically, as shown in Figure 15-18.

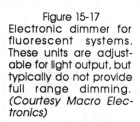

Figure 15-17
Electronic dimmer for
fluorescent systems.
These units are adjust-
able for light output, but
typically do not provide
full range dimming.
(Courtesy Macro Elec-
tronics)

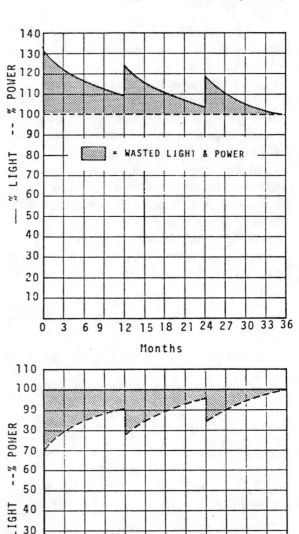

Figure 15-18 Graph at top shows the relative light produced by a system over a 3-year period as lamps depreciate and dirt builds up on fixture surfaces. The system produces more light than needed during most of its life. The shaded area represents wasted light and power. Automatic dimming controls reduce power input, as shown at bottom, to save energy while producing the design illuminance at all times. The shaded area represents saved power. The example assumes that fixtures will be washed and group relamped, as explained in Chapter 17.

HID DIMMING

High-intensity discharge lamps may be dimmed with transformers used for fluorescent systems, or with special electronic dimmers. The reductions in light output and power consumption will vary according to the lamp and ballast type. Excessive dimming will result in a color shift in the lamp.

Manufacturers' data should be consulted for information on reductions in light output and power consumption.

COMMUNICATION LINKS BETWEEN CONTROL AND SWITCHING DEVICES

HARDWIRING

Hardwiring consists of running wires from the control device to the switching device to provide a path for the signal to follow. This may be accomplished with line voltage, typically 120v or 277v, or with low voltage. The voltage will be dictated by the control and switching equipment selected.

Line voltage conductors are run in conduit, and treated as any other line voltage circuit. Low-voltage conductors for lighting control are usually contained in a multiconductor cable which may be run without conduit, subject to building code restrictions.

POWER LINE CARRIER SYSTEMS

The installation cost of control circuit wiring, either line or low voltage, may be prohibitive in many existing buildings which could benefit from remote or automatic switching. In these cases, power line carrier control may be the best solution

A simple power line carrier system consists of a relay and a remote switch (Figure 15-19). Unlike conventional relays and contactors, power line carrier systems do not require dedicated conductors between the switch and relay. The switch contains a small signal generator (transmitter) which imposes a coded signal on the existing building wiring system. The signal is typically a binary code, and numerous different codes can be generated by a single transmitter and received by a typical receiver. The signal travels over the electrical wiring to the relay, which contains a receiver to intercept the signal and instruct the relay to activate. When a coded signal is sent by a transmitter, it will travel to all relays connected to the wiring system, but only receivers which have been set to that specific code will respond.

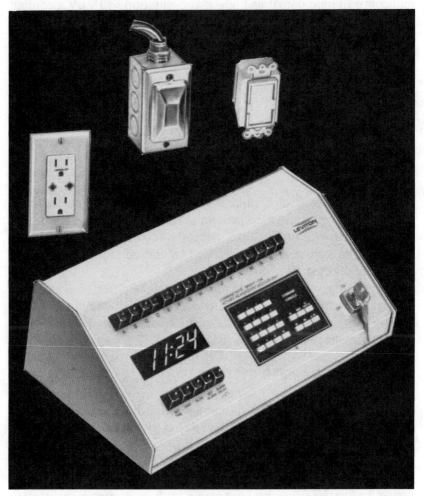

Figure 15-19
Components of a typical power line carrier control system.

Transmitters for power line carrier systems may be hard wired into the building electrical system, or may be plugged into a convenient wall receptacle. Transmitters are currently available in a variety of configurations ranging from a single switch capable of controlling hundreds of individual relays simultaneously, to microprocessors capable of controlling tens of thousands of relays.

Power line carrier systems are particularly applicable to existing hotels and motels where some form of control of lights and air conditioning loads in rooms is desirable to assure that vacant rooms are not being lighted or space conditioned. They are also useful in offices and other occupancies where remote or automatic control of electrical loads is desired but the installation cost of a hard wired system is prohibitive. In residential applications they can provide simple, easy to install control of security lighting.

TELEPHONE

Some computer-based systems may be provided with commands by telephone. A computer may provide global control for a building, based on normal working schedules, but some personnel may begin early while others work late. By simply picking up the telephone and dialing a code number, these workers can override the programmed schedule, and turn lights on when they are needed. Telephone lines may also be used to link outlying buildings to the central computer.

DETERMINING
AN APPROPRIATE CONTROL STRATEGY

There are a number of factors which govern the selection of a particular control strategy for a given building:

1. Operating schedule of building
2. Type of lighting system
3. Budgetary constraints
4. Dimming requirements
5. Space to locate controls
6. Availability of qualified personnel to operate and maintain the system

The relative importance of each factor will vary according to the specific job conditions, and several factors may require interactive consideration.

OPERATING SCHEDULE OF BUILDING

Many buildings operate on a fixed schedule. Employees come to work, go to lunch, and leave at a predetermined time, and maintenance and janitorial services are performed according to a schedule. In these cases scheduling of the lighting system operation to coincide with the occupancy schedule of the building will produce substantial savings in energy costs.

Scheduling may be accomplished with timeclocks, microcomputers, or large computer-based energy management systems. The sophistication of the system will depend on the needs of the client and the size of the building. In general, small buildings are candidates for programmable electronic timeclocks, with the larger, more complex computer-based systems reserved for large buildings. Note, however, that many large buildings have been successfully operated by simple timeclocks activating contactors or relays, and equipped with remote override controls.

Programmable electronic timeclocks or computer-based systems are indicated when several cycles or different switching functions per day are required, and provisions should be made for manual override or programming in the event it is necessary to operate lights outside of the normal schedule. Be aware that some timeclocks may be overridden only at the timeclock, and access to the inside of some mechanical timeclocks may present shock hazards to personnel who are unfamiliar with them. Manual override is almost always needed since workers can occasionally be expected to work outside normal working hours. The override should be activated by a remote device located at the point of need, typically the entrance to the building or room, or at individual workstations. Auxiliary override equipment may also be needed at guard stations. Override functions in some large systems may be performed by telephone.

If all of the lights in an area are to be switched simultaneously some provision must be made for warning personnel that the lights are about to be turned off. This may be accomplished by controlling the fixtures in two separate groups and turning off the first group, followed by a time delay of several minutes to allow overriding the control or vacating the room, before turning off the second group of luminaires. As an alternative, some systems can cause the lights to flicker several times as a warning prior to actually turning them off.

Scheduling needs may vary in different areas of a building, and sufficient control capability should be provided on a zone by zone basis to accommodate all necessary functions.

In some buildings, or small areas of large buildings, the operation does not follow a fixed schedule, so controls which are responsive to people using the space are indicated. Occupancy sensors, door switches, and twist timers may be considered.

TYPE OF LIGHTING SYSTEM

Automatic on/off control of incandescent systems imposes no special requirements on the control equipment. Discharge lamps, however, are electrically different, and specialized control equipment may be required. Collapsing electromagnetic fields in the ballasts used with discharge lamps may cause a high voltage across the switching device which exceeds the rating of some electronic switches used in inexpensive photocell controls, and cause premature failure of the device. On/off controls for discharge lamps should be rated for use with the specific lamp type.

When HID lamps are turned on and off by automatic daylighting controls, the control should include a suitable time delay to prevent frequent lamp cycling due to the restrike times of HID sources and the highly deleterious effects of frequent switching on metal halide lamps. Time delay features cause no switching to occur for some predetermined time period after the sensor has called for switching. They are of use in daylighting control systems on partly cloudy days when the illuminance from daylight can change dramatically over intervals of only a few minutes as clouds temporarily obscure the sun.

Dimming controls will also be dictated by the light source since some controls, as previously discussed, will operate only specific lamp types.

BUDGETARY CONSTRAINTS

It is an unfortunate fact that the budget for a project may simply be insufficient to do the most effective control job. It aesthetics are the primary concern and the desired effect cannot be obtained by other means, it is generally possible to obtain cost reductions in other areas to provide sufficient funds for the controls. When the controls are intended to reduce energy consumption, it is frequently a different story since the main concern is getting the building up and operating. The philosophy is that controls can be installed at a later date, although this probability is small.

When funds are simply not available to do the job right, consider a lower cost system which will realize at least part of the potential savings, and lends itself to conversion or upgrade to the desired system at a later date. A detailed analysis delineating the lost savings may be instrumental in changing priorities, but the time invested in the study may be lost if the effort is unsuccessful. Each job must be evaluated on an individual basis, and a course of action taken which is based on the odds of success.

DIMMING REQUIREMENTS

If the purpose of a dimming system is aesthetic, such as in intimate restaurants and lounges, theaters, and other spaces where dimming creates a mood, the control equipment must provide an adequate range of dimming. This eliminates transformer type dimmers for discharge lamps since the range of dimming is limited. Electronic dimmers will be required, and the desired range of dimming may be obtainable only with standard fluorescent lamps and magnetic or electronic dimming ballasts.

If the intent of a dimming system is to reduce energy usage, either transformers or electronic dimmers may be employed. If the building utilizes daylight, variable dimmers will be required. If it is possible to turn the electric lights completely off, as is normally the case in daylight industrial buildings, the control should incorporate on/off control.

If an existing building is not daylit, and it is desired to reduce lighting levels by 15% to 20%, a fixed reduction transformer may be the best choice.

SPACE TO LOCATE CONTROLS

While seemingly unimportant, an appropriate space to place control equipment must be available. Transformers and some electronic controls are bulky and may generate noise and heat that would be objectionable if they were located in or near work spaces. Computerized systems may require space for the equipment, plus provisions for cabling to connect the computer to the switching devices. Frequent access for programming, maintenance, or repairs may also be required.

Access to telephone equipment rooms and space for mounting interface equipment may also be required if telephone communication with the control system is used.

AVAILABILITY OF QUALIFIED PERSONNEL

Maintenance, repair, and reprogramming of complex control systems requires the services of personnel who are qualified to perform these functions. The most desirable option is to have them in-house, but this is not always possible. When outside services must be utilized, it is good practice to make sure that they are available before specifying a complex system. Some very expensive systems have been by-passed simply because the necessary expertise to operate or maintain them was not readily available. Some manufacturers have excellent after-sales service, while with others it is *"caveat emptor."* Verify the availability and reliability of service before specifying the installation of complex systems.

Chapter 16

ENERGY-SAVING LAMPS AND RETROFIT DEVICES

There are a large number of energy-saving lamps and retrofit devices which have been developed specifically to reduce the energy costs associated with operating a lighting system. This chapter will discuss some of the energy-saving products which are currently available, but cannot serve as an up-to-date guide for all of the new products since new products are introduced by manufacturers on an almost daily basis. Continuing contact with manufacturers is required to keep abreast of new energy-saving products.

When considering the installation of energy-saving products, it is easy to become carried away with the thought of conserving energy. It is imperative to remember that lighting systems are normally installed to provide light to allow people to see. If lighting levels are reduced below the required levels, the seeing process will be impaired, and the purpose of the lighting will be defeated. Some energy-saving products produce the same or more light than the original equipment, while other products reduce lighting levels. Verify sales claims, using technical data on the product, as opposed to sales literature, and make sure that both the quantity and quality of the illumination will be adequate for the visual task requirements.

INCANDESCENT SYSTEMS

Incandescent lamps used in the commercial sector are employed primarily for display or accent lighting. In some cases they are used for general lighting in retail operations where their color rendering qualities are desired. If the primary use of the space is merchandising, the requirement for color may be more important than energy efficiency. Incandescent sources offer important benefits such as light control–the ability to put light where it is needed. This permits the creation of high luminance ratios (5 to 1) required for highlighting feature displays. If the existing lighting system performs satisfactorily, the goal should be a reduction in energy consumption with no negative change in lighting levels or distribution. This can frequently be accomplished by simply installing a more efficient version of the lamp that is currently in use.

There are many energy-saving incandescent lamps that are direct replacements for standard lamps. In most cases they simply replace the existing lamp, with no fixture modification required. In fact, fixture modifications may void the UL listing, and should be approved by local building officials in advance if such approval is necessary.

REDUCED WATTAGE A LAMPS

Most lamp manufacturers offer a line of reduced wattage A lamps to replace standard wattages in standard sizes ranging from 40 watts to 150 watts. These lamps use improved manufacturing technology to achieve reductions in wattage of about 15%, with negligible reductions in light. There are slight variations in wattage ratings between manufacturers.

ELLIPSOIDAL REFLECTOR LAMPS

ER lamps (Figure 16-1) are designed for use in recessed baffle type fixtures in place of standard reflector lamps. As explained in Chapter 2, light reflected by the reflector in the lamp converges at a point in front of the lamp, so less light is lost to absorption by the baffle. This permits reductions in lamp wattage with no sacrifice in lighting levels. ER lamps are available in 50-watt, 75-watt, and 120-watt sizes to replace 75-watt, 150-watt, and 300-watt floodlamps, respectively.

ER lamps have a slightly more concentrated beam pattern than floodlamps, so levels directly below the fixture can be expected to increase. They are not spotlights, however, and should not be used as replacements for spots when concentrated beams are desired.

HALOGEN INFRARED REFLECTING LAMPS

The most dramatic new development in incandescent technology is the Halogen I/R. This lamp employs a dichroic coating that permits

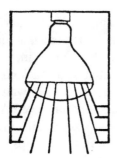

Figure 16-1
Conventional reflector lamp (left) in black baffle type fixture results in low efficiency since about 50% of the light is trapped by the baffle. Ellipsoidal reflector lamp (right) reduces loss since reflector design causes most light rays to cross in front of the lamp.

visible light to pass through, but reflects infrared energy back to the filament where it contributes to filament heating. The added heat causes the filament to produce more light without increasing power consumption. At the time of this writing the lamp is available in a 60-watt PAR version that produces about the same usable light as a standard 150-watt PAR lamp, and a 100-watt lamp will soon be introduced; however, this technology is certain to be applied to other wattages and lamp types. See Figure 16-2.

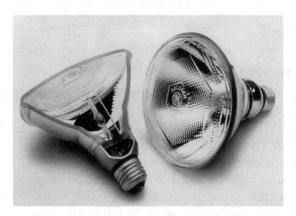

Figure 16-2
The newest incandescent lamp technology is the Halogen Infrared (HIR). Multi-layer thin film on capsule permits visible light to pass, yet reflects infrared energy back to the filament.

REDESIGNED PAR LAMPS

Despite the fact that standard PAR lamps deliver better light control than R lamps, there is still some light that is distributed at relatively high angles and is normally wasted, as illustrated in Figure 16-3. Redesigned reflectors redirect this wasted light back into the main beam. This permits nominal 13%-20% reductions in wattage with no visible change in performance. These high-performance PAR lamps are available in 65-watt, 85-watt, and 120-watt sizes as direct replacements for standard 75-watt, 100-watt, and 150-watt PAR lamps.

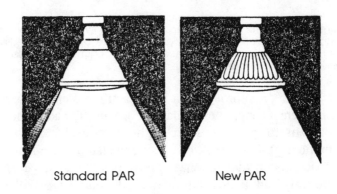

Standard PAR   New PAR

Figure 16-3
Reduced wattage PAR lamps use a redesigned reflector to redirect high angle light back into the main beam. Shaded area outside of main beam represents wasted light. *(Courtesy GE Lighting)*

HALOGEN CYCLE PAR LAMPS

The benefits of halogen cycle operation have been extended to PAR lamps with the introduction of reduced wattage halogen cycle PAR's, (Figure 16-4). These lamps provide a 13% to 20% reduction in power consumption and, as with the other previously discussed energy-efficient lamps, produce about the same usable light.

Halogen cycle PAR lamps are also available as replacements for some popular low-voltage lamps which are used for display and accent lighting. An example is the 12-volt, 36-watt PAR 36 narrow spot. This lamp replaces the standard 12-volt, 50-watt PAR and, in addition to saving watts, provides better color rendering.

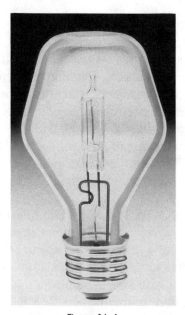

Figure 16-4

Halogen cycle lamps employ a small capsule containing the filament and halogen gas. *(Courtesy GTE Sylvania)*

ENERGY-SAVING TUBULAR QUARTZ LAMPS

These lamps use a dichroic coating which passes visible light, but reflects infrared energy back to the filament to raise filament temperature, as shown in Figure 16-5.

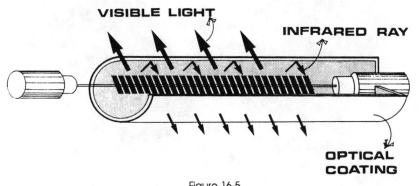

Figure 16-5

Tubular quartz lamps use dichroic coating to reflect infra-red energy back to the filament. These lamps were the first to receive the coating, which has recently been applied to PAR lamps.

A 900-watt lamp produces almost as much light as the standard 1500-watt lamp which it replaces, yet consumes 40% less energy, and a 350-watt version replaces a standard 500-watt lamp with a 30% reduction in power consumption. These lamps cost slightly more than the standard lamps they replace but, considering the operating cost savings, are certainly worth the added investment.

Reduced-wattage tubular lamps which utilize a krypton fill gas to provide reductions in wattage of 20% with only a 10% loss of light are also available. A 1200-watt version replaces the standard 1500-watt lamp, and a 400-watt version replaces the 500-watt lamp.

SCREW-IN, LOW-VOLTAGE CONVERSION KITS

The benefits of low voltage incandescent equipment are now available in the form of screw-in conversion kits to convert some existing line voltage luminaires to use low-voltage lamps. The kits consist of a transformer, equipped with a standard medium lamp base, and a low-voltage lamp. These units utilize the popular MR-16 and similar lamps, or standard low-voltage PAR lamps. They may normally be installed in recessed cans which are designed for R-30 or R-40 reflector lamps. Typical conversion units are shown in Figure 16-6.

Figure 16-6
Screw-in low voltage conversion kits can be used in some line voltage down lights.

MISCELLANEOUS

In addition to simple lamp changes there are several other energy-saving products for use with incandescent lamps. These products and/or ideas do not fit any of the above categories but should be considered when making recommendations for reducing power consumption of incandescent lighting systems.

SPECULAR REFLECTORS

Specular reflectors for use with ordinary A lamps can greatly improve system efficiency under certain conditions. The reflector/A lamp system can be used in lieu of an ER lamp in recessed deep baffle fixtures when the lamp is not deeply recessed in the fixture. It can also replace a reflector floodlamp when the surface of the lamp is flush with the ceiling or extends slightly below the ceiling surface. If aesthetics are not a concern, they can also be used in exposed, non-shielded applications. When used with an inside frosted lamp, the beam spread approximates that of a flood lamp. Clear A lamps produce a spot light effect.

Reflectors are available in a clip-on type which mounts directly on the lamp, and a screw-in version. Typical reflectors are shown in Figure 16-7.

Figure 16-7
Simple reflectors can be applied to standard "A" lamps for use in some down lights and track fixtures. Unit on left screws into fixture, and lamp screws into reflector. Unit on right clips onto lamp.

DIODES AND THERMISTERS

Diodes and thermisters are electronic components that are sometimes applied to incandescent lamps (Figure 16-8).

Figure 16-8
Diodes for use with incandescent lamps. Diodes extend lamp life, but at the
expense of efficacy.

Diodes are sometimes represented as energy-saving devices, which
is pure nonsense. Diodes are half-wave rectifiers. They cut off one half
of the 60-hertz AC power supply, so the lamp receives power during
only one half of each cycle. This means that the lamp is effectively
turned off half of the time. The results are the same as using a dimmer,
except that the level of dimming cannot be varied. Diodes typically
reduce power consumption by about 42 percent and light output by
about 70%. A 100-watt lamp operated with a diode will consume about
58 watts and produce less light than a conventional 40-watt lamp. In
addition, the spectral distribution of the lamp shifts into the red region
and colors are greatly distorted.

Lamp life will be extended, according to some sales literature, to
100 times normal, thus relamping is essentially eliminated. This is
highly doubtful since operation of incandescent lamps on diodes can

result in filament notching (small sawtooth types of irregularities on the filament), which can reduce life by up to 50% from the claimed life.

The use of diodes should be recommended with caution. While lamp life is greatly extended, the color shift will normally be unacceptable if color rendering is important. More importantly, the economics are generally very poor. A 300-watt PS lamp, equipped with a diode, will consume about 175 watts and replace a standard 100-watt A lamp (assuming that the lager envelope will fit in the fixture). Over the estimated 75,000-hour life of the lamp (according to some manufacturers' claims) it will consume an extra 5,600 kWh. The added power cost will usually be far greater than the relamping savings in all but the most difficult relamping situations.

Thermisters are electronic devices which limit the inrush current to the lamp. They also slightly reduce the voltage supplied to the lamp. This reduces power consumption by 2% to 4% and light output by 7% to 16%. Life is increased 2 to 2-1/2 times.

When diodes or thermisters are inserted in lamp sockets they raise the level of the metal lamp base and create a potential for electric shock since the metal base is exposed. Some manufacturers and distributors of these devices provide an insulator which slips over the neck of the lamp and provides protection. The insulator should always be used if the lamp base is exposed. Thermisters resemble diodes in appearance.

USE PAR LAMPS INSTEAD OF R LAMPS

PAR lamps possess better light control characteristics than R lamps and should be used whenever possible. Wattages can be reduced by one standard size with no visible reduction in performance, and the additional cost is quickly recovered in power cost savings.

QUICK REFERENCE CHART

The Quick Reference Chart, Figure 16-9, will serve as a reference for recommending incandescent lamp replacements. In most cases the recommended lamp will provide the same or more light than the lamp it replaces.

INCANDESCENT LAMP REPLACEMENT GUIDE

Existing Lamp	Fixture Type		
	Black Baffle	Specular Reflector	Bare Lamp
300R FLOOd	120ER40	Q250 PAR 38 FLOOD*	Q250 PAR 38 FLOOD*
200R FLOOD	120ER40	100 PAR/HIR FLOOD 150 PAR HALOGEN FLOOD	100 PAR/HIR FLOOD 150 PAR HALOGEN FLOOD
150R FLOOD	75ER30 75A W/CLIP-ON OR SCREW-IN REFLECTOR	60 PAR/HIR FLOOD 90 PAR HALOGEN FLOOD 120R FLOOD	60 PAR/HIR FLOOD 90 PAR HALOGEN FLOOD 120R FLOOD
150R SPOT	60 PAR/HIR SPOT 90 PAR HALOGEN SPOT 120 PAR SPOT	60 PAR/HIR SPOT 90 PAR HALOGEN SPOT 120 PAR SPOT	60 PAR/HIR SPOT 90 PAR HALOGEN SPOT 120 PAR SPOT
150 PAR FLOOD	60 PAR/HIR FLOOD 90 PAR HALOGEN FLOOD 120ER40	60 PAR/HIR FLOOD 90 PAR HALOGEN FLOOD 120 PAR FLOOD	60 PAR/HIR FLOOD 90 PAR HALOGEN FLOOD 120 PAR FLOOD
150 PAR SPOT	60 PAR/HIR SPOT 90 PAR HALOGEN SPOT 120 PAR SPOT	60 PAR/HIR SPOT 90 PAR HALOGEN SPOT 120 PAR SPOT	60 PAR/HIR SPOT 90 HALOGEN SPOT 120 PAR SPOT
100 PAR FLOOD	75ER30 75A W/CLIP-ON OR SCREW-IN REFLECTOR	80 PAR FLOOD	
100 PAR SPOT	80 PAR SPOT	80 PAR SPOT	80 PAR SPOT
75R FLOOD	50ER30	45 PAR HALOGEN FLOOD 50 PAR 30 HALOGEN FLOOD 55 PAR FLOOR 65 PAR FLOOD	45 PAR HALOGEN FLOOD 50 PAR 30 HALOGEN FLOOD 55 PAR FLOOD 65 PAR FLOOD
75R SPOT	45 PAR HALOGEN SPOT 50 PAR 30 HALOGEN SPOT 55 PAR SPOT 65 PAR SPOT	45 PAR HALOGEN SPOT 50 PAR 30 HALOGEN SPOT 55 PAR SPOT 65 PAR SPOT	45 PAR HALOGEN SPOT 50 PAR 30 HALOGEN SPOT 55 PAR SPOT 65 PAR SPOT
75 PAR FLOOD	50ER30	45 PAR HALOGEN FLOOD 50 PAR 30 HALOGEN FLOOD 55 PAR FLOOD 65 PAR FLOOD	45 PAR HALOGEN FLOOD 50 PAR 30 HALOGEN FLOOD 55 PAR FLOOD 65 PAR FLOOD
75 PAR SPOT	45 PAR HALOGEN SPOT 50 PAR 30 HALOGEN SPOT 55 PAR SPOT 65 PAR SPOT	45 PAR HALOGEN SPOT 50 PAR 30 HALOGEN SPOT 55 PAR SPOT 65 PAR SPOT	45 PAR HALOGEN SPOT 50 PAR 30 HALOGEN SPOT 55 PAR SPOT 65 PAR SPOT

* USE ONLY IN CERAMIC SOCKETS

Figure 16-9

Quick reference chart for incandescent lamp replacement. In most cases the replacement lamp will provide about the same illuminance as the existing lamp. When in doubt, try the replacement in a few fixtures.

FLUORESCENT PRODUCTS

The typical energy-conservation measure recommended for existing fluorescent lighting systems is the installation of energy-saving lamps. They are produced by most lamp manufacturers and marketed under a variety of trade names.

Energy-saving fluorescent lamps which replace most standard sizes of 30 watts and larger are available. They are produced in three basic generations, with the difference being in the efficiency, and types of phosphors used to produce light. The first-generation lamps use standard phosphors in cool white, cool white deluxe, warm white, warm white deluxe, daylight, and white colors. They are rated to produce about 12%-15% less light than standard lamps and consume about 15% less power. Second-generation lamps use a high-efficiency phosphor, "Lite-White," and are rated at about 7% less light while providing the same power reduction as the first-generation lamps. The third-generation energy-saving lamps use high-efficiency, high-color-rendering phosphors to produce about the same reductions in light and power as the second-generation lamps, but with substantial improvements in color.

Energy-saving fluorescent lamps are similar in construction and operation to standard lamps. The main difference is in the fill gas—krypton is added. This changes the electrical characteristics and results in a decrease in the power consumed by the lamp. This change also alters the ballast factor, as discussed in Chapter 6.

Energy-saving lamps also draw more current than standard lamps, and exhibit an increase in the voltage across the ballast capacitor. In pre-1978 ballasts this may exceed the dielectric capability of the capacitor, with resultant premature ballast failure. Since the typical life of a ballast in a troffer is 9 to 10 years, these older ballasts will soon be replaced with new ballasts designed to operate normally with the increased voltage, thus the problem will soon be solved through attrition.

Energy-saving fluorescent lamps are rated for use in ambient temperatures of 60° F or higher, and are not recommended for use on dimming circuits, or for operation on reduced-current ballasts.

Four-foot energy-saving lamps employing internal switches which disconnect the cathode heating circuits after the lamp has started are also available. See Figure 16-10. These lamps are rated at 32 watts, a two-watt savings over standard energy-saving lamps, and disconnection of the cathode heating circuit saves another 1/2 watt per lamp. The removal of cathode heat shortens rated lamp life by up to 25%; however, the added energy savings may more than offset the added relamping cost. Be aware that lamps may require up to a minute before the system will start up after being turned off, so it may be advisable to retain a few standard lamps to provide light if power is momentarily interrupted.

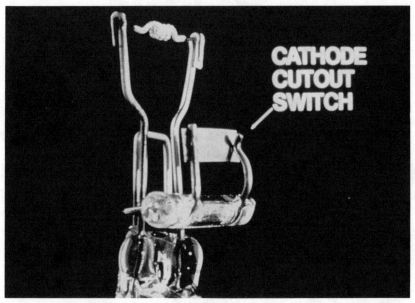

Figure 16-10
Thermal switch located in cathode heater lead turns cathode off after lamp
has started. *(Courtesy GE Lighting)*

FLUORESCENT CONVERSIONS
FOR INCANDESCENT LAMPS

The popular single ended compact fluorescent lamps in 5-watt, 7-watt, 9-watt, and 13-watt sizes are available for retrofit of incandescent fixtures. A wide variety of products exist for the conversion of table lamps and some recessed fixtures. New "quad" tube lamps in sizes up to 26 watts have recently been introduced and offer additional retrofit opportunities.

LAMP/BALLAST SYSTEMS

Some existing systems using standard lamps and ballasts may be candidates for conversion to more efficient systems by the installation of new lamp/ballast systems, particularly if the existing system is nearing end of ballast life. These systems employ special cathode cutout circuits which disconnect power to the lamp cathode once the lamp has started, or smaller diameter T-8 lamps which improve the optical and thermal characteristics of the fixture. These systems require the replacement of both the lamps and ballast when converting existing systems. They are available from most fixture manufacturers as extra cost options for new fixtures.

REACTIVE DEVICES

Reactive devices (Figure 16-11) reduce the power input to fluorescent systems by 20% to 50%, with slightly lower reductions in light output. There is a slight improvement in fluorescent lamp efficacy in the early stages of dimming, so the reductions in light output are slightly less than the reductions in power input. A 33% power reduction device can be expected to reduce light output by about 27%.

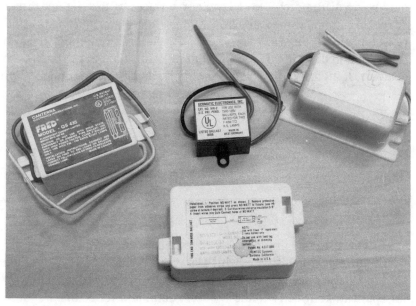

Figure 16-11
Reactive devices for use with fluorescent lamps. Some devices wire into circuit between ballast and lamp, while others simply clip onto lamp base.

Some devices attach to the end of one lamp in a two-lamp system, while others are mounted inside the fixture and wired directly into the lamp circuit. Lamps with built-in reactive devices are also available. The active element is a capacitor, which changes the impedance of the circuit to reduce current. They should contain a bleed-off resistor to reduce the possibility of electrical shock hazard. Reactive devices for rapid-start systems also include a small transformer.

The primary application of these devices is in systems where the illuminance may be reduced.

SPECULAR REFLECTORS

Specular retrofit reflectors are typically used in recessed troffers or bare lamp strip fixtures, although they can be fabricated for almost any fixture type. The installation of a reflector is generally accompanied by the removal of one-third or one-half of the lamps from the fixture. Reflectors are available in two basic types: semi-rigid reflectors which are secured in the fixture by mechanical means, and adhesive films which may be applied directly to the interior surfaces of the fixture. Either silver or aluminum may be used as the reflecting media. Semi-rigid reflectors are bent to some specific shape, as shown in Figure 16-12, to direct light in the desired manner. Film reflectors conform to the fixture contours, so light distribution characteristics can not be controlled.

Figure 16-12
Two of the many shapes used for reflectors. Design on left concentrates light downward, while complex bends on right create multiple images and slightly wider light distribution.

Silver reflectors are made by coating or impregnating a polyester film with elemental silver, which may be bonded to an aluminum substrate to produce a semi-rigid reflector, or coated with an adhesive for application directly to the fixture. Aluminum reflectors are made from specular anodized aluminum sheet.

Silver reflectors typically have reflectances in the range of 90% to 97%, while anodized aluminum products range from 70% to 90%. Anodized aluminum products employing a multi-layer optical coating achieve reflectances of about 95%. For comparison, white enamel paints used on most older fixtures and many currently manufactured fixtures range in reflectance from 80% to 85%. Modern white powder paints exhibit reflectances of about 90%.

The design of the reflector is critical to performance, and is usually more important than the reflectance of the material used. A properly configured reflector in a two-lamp troffer can be expected to improve fixture efficiency by 5% to 15% over a comparable two-lamp fixture without a reflector. When compared to a four-lamp troffer, the two-lamp reflectorized fixture will show an additional increase in efficiency of about 12% due to improvements in thermal factors and optical efficiency resulting from the removal of two lamps. A poorly designed reflector can actually reduce efficiency to less than that of an unreflect-ored fixture. Given comparable good design, a silver reflector can be expected to outperform an anodized aluminum reflector by about 10%. Information available at the time of this writing indicates that multi-layer optical coated reflectors provide about the same performance as silver products. Since silver and multi-layer optical coated reflectors are more costly than anodized aluminum reflectors, an economic analysis should be prepared, and the added cost weighed against the slight improvement in performance.

Reflectors can create multiple lamp images (Figure 16-13), thus eliminating the appearance of delamped fixtures. This benefit alone may be sufficient justification for their installation.

Figure 16-13
Photo shows multiple lamp images, but fixture contains only two lamps. This eliminates the visual appearance of a delamped fixture.

ALTERNATIVES TO REFLECTORS

When considering the installation of reflectors, several other alter-natives should also be considered to assure that the action taken provides the desired results. New lamp/ballast systems, previously discussed,may be viable options and should be investigated. Another, often overlooked, is simply the removal of one or two lamps, and the implementation of a routine program of fixture washing and group lamp replacement. If the reductions in light levels which result from simple delamping and routine washing are slightly greater than acceptable, consider leaving all lamps in the fixture and installing reactive devices in conjunction with a maintenance program. It is a sad but true fact that many lighting systems are not maintained in the mistaken belief that fixture washing is not necessary and the money spent on this function is wasted.

Many older office lighting systems consist of four-lamp recessed troffers mounted on 8-foot centers, which produces an average of about 100 footcandles if the system is maintained. These systems, if not maintained, can be expected to produce about 65-70 footcandles after 10 years. A new reflectorized system using two lamps will generally produce the same or higher illuminance. If the reflectorized system is not maintained, if can be expected to produce 35-45 footcan-dles after 10 years. Figure 16-13 shows a comparison of the calculated illuminance for several types of semi-rigid reflectors, and fixtures without reflectors. Note that film reflectors have been omitted due to the lack of reliable photometric data for these products. The comparison is based on the following parameters:

1. The installation is in a 10-year-old office building which operates 12 hours per day, 5.5 days per week. The RCR is 1, and reflectances are 80% ceiling, 50% wall, and 20% floor. The dirt condition is "clean."

2. The system was originally designed to produce a maintained illuminance of 100 footcandles using four-lamp troffers mounted on 8-foot centers with standard F40CW lamps. The assumed maintenance schedule called for washing every 12 months, and group lamp replacement every 3 years.

3. Light loss factors are based on data contained in Chapter 6.

Eight different scenarios are shown, each with two different maintenance schedules, for initial and end-of-year illuminance for a 10-year period. The column headed "No-Maint." assumes that fixtures are not washed, and lamps are replaced only when they fail. The "Maint."

heading assumes washing every 12 months, relamping every 3 years, and prompt replacement of early failures.

Condition 1 represents the base case, an installation consisting of four-lamp fixtures with standard lamps.

Condition 2 is similar to Condition 1, except energy-saving lamps are used.

Condition 3 represents the most efficient silver reflector for which photometric data are available at the time of the analysis. A concentrating light distribution is used to direct light downward and maximize coefficients of utilization.

Condition 4 is a wider distribution-design silver reflector. Most silver products will perform somewhere in between Conditions 3 and 4.

Condition 5 is an aluminum reflector with a configuration similar to the silver reflector used in Condition 3. Reliable photometric data for this configuration are not available, and the coefficient of utilization is based on a comparison of other silver and aluminum reflectors of known characteristics. While not precise, the estimate is believed to be sufficiently accurate for purposes of comparison.

Condition 6 is a wider distribution aluminum reflector, similar to the silver reflector in Condition 4.

Condition 7 represents a system of unmodified new two-lamp fixtures.

Condition 8 represents a 10-year-old fixture which has been cleaned and delamped to two new standard lamps. Failure to clean the fixture periodically has resulted in a permanent loss of reflectivity, and performance has been degraded by 5%. Note, however, that luminaire surface deterioration can vary widely between installations, depending upon the severity and corrosiveness of contaminants.

The predictions of illuminance in Figure 16-14 are based on the average illuminance within the space. Reflectors normally concentrate light downward, so the uniformity of illuminance will normally show wider variations with reflectors. Figure 16-15 shows a comparison of the calculated illuminance at a series of points surrounding a fixture. The values include illuminance from adjacent fixtures and the reflected component.

Since reflectors typically concentrate light downward, an increase in veiling reflections may also be anticipated, with a commensurate degradation of the quality of the illuminance. These effects may be

End of Year Footcandles
Typical 2' x 4' troffer in clean office environment
Luminaire Spacing: 8' x 8' grid pattern 64 sq. ft/fixt.
Ballast Factor: 0.94 std lamps, 0.87 energy -saving lamps

	Condition 1 4 Std CW lamps No Reflector CU = 0.69		Condition 2 4 E/S CW lamps No Reflector CU = 0.73		Condition 3 2 Std white lamps Silver Reflector #1 CU = 0.91		Condition 4 2 Std white lamps Silver Reflector #2 CU = 0.76	
	Footcandles		Footcandles		Footcandles		Footcandles	
Year	No-Maint.	Maint.	No Maint.	Maint.	No-Maint.	Maint.	No-Maint.	Maint.
0	128	128	111	111	84	84	70	70
1	103	103	90	90	68	68	57	57
2	92	98	80	85	61	64	51	54
3	86	94	75	82	57	62	47	52
4	81	103	70	90	53	68	44	57
5	77	98	67	85	51	64	43	54
6	75	94	66	82	50	62	42	52
7	73	103	64	90	48	68	40	57
8	71	98	62	85	47	64	39	54
9	69	94	60	82	46	62	38	52
10	68	103	59	90	45	68	37	57

	Condition 5 2 Std CW lamps Alum.Reflector #2		Condition 6 2 Std CW lamps Alum.Reflector #2		Condition 7 2 Std WH lamps New 2-lamp Fixture CU = 0.73		Condition 8 2 Std CW lamps Deteriorated Fixture CU = 0.69	
	Footcandles		Footcandles		Footcandles		Footcandles	
Year	No-Maint.	Maint.	No Maint.	Maint.	No-Maint.	Maint.	No-Maint.	Maint.
0	77	77	68	68	68	68	64	64
1	62	62	55	55	55	55	52	52
2	55	59	49	52	49	52	46	49
3	52	57	45	50	45	50	43	47
4	48	62	43	55	43	55	40	52
5	47	59	41	52	41	52	39	49
6	45	57	40	50	40	50	38	47
7	44	62	39	55	39	55	37	52
8	43	59	38	52	38	52	36	49
9	42	57	37	50	37	50	35	47
10	41	62	36	55	36	55	34	52

Figure 16-14
Comparison of average illuminance produced by various fixture and reflector options. See text for explanation.

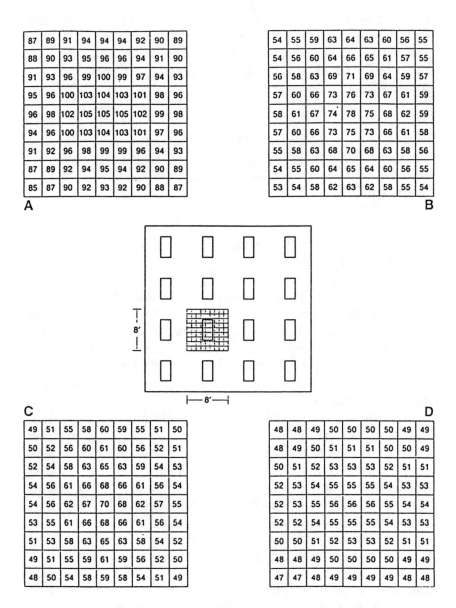

Figure 16-15
Illuminance at a series of points below fixtures with and without reflectors. The numbers represent the illuminance at points within the illustrated grid pattern. (a) is a typical unmodified four-lamp lensed troffer; (b) is the same fixture with a concentrating design silver reflector; (c) uses a wider spread multiple imaging silver reflector, and (d) is an unmodified two-lamp troffer.

estimated by predicting the ESI (Chapter 9) for each system. Comparisons of the raw and ESI illuminance for a point directly below a fixture and another point midway between fixtures for the layout in Figure 16-14 for concentrating and spread silver reflectors, and a new two-lamp fixture without a reflector are:

	Nadir		Midpoint	
	Raw FC	ESI FC	Raw FC	ESI FC
Concentrating	78	25	58	57
Spread	70	21	54	52
No reflector	56	23	53	52

From the analysis, it can be seen that all three options are about equal in terms of veiling reflections.

Before the suitability of reflectors (or any other retrofit device or system) can be reached, an analysis of the lighting requirements must be performed. Many older buildings were designed in an era of higher design illuminance specifications, so some reductions in lighting levels may be acceptable. Once the required quantity and quality of light have been determined, a site-specific analysis may be prepared to determine the appropriate course of action.

Maintenance costs should be thoroughly investigated. The cost of washing reflectors is unknown at this time; however, it may be higher than the cost of washing a painted fixture surface. It is obvious, however, that if reflectors are installed, a routine program of fixture maintenance must be implemented or the investment will be wasted. Most reflectors of current design require the removal of the reflector for ballast replacement which can be expected to increase ballast replacement labor cost. One manufacturer is working on a design to facilitate access to the ballast compartment to overcome this problem.

It is also very important to verify the durability of the product to be used. Sheet aluminum has been used in the lighting industry for several decades, and the durability is well known. Films, however, are relative newcomers, and there have been reported cases of film separation, flaking, and discoloration. Some films have also shown little abrasion resistance, and may scratch when cleaned.

ENERGY-SAVING HID RETROFIT LAMPS

High-intensity discharge lamps, as previously discussed, require proper ballasting. They are not usually interchangeable between wattages and types. There is one notable exception to this rule: mercury vapor lamps may be operated on metal halide ballasts as long as the wattage ratings of both the lamp and ballast are the same. I frankly don't know why anyone would want to do this, but it works.

Some metal halide and high-pressure sodium lamps are made specifically for operation on mercury vapor ballasts as an energy conservation measure. Since the lamp is usually operating under less than ideal conditions there may be trade-offs in performance when compared to a standard lamp operating on a standard ballast. Most retrofit HID lamps operate only on specific mercury vapor ballast types, and there are wide variations in performance between various lamp manufacturers' products, so a thorough investigation of the available products should be conducted before making a decision.

The use of retrofit lamps can frequently be justified on an economic basis in cases where the installation of new luminaires is too costly or payback periods are unacceptable due to short operating hours.

Chapter 17

LIGHTING SYSTEM MAINTENANCE

This chapter is on lighting maintenance, but before the discussion of maintenance begins, consider the following:

You are buying a new house. The bank offers you two options, A and B. If you select option A you will make a downpayment of $25,000, and payments will be $1,500 per month. Option B has a downpayment of $20,000, payments of $1,200 per month, and an additional payment of $100 at the end of each year. Either option will retire the loan in 20 years. Which option do you choose? If you have selected option A, which required an additional immediate outlay of $5,000, plus additional annual payments of $3,600 to avoid a payment of $100 once per year and are convinced that this is the right choice, read no further, put this book on the shelf, and investigate the purchase of stock in a company which purports to have developed a perpetual-motion machine. If you selected option B, or are now questioning the selection of option A, read on.

The proper maintenance of a lighting system is the most neglected, and frequently the most cost-effective, means of reducing the overall cost of light. Most people change oil and lubricate their cars on a regular (or somewhat regular) basis, yet will not even consider maintaining a lighting system. The end result of failure to follow a maintenance schedule for either a car or a lighting system has the same ultimate effect: the bottom line cost is higher if maintenance is not performed. The results with the car are more dramatic: catastrophic failure of the engine or other moving parts, and an immediate large cash outlay to fix the problem. The results of failure to maintain the lighting system are more subtle since they result in a moderate added monthly cost in the form of a higher electric bill which continues over the life of the system and, over time, may cost many times the cost of car repairs.

Without proper maintenance a lighting system may perform at a level of only 50% to 60% of its capability. There are many office lighting systems installed in the 1960's and 70's which were designed to provide 100 or more footcandles, but now provide 50-70 footcandles. These systems are consuming the energy required for the original design level, but have dropped dramatically in efficiency due to lack of maintenance. In many cases the operating costs of many of these systems could be reduced by 40% to 50% simply by delamping and instituting a routine maintenance program. It is amazing that so many otherwise intelligent people in decision-making positions consider lighting system maintenance to be a cost which can be avoided simply by not maintaining the system, and believe that they have saved money. They have a choice: they can pay for maintenance, or they can pay equipment manufacturers and utilities. The cost for equipment and utilities is much higher.

WHAT IS LIGHTING SYSTEM MAINTENANCE?

Routine maintenance consists of washing the interior reflective surfaces of the fixture, both sides of the lens or protective cover (if the fixture has one), and the lamps, at some regular time interval, Figure 17-1. In some cases lamps will be replaced on a group basis when washing is performed. Broken lamp holders, loose wires, and failed ballasts may also be repaired or replaced at this time if they have not been taken care of at the time of failure.

Maintenance may be performed by in-house maintenance personnel or by a lighting maintenance company which specializes in this type of work. Unless a facility is either very small and the lighting equipment easy to maintain, or large enough to employ a full-time maintenance staff with the time and equipment required to do the job, it is generally more economical to contract the job out to a lighting maintenance company. They have the equipment and experience to do the job quickly, efficiently, and properly.

Figure 17-1
Typical fixture washing; (top) parabolic troffer, (bottom) metal halide with protective glass cover.

EQUIPMENT REQUIREMENTS

The equipment necessary to perform maintenance depends upon the type of lighting system, the mounting height and accessibility of fixtures, and the number of luminaires.

The first requirement is gaining access to the fixture, which normally is done on ladders, scaffold, lifts, or bucket trucks. Ladders, scaffolds, and lifts are normally used indoors, with bucket trucks, Figure 17-2, reserved for outdoor installations. The choice is dictated by the physical conditions at the site.

For safety reasons, ladders should be made of fiberglass or other electrically nonconductive material. Ladders are normally used for systems with low mounting heights; however, they are sometimes used for heights of up to 22 feet when other means are not available. Lifts are used for industrial and warehouse applications when there is sufficient space to permit their use, and a large volume of work is to be done. Scaffolds are used when there are obstructions on the floor, and ladders or lifts are impractical due to the obstructions.

Washing of fixture interiors is accomplished with cloth rags or towels and mild cleaning agents such as dishwashing detergent or very mild solutions of TSP. Avoid the use of harsh chemicals which may damage the reflective surfaces and degrade fixture performance. Glass lenses or covers, such as those used on some high-intensity discharge or incandescent luminaires, are cleaned with mild detergents or other non-abrasive cleaners. Plastic lenses are also cleaned with mild detergents. Both fixture interiors and lenses should be thoroughly rinsed with clean water after washing. Lenses should be allowed to air dry as opposed to drying them with cloth to avoid the build up of a static charge which will attract dirt particles.

Flat plastic lenses (those without bends or angled surfaces) are best washed by machine when a large number of lenses are to be cleaned. These machines are available from several manufacturers and are used by lighting maintenance companies. They normally do a better job than can be done manually, and do the job in much less time.

Parabolic wedge louvers, called paracubes, may be cleaned with an ultrasonic cleaner. These machines use ultrasonic waves and solvents to remove dirt from the interior surfaces, and do an excellent job.

Spot relamping, the replacement of lamps as they fail, may be performed by climbing a ladder or using a lift, or in some cases may be

Figure 17-2
Access to fixtures from: (top left) ladder, (top right) scissors lift, (bottom) bucket truck.

accomplished faster, easier, and in a safer manner by using a pole-mounted device which grips the lamp and allows removal or insertion from the ground. These devices may be used for open-bottomed fixtures, or enclosed fixtures which incorporate a latching mechanism on the lens which permits opening the fixture by simply pressing on the cover.

DISPOSAL OF OLD LAMPS AND BALLASTS

Ballast capacitors manufactured prior to 1978 contain polychlorinated byphenols (PCB's), and are classified as hazardous waste. Fluorescent, mercury vapor, metal halide, and high-pressure sodium lamps contain mercury, and may be classified as hazardous waste by some states. A check with local authorities to determine the required disposal procedure is recommended before disposing of old lamps, and ballasts or capacitors which contain PCB's.

DETERMINING AN
ECONOMIC MAINTENANCE INTERVAL

The determination of a time interval for fixture washing and lamp replacement is simply an economic choice. Maintenance procedures represent a cost which must be weighed against the savings which accrue from a reduction in the number of fixtures installed, reductions in relamping costs, and savings from reduced energy costs. There is no single maintenance interval that is right for all systems, since several variables might be considered: the installed fixture cost, the maintenance cost, and the energy cost. For new installations all three components are considered. When evaluating existing systems, only the elements of maintenance and energy are of concern since the number of luminaires is fixed and they are already in place.

When designing a new lighting system there are a finite number of layouts which will fit the room, and the calculated illuminance produced by each layout will depend, in part, on the light loss factors that were used in the calculation of the required number of fixtures. Changing the luminaire dirt and lamp lumen depreciation factors will alter the required number of luminaires, with frequent washing and relamping resulting in the least number of fixtures. The installation using the fewest fixtures or lamps will normally be the most cost effective.

    As discussed in Chapter 10, Design Methods, the number of luminaires which will fit into grid ceilings such as the popular 2' x 4' T Bar system is somewhat fixed. There may be occasional opportunities to use a layout with wider fixture spacing by maintaining the system, but in many cases reducing the number or wattage of lamps in each fixture will be more feasible.

Example

    A 40' x 40' office is to be lighted to 50 footcandles. the RCR is 1.9, and room surface reflectances are 80% ceiling, 50% wall, and 20% floor, which results in coefficients of utilization of 0.73 for the selected fixture in a two-lamp configuration and 0.68 with three lamps. Lights will operate 3500 hours per year, and the dirt condition is clean.

    Using the layout procedures outlined in Chapter 10, layouts of either 25 fixtures on 8' x 8' centers or 35 fixtures on 6' x 8' centers will fit the space.

    If a relamping interval is assumed, the equation used to calculate the required number of fixtures can be rearranged to solve for a luminaire dirt depreciation factor, and the required maintenance interval can then be read directly from the LDD graphs.

$$LDD = \frac{(FC)\,(Area)}{(\# \text{ Fixt.})\,(Lumens/Fixt.)\,(CU)\,(BF)\,(LLD)\,(RSDD)}$$

The light loss factors for the example are, from Chapter 6:
    Ballast Factor = 0.94 (Standard 40-watt lamps)
    Lamp Lumen Depreciation: 0.85 (3-year relamping)
                                    0.82 (Spot relamping)
    Room Surface Dirt Depreciation: 0.96

    At this point some assumptions must be made. They can be easily changed if they are incorrect. Since the design illuminance is a moderate level, 50 footcandles, assume that 25 two-lamp fixtures will be used.

    The LDD, assuming a 3-year relamping interval, is:

$$LDD = \frac{(50 \text{ FC})\,(1600 \text{ Sq. Ft.})}{(25 \text{ Fixt.})\,(6300 \text{ Lm.})\,(.73)\,(.94)\,(.85)\,(.96)}$$

$$LLD = 0.91$$

The answer should be near the 0.85-0.90 range for offices, and 0.91 is sufficiently close, so a layout consisting of 25 two-lamp fixtures is probably the most economical installation. From Figure 17-2, a washing interval of 7 months is indicated to maintain the LDD at 0.91. This is not practical in industry, so a 12-month interval should be considered. The LDD at 12 months is 0.88, and the maintained illuminance will be:

$$E = \frac{(\#\ Fixt.)\ (Lm/Fixt.)\ (CU)\ (BF)\ (LLD)\ (LDD)\ (RSDD)}{Area}$$

$$= \frac{(25\ Fixt.)\ (6300\ Lm)\ (0.73)\ (0.94)\ (0.85)\ (0.88)\ (0.96)}{1600\ Sq.\ Ft.}$$

$$= 49\ Fc$$

This is sufficiently close to 50 Fc, and can be expected to provide adequate illumination.

The alternative is to use three-lamp fixtures on the same spacing, or a layout of 35 two-lamp fixtures on 6' x 8' spacing, replace lamps only on burnout, and omit washing. For purposes of comparison, the illuminance produced by each system over a 10-year period is shown in Figure 17-3. Note that the alternatives produce illuminances which are considerably higher than the design level during the first 4-6 years, but ultimately provide about the same performance as proposed system, and at a higher first cost and ongoing operating and maintenance expense, as shown in Figure 17-4.

Layouts of surface- or pendant-mounted fixtures on ceilings which do not impose constraints on fixture locations provide opportunities for reducing the number of fixtures. In these cases the required number of fixtures is calculated using several maintenance intervals, and the layout using the least number of fixtures which provides acceptable uniformity is selected.

Existing buildings can provide excellent opportunities for reductions in energy cost through a routine maintenance program if one does not already exist, particularly in older buildings that were designed for lighting levels in excess of current standards. The procedure for determining the maintenance interval consists of three steps: 1) Determine the required illuminance using procedures in Chapter 9; 2) determine what the existing system is currently producing by measuring the illuminance, also using procedures from Chapter 9; 3) determine the number of lamps per fixture required to produce the desired illuminance after a maintenance program has been implemented, using procedures from Chapters 6 and 7.

End of Year Footcandles

	25 Fixtures 2 Std. CW Lamps 12 month wash 36 month relamp CU = 0.73	35 Fixtures 2 Std. CW Lamps No washing Relamp @ Burnout CU = 0.73	25 Fixtures 3 Std. CW Lamps No washing Relamp @ Burnout CU = 0.68
Year	Footcandles	Footcandles	Footcandles
0	65	91	91
1	52	73	73
2	50	66	65
3	49	61	61
4	52	57	57
5	50	55	55
6	49	54	53
7	52	52	52
8	50	51	51
9	49	49	49
10	52	48	48

Figure 17-3
Comparison of end-of-year average illuminance provided by three systems with different maintenance schedules. See text for explanation.

	25 Fixtures 2 Lamp 12 Mo. Wash 3 Yr. Relamp	35 Fixtures 2 Lamp No Wash Spot Relamp	25 Fixtures 3 Lamp No Wash Spot Relamp
First Cost			
Fixtures, installed	$3500.00	$4900.00	$3750.00
Operating & Maint. Cost			
Power Cost @ $0.08/kWh	$602.00	$842.80	$903.00
Replacement Lamp Cost (Yr)	25.00	18.38	19.69
Relamping Labor Cost (Yr)	5.50	61.25	65.63
Washing Cost (Yr)	75.00	-0-	-0-
Total Annual O & M Cost	$707.50	$922.43	$988.32

Figure 17-4
Economics of the three systems shown in Figure 17-3.

Note that some older fixtures may have undergone a permanent loss of reflectivity due to the corrosive action of some contaminants if fixtures have not been cleaned for several years. When in doubt it is a good idea to try cleaning and relamping a small test area in the building, measuring the average illuminance, and comparing the measured values to calculations. The extent of permanent loss may then be estimated based on the differences. A rule of thumb is that the maintained illuminance in an average office will be about 75% of the initial measured illuminance if fixtures are washed annually and group relamped at 50% rated lamp life. Note that this is only a rule of thumb, but it may be useful in evaluating the effects of cleaning. If a maintained level of 50 fc is desired, the illuminance after washing and relamping should be in the 65-70 fc range.

Calculations for existing systems may be simplified by performing them on a "per fixture" basis. For example, in an installation of troffers on 8' x 8' centers, each fixture covers an average area of 64 square feet, so calculations may be based on one luminaire, and 64 square feet. The average illuminance is calculated from:

$$E = \frac{(\# \text{ lamps/fixt.}) \ (\text{lumens/lamp}) \ (\text{CU}) \ (\text{LLF})}{\text{Area per fixture}}$$

The coefficient of utilization is based on the RCR for the room, and light loss factors obtained from methods in Chapter 6.

Example

An RCR 1 office is lighted with four-lamp lensed troffers using four F40CW energy-saving lamps. Fixtures are on 8' x 8' centers. The system is 8 years old, and has not been maintained. Lights operate 3500 hours per year. Meter readings at various work stations indicate the average illuminance is about 60 footcandles. The recommended illuminance for the task is 50 Fc. What maintenance intervals and lamps might be used to reduce the number of lamps and still provide adequate illuminance?

Solution

Since the existing system uses four lamps per fixture, the only simple delamping option is to go to two lamps.

1. Obtain photometrics for the fixture in both four- and two-lamp configurations, and determine the CU's. If photometrics for the two-lamp fixture are not available, a reasonable estimate may be made by

multiplying the four-lamp CU by 1.1. The CU's for the example are, from Figure 17-5, 0.77 for four energy-saving lamps, and 0.80 for two lamps.

CU Table
Example Troffer

ρ_{FC}	20		
ρ_{CC}	80		
ρ_W	70	50	30
1	77	(73)	70
2	71	65	61
3	66	60	54
4	61	54	50
5	56	49	45
6	52	45	39
7	49	40	35
8	45	34	31
9	42	32	28
10	39	30	25

Four Standard F40 Lamps

Figure 17-5
CU table for the example in text. CU is 0.73 for standard lamps. This is adjusted by a luminaire thermal factor of 1.06 for four energy-saving, 34-watt lamps, so a CU of 0.77 is used for this configuration. For two standard lamps the CU can be approximated by multiplying the base CU by 1.1 to obtain the 0.80 value.

2. Determine the maintenance intervals and lamps to be used for the study. For the example we will use washing intervals of 12 months and 24 months, and relamping intervals of 3 years, 4 years, and spot relamping only. Lamps will be F40CW standard and energy saving. This yields the following possible combinations:

Option
1   Standard lamp, 12-month wash, 36-month relamp
2   Standard lamp, 12-month wash, 48-month relamp
3   Standard lamp, 12-month wash, spot relamp
4   Standard lamp, 24-month wash, 48-month relamp
5   Standard lamp, 24-month wash, spot relamp
6   Energy-saving lamp, 12-month wash, 36-month relamp

7   Energy-saving lamp, 12-month wash, 48-month relamp
8   Energy-saving lamp, 12-month wash, spot relamp
9   Energy-saving lamp, 24-month wash, 48-month relamp
10 Energy-saving lamp, 24-month wash, spot relamp

Note that in reality the number of options would be limited to only three or four, based on experience. For purposes of illustration, all 10 options will be analyzed.

3. Determine the light loss factors for each option. Light loss factors are determined using the procedures discussed in Chapter 6. For existing buildings only the ballast factor, lamp lumen depreciation factor, and luminaire dirt depreciation factors are normally used.

The following light loss factors apply to the example problem:
Ballast Factor: 0.94 (Standard Lamps)
                    0.87 (Energy-Saving Lamps, standard ballasts)
Lamp Lumen Depreciation Factor (All Lamps):
                    0.85 (3-year relamp)
                    0.82 (4-year relamp)
                    0.79 (Spot Relamping)
Note that the spot relamping factor is based on output at 70% of rated lamp life. Rated life at 12 hours per start is 26,000 hours, so the LLD at 18,200 hours is used.

Luminaire Dirt Depreciation Factor:
    0.88 (12-month wash cycle)
    For no washing

Year	LDD	Year	LLD
1	0.88	6	0.72
2	0.83	7	0.70
3	0.80	8	0.68
4	0.77	9	0.66
5	0.74	10	0.65

LDD factors are calculated from the equation given in Chapter 6 since the time span exceeds the scope of the LDD graphs.

4. Calculate the average maintained illuminance, annually, for each of the options. The results are shown in Figure 17-6.

End of Year Footcandles

OPTION

Year	Existing	1	2	3	4	5	6	7	8	9	10
1	93	60	60	60	60	60	48	48	48	48	48
2	84	57	57	57	54	54	46	46	46	44	44
3	78	55	55	55	55	55	45	45	45	45	45
4	73	60	53	53	50	50	48	43	43	41	41
5	67	57	60	51	60	51	46	48	42	48	42
6	65	55	57	51	54	49	45	46	42	44	39
7	64	60	55	51	55	51	48	45	42	45	42
8	62	57	53	51	50	49	46	43	42	41	39
9	60	55	60	51	60	51	45	48	42	48	42
10	59	60	57	51	54	49	48	46	42	44	39

Figure 17-6
Average end of year illuminance produced by the 10 options in the example problem.

5. Determine the options which are feasible from a performance standpoint, and perform an economic analysis to determine the most economical maintenance schedule. A 50-footcandle design illuminance was specified for the example, so Options 1 through 5 are viable. Note that Option 5 periodically drops to 49 footcandles, but in practice this is only 2% low, and is considered acceptable. Options 6 through 10 produce illuminances which are 10% to 22% below the desired level, and are not considered adequate.

The economic analysis is performed using procedures described in Chapter 18, and a comparison of the average annual operating and maintenance costs is shown in Figure 17-7. For the conditions stated in the analysis, Option 4 has the lowest annual cost, $2675.65, a savings of $2280.08 when compared to the original system, or a reduction in O & M costs of 46%. Note that the actual savings will vary according to site conditions and costs for power, lamps, and labor, so other maintenance intervals may be more economically advantageous. A site-specific analysis should be prepared on a case-by-case basis.

Average Annual Operating and Maintenance Costs (Per 100 Fixtures)

	Existing	Option 1	Option 2	Option 3	Option 4	Option 5
kW electrical load	16.4	8.6	8.6	8.6	8.6	8.6
Power cost @ $0.08						
per kWh	$4,592.00	$2,408.00	$2,408.00	$2,408.00	$2,408.00	$2,408.00
Replacement lamp cost	94.50	100.00	81.00	40.38	81.00	40.38
Relamping labor cost	269.23	22.23	36.65	174.62	36.65	134.62
Washing cost	—	$300.00	300.00	300.00	150.00	150.00
Total Annual						
Operating Cost	$4,955.73	$2,830.23	$2,825.65	$2,923.00	$2,675.65	$2,733.00

Assumptions: (1) Cost per lamp: $1.75 (Energy Saving), $1.50 (Standard)
                      (2) Group relamping is performed when fixtures are washed, and requires an
                         additional two minutes per lamp.
                      (3) Spot relamping time is 30 minutes per lamp
                      (4) Relamping labor rate is $10.00 per hour.

Figure 17-7
Annual operating and maintenance cost for the five systems and maintenance schedules described in the text.

TROUBLESHOOTING

Problems can be expected to occasionally develop in any lighting system. The most common complaint is short lamp life. On rare occasions it is a valid problem, but this is the exception, not the rule. The following factors should be considered when evaluating complaints of short lamp life.

1. All lamps will not burn for the rated period. Normal failures will begin to occur at 40%-45% rated life, and will increase in frequency until 50% of the lamps have failed at the end of rated life. Remember though, that for each lamp that fails early, another lamp will burn for a corresponding period beyond rated life.

2. In a large installation, it is human nature to forget which lamps have been replaced, and normal failures in adjacent sockets may be remembered as having been recently replaced when in fact the replacement lamp was installed in another socket.

This problem can be resolved by writing the installation date on all lamps, using a fine-point, felt-tipped "permanent—writes on all surfaces" marking pen on the bases of incandescent and fluorescent lamps, or using the "date code" stamping on the bases of HID lamps. If a large-scale problem does in fact exist, it can then be readily identified.

Most problems involve a single fixture, and the guidelines in Figure 17-8 will assist in troubleshooting both large-scale problems and failures involving single lamps or fixtures.

FIGURE 17-8. GUIDE FOR TROUBLESHOOTING
INCANDESCENT SYSTEMS

SYMPTOM	POSSIBLE CAUSE	CORRECTIVE ACTION
Lamp fails immediately Bulb coated with white or gray powder	Air in lamp	Install new lamp
Lamps fail after short operating period. Replacement lamps also fail.	High voltage	Check voltage. If high, contact utility. If voltage cannot be reduced, use higher voltage rated lamp.
	Lamp subjected to vibration	Use vibration service lamp
Lamps in recessed fixtures cycle on and off.	High temperature is activating thermal protector in fixture	Make sure fixture is not covered with insulation, and ventilation is not obstructed. Install lower wattage lamps.
Early failure of many lamps in the installation	High voltage	Check voltage at lamp. Contact Utility if high voltage is encountered. Use higher voltage rated lamp.
Glass envelope separates from base.	Excessive lamp temperature	Make sure lamp wattage does not exceed rated fixture wattage. Make sure that fixture is not covered with insulation, and adequate ventilation is provided.
Lamps in outdoor fixtures are difficult to remove from socket.	Aluminum base on lamp	Use lamp with brass base. If problem is severe coat base of replacement lamp with a thin layer of petroleum jelly or other non-conductive grease.
Glass envelope blisters or bulges.	Excessive lamp temperature	Make sure lamp wattage does not exceed rated fixture wattage. Make sure fixture and lamp type are compatible. Some poorly designed reflectors may reflect heat back to the bulb, raising bulb wall temperature. If this is the cause, replace fixture.
Lamps "hum" when operated on a dimmer.	Normal condition	Install filter. See Chapter 2, section on dimming.

FLUORESCENT SYSTEMS

SYMPTOM	SYSTEM	POSSIBLE CAUSE	CORRECTIVE ACTION
Lamps fail to light	Any	Failed lamps	Replace with new lamps
		Poor contact between lamp and lampholder	Seat lamps properly in lampholders. Make sure contact surfaces are clean. Check spacing between lampholders to assure that contact is made with lamp. Adjust spacing if necessary.
		Abnormally low ambient temperature	Make sure ballast is rated to start lamps at ambient temperature. See Chapter 2 for lamp and ballast temperature ratings. Lamps may flicker in an attempt to light, which will shorten lamp life.
	Rapid start	Fixture not properly grounded	Rapid and trigger start lamps require a grounded metal surface in close proximity to the lamp. The surface must extend the full length of the lamp, and serves as a starting aid. Provide a good electrical ground.
	Trigger start	Loss of cathode heat	Make sure lamps are properly seated in lampholders, and electrical contact surfaces on lamps and lampholders are clean and tight. Check cathode voltage. If no cathode voltage, check for loose connections at lampholders. If connections are good, replace ballast.
Lamps cycle on and off	Any system with thermally protected ballasts	Ballast has failed	Replace ballast
		Fixture covered with building insulation material	Remove material to prevent heat retention in fixture.
Short lamp life	Any	Short operating hours per start	Lamp life ratings are based on operation for 3 hours per start. Shorter operating schecules will result in shorter lamp life.
		High or low voltage	Either over or under voltage may shorten lamp life. Actual voltage at the fiixture should be within 10% of rated ballast voltage.
		Ballast has failed	Check ballast secondary voltage. If low, replace ballast.

FLUORESCENT SYSTEMS (CONT'D)

SYMPTOM	SYSTEM	POSSIBLE CAUSE	CORRECTIVE ACTION
Short lamp life (Cont'd)	Rapid start	Loss of cathode heat	Loss of cathode heat frequently results in a dense blackening of one or both ends of the lamp. If only one end is blackened the problem is usually a bad connection or improper seating of the lamp. Make sure lamps are properly seated in lampholders, and electrical contact surfaces on lamps and lampholders are clean and tight.
	Trigger start		Check cathode voltage. If no cathode voltage, check for loose connections at lampholders. If connections are good, replace ballast.
	HO/VHO	Wrong lamp	HO and VHO lamps, while electrically different, are physically interchangeable. Use of HO lamps with VHO ballasts or VHO lamps with HO ballasts will result in premature lamp failure. Ballast life may also be shortened.
Flickering or swirling	Any	Low bulbwall temperature	Low ambient temperature may cause pulsating until lamp warms up. If condition continues, enclose lamp or use lamp with lower temperature rating.
			Lamps located near air conditioning outlets may be affected by cold drafts. Redirect air, or shield lamp. Energy saving lamps are particularly susceptible to this problem.
		Impurities or loose phosphor in lamp.	Impurities or loose phosphor which has been dislodged during shipping may enter the arc stream and swirl. Turn lamps on and off several times. If condition persists, remove lamp from fixture and gently tap one end on the floor to settle particles to one end. Replace lamp in fixture. If the condition still exists, operate lamps for several days, during which time the problem normally disappears. If it does not, replace the lamp.
		Poor contact between lamps and lampholders.	Check contact points. Clean if necessary.
Dark ends on lamps (Continued)	Any	Normal condition	Lamp ends will gradually darken over life. If is not usually a concern unless accompanied by short lamp life. If one end becomes very dark after a short burning period, check cathode voltage on rapid start systems.

FLUORESCENT SYSTEMS (CONT'D)

SYMPTOM	SYSTEM	POSSIBLE CAUSE	CORRECTIVE ACTION
Dark ends on lamps (Cont'd)	VHO	Normal condition	VHO lamps have pressure control chambers at the ends of the lamp. Ends may appear to be dark as a result. This is normal and does not affect performance.
Ballast noise	Any	Normal condition	All ballasts produce some audible noise as as result of electromagnetic fields within the ballasts. If noise is objectionable, check the manufacturer's sound rating, and change to a quieter ballast if possible. Loose parts on the fixture may rattle or hum, and loose mounting of ballasts may amplify the problem.

HIGH INTENSITY DISCHARGE SYSTEMS

SYMPTOM	SYSTEM	POSSIBLE CAUSE	CORRECTIVE ACTION
Lamp will not start	Any	Failed ballast	Check for proper secondary voltage. Visually examine ballast. Burned insulation indicates excessive heat, possibly from shorted windings. Bulged capacitors may be internally shorted. Replace components as necessary.
		Wrong lamp	Make sure lamp type and wattage are compatible with the ballast.
		Use of wrong ballast tap	Many ballasts are tapped for several primary voltages; i.e., 120v, 208v, 240v, 277v. Connection of the wrong tap to the power supply may result in failure to start lamps, or damage to lamps.
	Mercury Metal halide	Normal end of life	Normal failure mode for mercury and metal halide lamps is the inability to start. The lamp may try to start, and arcing may be observed, but the lamp will not develop full light output.
	High pressure sodium	Igniter failure	Remove the sodium lamp and install a mercury vapor lamp of about the same wattage. If the mercury lamp starts, the igniter has failed. If the mercury lamp does not light, the problem is usually in the transformer.
		Wrong lamp	150 watt high pressure sodium lamps are available in both 55 volt and 100 volt ratings, referred to as S55 and S56, respectively. Make sure the proper voltage lamp is used.

HIGH INTENSITY DISCHARGE SYSTEMS (CONT'D)

SYMPTOM	SYSTEM	POSSIBLE CAUSE	CORRECTIVE ACTION
Lamp cycles on and off	High pressure sodium	Normal failure	High pressure sodium lamps cycle on and off at normal end of life. Replace the lamp as soon as possible to minimize damage to igniter.
Short lamp life	Any	Wrong ballast	Operation on the wrong ballast type may cause premature failure. Lamps operated on higher wattage rated ballasts will appear brighter. Overwattaged high pressure sodium lamps will have a whiter color than lamps operated at design wattage.
		Defective lamp	Replace Lamp
Arc tube swollen or blackened shortly after installation	Any	Reflector design	Poorly designed reflectors may reflect heat back to the arc tube and cause swelling or premature darkening of the arc tube. The problem may also occur when using high pressure sodium lamps in reflectors designed for mercury lamps. Replace reflector.
		Wrong ballast	Make sure lamp wattage and ballast wattage match. Overwattage operation may cause premature blackening and short lamp life.
Lamps "flicker"	Any	Stroboscopic effect	Strobe is normal with HID lamps since an arc strikes and extinguishes 120 times per second. The effects are reduced by the use of phosphor coated mercury or metal halide lamps. High pressure sodium lamps are most likely to cause problems. Strobe is more easily noticed when lamps are connected to a single phase power supply system. Staggering lamps between phases on a three-phase system generally eliminates problems.
	Metal halide	Arc swirl	The arc in some metal halide lamps tends to swirl inside the arc tube. This is not actually a flicker problem, but may be perceived as such. If the problem is severe, try changing lamps. The use of multifaceted reflectors may also reduce the severity.
Low light output (Continued)	Any	Normal condition	All HID lamps depreciate in light output as they age. Losses may be as high as 50%-60% in extreme cases. Date code all lamps upon installation to verify burning hours. Replace old lamps which have depreciated beyond a usable level.

HIGH INTENSITY DISCHARGE SYSTEMS (CONT'D)

SYMPTOM	SYSTEM	POSSIBLE CAUSE	CORRECTIVE ACTION
Low light output (Cont'd)	Any	Low voltage	Check primary voltage at fixture and compare to ballast primary voltage rating. If over 5% low, check voltage at electrical service. If low at service, contact utility. If voltage at service panel ok, check for excessive voltage drop in electrical wiring. A ballast with good regulation may be required.
		Defective ballast	Check ballast output. Replace if necessary

Chapter 18

ECONOMICS

An economic analysis is simply a means of comparing the financial consequences of alternative investments. The fundamental questions are whether or not the investment will be recovered, plus a return commensurate with the risk, and how does that return compare with other possible investments.

The basic question in the lighting industry generally takes one of two forms: 1) a lighting system is to be installed in a new building; should we use System A, which has a low first cost and a high operating cost; or 2) should an existing system with high operating costs be replaced with a new, more efficient system with low operating costs? There may be slight variations of the themes but, in general, lighting installations fall into one of the two categories.

There are two common types of analyses which are performed: the simple payback analysis, and the life cycle cost analysis. The simple payback method asks only "What does it cost, what does it save, and how soon do I get my money back?" The answers give an indication of the attractiveness of a proposal, but if interpreted incorrectly they may lead to poor decisions. The life cycle cost method is more complex, and provides much more accurate information. It recognizes the value of money changes over time, and that a dollar today is worth more than a dollar at this time next year. To illustrate this point, assume that you have purchased a lottery ticket and just learned that you have won $1 million. The money is to be paid in 20 equal annual installments of $50,000. Did you really win $1 million. The answer is no, you did not. The amount which you have actually won, in today's dollars, will vary as the inflation rate changes. If we assume a rate of five percent per year, you have won $623,100. A substantial amount, but far less than $1 million.

The life cycle cost method recognizes that expenditures and savings will occur over a period of time, and expresses them in today's dollars. It will also include the impact of taxes.

Despite its lack of sophistication the simply payback method has some advantages over the life cycle cost analysis and is widely used by industry to evaluate potential investments. It is easy to prepare and requires only minimal information: cost and savings. The life cycle cost method requires more time to prepare since the magnitude of individual expenditures and the future time at which they will occur are needed. Most importantly, the minimum rate of return which the company will accept on an investment must also be known, and industry is not anxious to share this information. Some companies that use sophisticated accounting methods have developed guidelines by which investments can be compared on a simple payback method. These guidelines are frequently used by plant managers and facilities engineers to evaluate some expenditures.

It must be stressed that an economic analysis may be used only to evaluate the financial implications of an investment. It says nothing about the suitability of the equipment, nor does it recognize intangibles such as comfort, convenience, status symbols, or pride of ownership. The value of the intangible factors cannot be included in an economic analysis, but there may be occasions when they are important enough to override the economic factors. The ultimate decision is up to the individual who is paying the bill.

SIMPLE PAYBACK ANALYSIS

Figure 18-1 is a form which may be used for simple payback analyses. The following instructions are keyed to the form.
(1)   Enter the number of lamps in the existing system.
(2)   Enter the watts per lamp for the existing system. Obtain from ballast manufacturer's data or test reports for fluorescent or HID lamps. Be sure to apply thermal factors if applicable. Use rated lamp watts for 120-volt incandescent systems. Transformer losses for low-voltage incandescent systems must be included if applicable. Obtain from fixture manufacturer. If loss data are not available it may be approximated by adding 10% to lamp watts.

ECONOMIC ANALYSIS - NEW FIXTURES

POWER COST SAVINGS (Annual)

Existing System _____ Lamps x _____ Watts = _____ Watts

Proposed System _____ Lamps x _____ Watts = _____ Watts

                              Watts Reduced = _____/1000 = _____kW

_____kW x $ _____/kW Demand Charge = $ _____

_____kW x _____Hrs/Mo. x $ _____/kW = _____

                 Total Monthly Savings = $ _____ x 12 = $ _____/Yr

INSTALLED COST

_____ Fixtures X $ _____/Fixt. = $ _____

_____ Lamps x $ _____/Lamp   = $ _____

                     subtotal $ _____ + _____ = $ _____
                                                   (tax)

Labor

_____ Fixtures x $ _____/Fixture = $ _____

                          Total Installed Cost $ _____

REPLACEMENT LAMP COST

Existing Lamp _____ Net Cost $ _____ + _____ (tax) = $ _____

                     Labor Cost to Replace 1 Lamp $ _____

                     Total Replacement Cost per Lamp $ _____

_____ Burning hrs/yr x _____ x $ _____/Lamp = $_____/Yr
        Rated Lamp Life   (# Lamps)

Proposed Lamp _____ Net Cost $ _____ + _____(tax) = $ _____

                     Labor Cost to Replace 1 Lamp $ _____

                     Total Replacement Cost per lamp $ _____

_____ Burning Hrs/Yr x _____ x $ _____/Lamp = $ _____
        Rated Lamp Life   (# Lamps)

                     Added Savings / Cost $ _____

TOTAL ANNUAL SAVINGS

$ _____ Power Cost Savings (+ -) $ _____ Relamping Cost/Savings

                          $ _____ Net Savings

SIMPLE PAYBACK

$ _____ Installed Cost/ $ _____ Net Savings = _____ Yrs

Figure 18-1 (a)

Economic analysis forms for determining simple payback of a new lighting system. (a) is keyed to the instructions in the text, (b) is a blank form that may be reproduced for use.

ECONOMIC ANALYSIS - NEW FIXTURES

POWER COST SAVINGS (Annual)

Existing System __(1)__ Lamps x __(2)__ Watts = __(3)__ Watts

Proposed System __(4)__ Lamps x __(5)__ Watts = __(6)__ Watts

Watts Reduced = __(7)__ /1000 = __(8)__ kW

__(9)__ kW x $ __(10)__ /kW Demand Charge = $ __(11)__

__(12)__ kW x __(13)__ Hrs/Mo. x $ __(14)__ /kW = __(15)__

Total Monthly Savings = $ __(16)__ x 12 = $ __(17)__ /Yr

INSTALLED COST

__(18)__ Fixtures X $ __(19)__ /Fixt. = $ __(20)__

__(21)__ Lamps x $ __(22)__ /Lamp = $ __(23)__

subtotal $ __(24)__ + __(25)__ = $ __(26)__
                                    (tax)

Labor

__(27)__ Fixtures x $ __(28)__ /Fixture = $ __(29)__

Total Installed Cost $ __(30)__

REPLACEMENT LAMP COST

Existing Lamp __(31)__ Net Cost $ __(32)__ + __(33)__ (tax) = $ __(34)__

Labor Cost to Replace 1 Lamp $ __(35)__

Total Replacement Cost per Lamp $ __(36)__

__(37)__ Burning hrs/yr x __(39)__   x $ __(40)__ /Lamp = $ __(41)__ /Yr
__(38)__ Rated Lamp Life   (# Lamps)

Proposed Lamp __(42)__ Net Cost $ __(43)__ + __(44)__ (tax) = $ __(45)__

Labor Cost to Replace 1 Lamp $ __(46)__

Total Replacement Cost per lamp $ __(47)__

__(48)__ Burning Hrs/Yr x __(50)__   x $ __(51)__ /Lamp = $ __(52)__
__(49)__ Rated Lamp Life   (# Lamps)

Added Savings / Cost $ __(53)__

TOTAL ANNUAL SAVINGS

$ __(54)__ Power Cost Savings (+ -) $ __(55)__ Relamping Cost/Savings

$ __(56)__ Net Savings

SIMPLE PAYBACK

$ __(57)__ Installed Cost/ $ __(58)__ Net Savings = __(59)__ Yrs

Figure 18-1 (b)

(3) Enter total watts for existing system. Multiply (1) by (2).

(4) Enter the number of lamps in the proposed system.

(5) Enter the watts per lamp for the proposed system. See instructions for Step (2).

(6) Enter total watts for proposed system. Multiply (4) by (5).

(7) Enter watts reduced. Subtract (6) from (5).

(8) Enter total kW reduced. Divide (7) by 1000 to convert to kW. This is done since power cost is based on kilowatts and kilowatt-hours instead of watts and watt hours.

(9) Enter the kilowatt reduction from step (8). Many utilities base their charges to customers on two components: demand, measured in kilowatts, and energy, measured in kilowatt hours. The demand is the average kilowatts drawn during some period of time, usually 15 or 30 minutes. For example, if ten 100-watt lamps were operated continuously for 15 minutes, the average demand for that 15-minute interval would be 1000 watts, or 1 kilowatt. If the ten lamps were operated for 7-1/2 minutes, or one half of the 15-minute interval, the average demand for the interval would be 1/2 kilowatt. Information on demand charges may be obtained from the utility. Note that some utilities have time-of-use rates for customers, typically the large ones, and rates vary according to the time of day, day of the week, and season when the energy is used. These rates are highly complex, and the assistance of the utility should be requested in calculating energy savings. If no demand charge is assessed, go to Step (12).

(10) Enter demand charge per kW. Obtain from utility.

(11) Enter reduction in demand charge. Multiply (9) by (10). Include demand charge only if peak billing demand is established at a time when lights are operating.

(12) Enter kW reduced from Step (8).

(13) Enter hours per month lights operate.

(14) Enter cost per kWh. Obtain from utility.

(15) Enter reduction in energy charge. Multiply (12) by (13) by (14).

(16) Enter the total monthly power cost savings. Add (11) and (15).

(17) Enter total annual power cost savings. Multiply (16) by 12 months.

(18) Enter the number of new fixtures in the proposed system.

(19) Enter the cost per fixture. Do not include cost of lamp unless lamp is provided with fixture and is included in fixture price.

(20) Enter the total cost for new fixtures. Multiply (18) by (19).

(21) Enter number of new lamps if not included with fixture. If lamp is included with fixture go to Step (24).

(22) Enter cost of new lamp.

(23) Enter total cost of new lamps. Multiply (21) by (22).

(24) Enter total cost of new fixtures and lamps. Add (20) and (23).

(25) Enter State and local sales tax. Multiply (24) by combined tax rate.

(26) Enter total cost of new fixtures and lamps. Add (24) and (25).

(27) Enter number of new fixtures. Obtain from (18).

(28) Enter labor cost to install one new fixture.

(29) Enter total cost for labor. Multiply (27) by (28).

(30) Enter total installed cost of new system. Add (26) and (29).

(31) Enter a description of the existing lamp (e.g., 75ER30, 34-watt fluorescent).

(32) Enter the net cost of the existing lamp. This information may be obtained from local vendors.

(33) Enter amount of State and local sales tax, if applicable. Multiply (32) by the combined tax rate.

(34) Enter the net cost per existing lamp. Add (32) and (33).

(35) Enter labor cost to change one lamp. This entry will vary according to the difficulty of lamp replacement, the labor rate of the person doing the work, and the number of lamps being replaced at the same time. Costs typically range from 25 cents per lamp for group replacement of fluorescent lamps, to $10 or more for hard to reach HID lamps in high ceilings. Lamps in pole-mounted outdoor fixtures may cost $50 or more to replace.

(36) Enter total cost to replace one lamp. Add (34) and (35).

(37) Enter total burning hours per year for the existing lamp. Multiply (13) by 12 months/year.

(38) Enter rated life for the existing lamp. Obtain from lamp manufacturer's catalog. Note that life ratings for fluorescent lamps are typically based on operation for 3 hours per start, and longer burning hours extend life. See text on fluorescent lamps to obtain rated life for longer burning cycles.

(39) Enter number of lamps in the existing installation. Obtain from (1).

(40) Enter cost per lamp for existing installation. Obtain from (36).

(41) Enter average annual cost for replacement lamps for the existing system. Divide (37) by (38) to obtain the average number of replacement lamps per socket per year, multiply by (39) to obtain the total average number of replacement lamps per year for the installation, and multiply by (40) to obtain the average cost per year for replacement lamps. Enter this value in (41).

(42) Enter a description of the proposed lamp.

(43) Enter the net cost of the proposed lamp.

(44) Enter the State and local sales tax for the proposed lamp. Multiply (43) by the combined tax rate.

(45) Enter the total cost per lamp for the proposed system. Add (43) and (44).

(46) Enter the labor cost to change one lamp. See Step (35).

(47) Enter the total installed replacement cost, per lamp, for the proposed lamp. Add (45) and (46).

(48) Enter the burning hours per year for the proposed system. This is normally the same as entry (37).

(49) Enter the rated life of the proposed lamp. See Step (38).

(50) Enter the number of lamps in the proposed installation. Obtain from (4).

(51) Enter the installed cost per lamp for the proposed system. Obtain from (47).

(52) Enter the average annual cost for replacement lamps for the proposed system. Divide (48) by (49) to obtain the average number of replacement lamps per socket per year, multiply by (50) to obtain the total number of replacement lamps per year for the installation, and multiply by (51) to obtain the average cost per year for replacement lamps.

(53) Enter the added cost or savings of the proposed system. Obtain from the difference between (41) and (52). If the cost of the proposed system is higher than the existing system, circle "cost." If the proposed system is less costly than the existing system, circle "savings."

(54) Enter annual power cost savings. Obtain from (17). If the annual relamping cost of the proposed system is less than the annual relamping cost of the existing system, circle "+." If the existing system is less costly, circle "–."

(55) Enter the annual relamping cost or savings. Obtain from (53).
(56) Enter the net savings per year. If the annual relamping cost produces a savings (+), add (54) and (55). If the proposed system is more costly (–), subtract (55) from (54).
(57) Enter the installed cost of the proposed system. Obtain from (30).
(58) Enter the net savings. Obtain from (56).
(59) Enter the simply payback, in years. Divide (57) by (58).
This completes the simple payback analysis.

LIFE CYCLE COST ANALYSIS

The life cycle cost analysis includes the calculation of interest. Most of us think of interest in terms of the interest we pay on a loan, or the interest a bank pays on our deposits. In other words, money paid for the use of money. In a much broader sense, interest is also defined as the return on an investment. In this chapter the terms "interest" and "return on investment" will be considered synonymous.

When interest is considered in an economic analysis the concept of equivalence is introduced: a present sum of money, plus interest, is equivalent (equal to) a future sum. This concept is used to express all expenditures, both present and future, in today's dollars.

The following symbols are used in interest calculations:

i = The interest per period. A period may be any mutually agreed upon period of time, such as a day, a month, or a year, and is the interval at which payments are made.

n = The number of periods occurring during the term of the loan or investment. If a loan has a 3-year term and payments are made at monthly intervals, the number of periods, n, is 36.

P = A present sum of money; the amount originally loaned or invested, or the present value of a sum to be paid or collected at a future time.

F = The sum of money at some future time. It is equal to a present sum, P, plus interest, i, at the end of a number of intervals, n.

A = The amount of a payment or receipt in a uniform series of payments or collections continuing over n periods.

There are six basic interest formulae used to calculate the present value or future value of an investment, or the amount of a payment. The desired information is obtained by multiplying the present or future value by a "factor" which may be obtained from tables published by financial institutions, or found in engineering economy or some finance textbooks. Since the interest rates and time periods published in the tables do not always coincide with the desired ones, the equations for calculating the factors will be given, along with explanations of their use.

SINGLE PAYMENT COMPOUND AMOUNT FACTOR (SPCAF)

Used to determine the future value of a present amount of money invested or borrowed at the beginning of a period and withdrawn at the end of some number of periods, n. The future value is equal to the present value, P, times the factor.

$$SPCAF = (1 + i)^n$$
and
$$F = P (1 + i)^n$$

Example

$1000 is to be invested today at a rate of 1% per month, and withdrawn with interest at the end of 12 months. How much money will be withdrawn?

P = $1000
i = .01
n = 12
F $= P (1 + i)^n = 1000 (1 + .01)^{12} = \$1126.82$

SINGLE PAYMENT PRESENT WORTH FACTOR (SPPWF)

Used to determine the present worth of a sum of money to be paid or withdrawn at some future time.

$$SPPWF = \frac{1}{(1 + i)^n}$$

Example

You have decided to buy a new car for $12,000. The salesman has assured you that the car will be worth $6,000 in 4 years, so it will only cost you $6,000 to drive it for that period. What is the present worth of the $6,000 salvage value? Assume inflation at 6% per year.

$$F = \$6,000$$

$$i = .06$$

$$n = 4$$

$$P = F \left(\frac{1}{(1 + i)^n}\right) = 6000 \left(\frac{1}{(1 + .06)^4}\right) = \$4752.56$$

In 4 years the car will have a value of $4752.56 in today's dollars, so the cost is $7,247.44, not $6,000.

SINKING FUND FACTOR (SFF)

Used to determine the amount of a periodic payment necessary to produce a specific amount of money at a specific future time. It is frequently used by industry to determine the amount which must be allocated each year towards the future replacement of equipment with a fixed life and high replacement cost.

$$SFF = \frac{1}{(1 + i)^n - 1}$$

Example

Your existing car has several years of remaining life. You have decided to drive it for another 3 years, and then buy a new car at an estimated future cost of $14,300. You want to pay cash for the new car, so you will make monthly deposits in a savings account which earns a nominal 6% per year, compounded monthly. How much must you deposit each month to have $14,300 at the end of 3 years?

$F = \$14,300$

$i = 6\%$ per year. Since deposits are made monthly the interest rate must be converted to a monthly rate of .06/12, or .005.

$n = 36$ months

$$A = F \left(\frac{1}{(1 + i)^n - 1} \right) = 14{,}300 \left(\frac{1}{(1 + .005)^{36} - 1} \right) = \$363.53$$

CAPITAL RECOVERY FACTOR (CRF)

The capital recovery calculation is one of the most common and useful economic analysis tools. It is used to calculate car payments, house payments, and most other installment loans.

$$CRF = \left(\frac{i (1 + i)^n}{(1 + i)^n - 1} \right)$$

Example

After thinking it over you have decided that you really want to drive a new car instead of your old car, as cited in the previous example, and have made a decision to buy now. The quoted price of $14,300 in the previous example included anticipated cost increases over the next 3 years, so you can buy it today for $12,000. You will finance the entire amount, at 12% nominal interest, for 36 months. How much is the payment?

P = $12,000
I   = 12% (1% per month)
n = 36

$$A = P \left(\frac{i (1 + i)^n}{(1 + i)^n - 1} \right) = 12{,}000 \left(\frac{.01 (1 + .01)^{36}}{(1 + .01)^{36} - 1} \right) = \$398.57$$

Note the difference between this example and the previous example. Purchasing the car today with borrowed money will cost an additional $35.04 per month over the amount required if you make deposits and pay cash for the car in 3 years. The difference is based on the fact that in the first example the bank will pay you interest for the use of your money, but if you buy the car today you will pay the bank interest for the use of their money. The decision is simply whether or not driving a new car now is worth the additional $35.04 per month.

UNIFORM SERIES COMPOUND AMOUNT FACTOR (USCAF)

Used to determine the future value of a series of equal payments or deposits over some period of time.

$$\text{USCAF} = \frac{(1 + i)^n - 1}{i}$$

Example

You are considering opening an individual retirement account (IRA), and plan to deposit $2000 per year, in 12 equal monthly install-ments of $166.67. The account pays 6% per year, compounded monthly. How much money will be in the account at the end of 20 years?

A = $166.67
i = 6% per year, or .005 per month
n = 240 months

$$P = A\left[\frac{(1 + i)^n - 1}{i}\right] = 166.67\left(\frac{(1 + .005)^{240} - 1}{.005}\right) = \$77,008.36$$

UNIFORM SERIES PRESENT WORTH FACTOR (USPWF)

Used to determine the present value of a uniform series of deposits over some period of time.

$$\text{USPWF} = \frac{(1 + i)^n - 1}{i\,(1 + i)^n}$$

Example

Find the present worth of the $77008.36 in the IRA used in the previous example. Assume inflation at 5% per year.

A = 166.67
i = 5% per year, or .004167 per month
n = 240

$$P = A\left[\frac{(1 + i)^n - 1}{i\,(1 + i)^n}\right] = 166.67\left(\frac{(1 + .004167)^{240} - 1}{.004167\,(1 + .004167)^{240}}\right)$$

$$= \$25,253.88$$

The $77,008.36 which will be available in 20 years is equivalent to $25,253.88 in today's dollars.

USE OF INTEREST TABLES

When available, interest tables will provide the same information as the calculations just discussed, and much faster and easier.

Example

You are buying a new car and wish to finance $12,000 for 3 years at a nominal interest rate of 12% per year.

    P   = $12,000
    i   = 1% per month
    n   = 36

The capital recovery factor (CRF) is .033214, and the amount of the payment is:

    A = 12,000 (.033214) = $398.57

Note that this is the same answer as was obtained using the interest formula in the example calculation of an application of the CRF.

ECONOMIC COMPARISON METHODS

There are two commonly used methods for comparing alternative investments: present worth, and equivalent uniform annual cost. Each considers the time value of money at some interest rate or minimum attractive rate of return.

The present worth method relates all expenditures or receipts to today's dollars, regardless of the time at which they actually occur. Alternative investments which have differing costs or returns that occur at different intervals may then be compared as if the entire required investment or realized return were to be made or received today. The present worth method is widely used to determine the rate of return on a prospective investment.

The equivalent uniform annual cost method is widely used in engineering studies. It is inevitable that some costs will vary from year to year. Maintenance costs tend to increase as a lighting system ages, energy costs tend to increase, taxes may fluctuate, and labor costs escalate over time. In lighting systems, relamping costs can be expected to be very low during the first year of operation, slightly higher during the second or third years, and very high during the next year or two. Equivalent uniform annual cost methods permit a comparison of two or more systems with annual expenditures which may vary from year to year, by expressing the costs associated with each alternative as a series of uniform annual expenditures.

Business operates with limited capital. Those in the business of selling lighting systems should underline and highlight the previous sentence and commit it to memory. A lighting system change may pay for itself in energy and maintenance savings in a year or less, yet the prospective client frequently turns down the proposal. There are several reasons why this might happen, but in most cases it is the result of limited capital. The proposal for a new lighting system is not the only proposal competing for limited funds. The manufacturing department may want a new machine to increase production or reduce manpower, engineering may need a new computer for product design, or a new truck may be needed to deliver orders to customers. In each case the economic benefit to the company may be different. Management, desiring to maximize profit, will rank projects based on their potential return on investment, and allocate funds to the projects with the highest return. When the available funds have been exhausted, no further expenditures are possible. If other projects produce higher rates of return, the lighting system will not be installed.

COMPONENTS OF COST AND SAVINGS

The costs associated with a lighting system must be identified when preparing an economic analysis. If one or more of the costs are omitted the analysis will be flawed, and the results may lead to erroneous conclusions. The costs normally associated with a lighting system, over its usable life, are:

1. Fixture cost
2. Lamp cost
3. Installation labor and materials cost
4. Electrical energy cost
5. Relamping cost (lamps and labor)
6. Luminaire cleaning cost
7. Ballast replacement cost (labor and material)

8. Miscellaneous maintenance costs (lampholders, etc.)
9. Heating and air conditioning equipment cost
10. Heating and air conditioning energy costs
11. Extra income taxes on savings
12. Property taxes
13. Insurance costs
14. Salvage value

Fixture, lamp, and installation costs are one-time expenses which occur when the system is initially installed, and are easily estimated. Other costs which occur over time, either on a regular basis or intermittently, are more difficult to estimate.

ELECTRICAL COSTS

Electrical energy rates have become increasingly complex in recent years due to the introduction of time of use pricing and general restructuring of rates. The days of simply dividing the total bill by the number of kilowatt-hours to arrive at an average cost per kWh for use in economic studies are over for all but a few exceptional cases. Most utilities, as part of their service, will not only explain their rates, but will assist in the calculation of energy costs and the savings which may be realized by reducing the usage. They may also be able to provide estimates of rate increases or decreases which are expected in future years. Be aware, however, that these are only estimates. Electrical energy costs are greatly influenced by fluctuating costs for oil, gas, coal, and other fuels used for generation, as well as the cost of programs which are mandated by governmental regulatory agencies and must be borne by the utility's customers.

RELAMPING COSTS

The number of lamps which can be expected to be replaced each year is based on statistical averages. In a large group of lamps, a predictable number can be expected to fail during any given time interval. There are three common schedules by which failed lamps are replaced: individual replacement, also called spot replacement, where lamps are replaced individually on an as-they-fail basis; group replacement with spot replacement of early failures, where all lamps in the facility are replaced at one time, usually at 50% to 70% of rated lamp life, and early failures are spot relamped; and group replacement with no replacement of early failures, where all lamps in the facility are replaced at one time, and early failures are not replaced until the replacement of all lamps.

Spot replacement is generally the most cost effective method for systems using high-intensity discharge lamps and other lamps with relatively high lamp cost. The exception is the case when the labor cost to replace a single lamp is high, but the cost of gaining access to subsequent lamps is relatively low in comparison. Examples are small parking lots, where the cost of bringing the necessary relamping equipment such as bucket trucks or man lifts to the site is high, but the time required for the actual replacement of a lamp is low, and relamping of installations requiring the erection of scaffolding.

Group replacement with spot replacement of early failures is recommended for fluorescent and incandescent systems, all high-pressure sodium systems falling into the exceptions category in the previous paragraph, and some other systems with high labor costs for replacing a single lamp and substantially lower costs for subsequent lamps replaced at the same time.

Group replacement with no replacement of early failures is seldom desirable under current practice since it requires the installation of additional luminaires to make up for the light which is lost when a lamp fails and is not promptly replaced. It should never be specified for high-pressure sodium systems since failure to replace failed lamps may result in premature igniter failure. The only economical application of this method is in cases where the replacement of an individual lamp is cost prohibitive. For example, a local church has a ceiling which is about 50' high. The only way to replace lamps is to manually erect a scaffold since the fixture cannot be accessed from the top for maintenance. The cost of replacing a single lamp is prohibitive, so the system is group relamped when 25% of the lamps have failed. Note that 25% is not a magic number. It is economical for the stated case. Each installation must be judged individually, and an economically feasible maintenance schedule derived, as discussed in Chapter 17.

The average annual relamping cost for the three methods may be calculated from:

Spot Replacement

$$\text{Annual Cost} = \left(\frac{B}{R}\right) (C + I) \ (N)$$

Group Replacement with Spot Replacement of Early Failures

$$\text{Annual Cost} = \left(\frac{B}{A}\right) ((K) \ (C) + (K) \ (I) + C + G) \ (N)$$

Group Replacement with No Replacement of Early Failures

$$\text{Annual Cost} = \left(\frac{B}{A}\right) (C + G) \ (N)$$

where

B = Burning hours per year
R = Rated lamp life
A = Burning time in hours between replacements
C = Lamp cost, per lamp, including sales tax
I = Labor cost to replace one lamp on a spot basis
G = Labor cost to replace one lamp on a group basis
K = Percentage of lamps failing before group replacement
N = Number of lamps in the installation

Note that C, the lamp cost, may vary according to the number of lamps which are purchased at one time. Large quantity purchases will normally result in a lower cost per lamp than the purchase of only one or several cases of lamps. The appropriate cost should be used for each individual calculation.

LUMINAIRE WASHING COST

Periodic washing of luminaires can provide substantial reductions in both the first cost and the ongoing operating costs of a lighting system. This topic was discussed in detail in Chapter 17, and guidelines provided for determining an economic washing interval for a lighting system. When the washing interval has been determined, the washing cost may be calculated from:

$$\text{Cost} = (N) \ (H) \ (R)$$

where

N   = Number of luminaires
H   = Hours to wash 1 fixture
R   = Hourly labor rate, including fringe benefits

HEATING AND AIR CONDITIONING COSTS

Changes in lighting loads may impact the heating and air conditioning loads in a building. One kilowatt-hour of electricity produces 3413 BTU's of heat, which will either contribute to heating a building or must be removed from the building. Reduced lighting loads in new buildings may result in reductions in the total capacity of cooling equipment, but a higher heating capacity may be required in areas with cold climates. Reductions in lighting loads in existing buildings may also require modifications to the space conditioning equipment.

The determination of the effects of lighting on space conditioning equipment loads and energy consumption is a discipline within itself, and is beyond the scope of this text. If major changes in lighting loads are contemplated in space conditioned buildings the services of a mechanical engineer may be required.

TAXES AND INSURANCE

The impact on taxes resulting from the installation of a lighting system should be assessed by a qualified individual. Prior to the Federal Income Tax revisions in 1987 it was common practice to use a 50% combined tax rate for most analyses. The advisability of continuing this practice is now questionable, and professional advice should be obtained for specific cases.

Insurance rates are seldom affected when a new lighting system is installed, but it may be advisable to check with an insurance advisor.

SALVAGE VALUE

Old lighting equipment seldom has an appreciable salvage value. In most cases the cost of removing old fixtures is more than they are worth. For this reason the salvage value is normally omitted from lighting studies. It if is included, it is normally a cost item rather than an income amount.

SUMMARY

The simple payback method of economic analysis is the most common, and is widely used by both large and small companies for evaluating proposals for many expenditures.

Detailed economic studies based on engineering economy or established accounting procedures provide a more accurate method of evaluating prospective investments, but are much more complex. Their preparation is better left to financial experts who have spent years studying and learning their discipline.

The discussion of interest, present worth, and equivalent uniform annual cost is intended only to familiarize the reader with the basic principles of life cycle cost analysis. A course in engineering economy is highly recommended for those who wish to increase their knowledge of economic analysis. The interest formulae are precise, and will yield exact answers. Those wishing to calculate car, house or installment loan payments, or the interest earned on an investment with a known interest rate, may use the methods described with full confidence. If the arithmetic is performed correctly, the amount of the payment or interest earned will be precise.

THE BOTTOM LINE

Lighting accounts for a major portion of the electrical energy consumed by commerce and industry. Percentages vary, but the electrical load from lighting accounts for 40% to 50% in most offices and stores, and 2% to 10% in most industrial facilities. In some specialized occupancies such as warehouses, lighting may be as much as 80% to 90% of the electrical load. It is obvious that efficient lighting systems provide major opportunities for reductions in operating costs.

But what is an efficient lighting system? Efficiency cannot be based on simple metrics such as lumens per watt or fixture efficiency since they ignore the basic purpose of a lighting system: to allow people to see in order to perform a visual task. An efficient system is one which provides illumination of sufficient quantity and quality for the task being performed, at the lowest cost. The elements of quantity, quality, and cost have been discussed in this text and may be summarized as follows: quantity is simply lumens per unit area, footcandles or lux; quality is a much broader topic which includes elements of glare, color, contrast, luminance ratios and uniformity ratios, and the aesthetics of both the lighting equipment and the visual impact of the lighted environment; cost refers to the installation, owning, operating, and maintenance costs of the system.

It is easy to become over zealous in our attempts to reduce lighting energy consumption. When designing or specifying a lighting system we must remember that with few exceptions, lighting systems are installed to permit a worker to perform a visual task. Failure to provide adequate illumination will result in impaired visual performance which may be accompanied by decreased productivity, increased errors,and dissatisfied workers. When the individual components of the total cost of production of either goods or services is examined, the cost of lighting is miniscule when compared to the cost of workers, space, machinery, and raw materials. A few dollars saved on power costs may be lost many times over if lighting levels are reduced below those required for effective seeing.

Conversely, the philosophy of "more light, better sight" may result in increased lighting cost with no offsetting benefits and, if the added light is of poor quality, may actually reduce the ability to see. Unfortunately, many lighting systems are "designed" by individuals with little training in lighting who compensate for lack of knowledge with brute force, and frequently specify far more light than is actually needed, just to be "safe."

Quality in a lighting system doesn't just happen. It requires careful evaluation of the design objective for the project, and matching the equipment and layout to the needs. The owners of industrial buildings, warehouses, and similar structures need systems which enhance production and create a pleasant work environment, but the aesthetic concerns are secondary when compared to operations such as restaurants and stores, where the primary objective is to create an environment which attracts customers and positively influences their feeling of comfort and mental attitude.

Over the past decade and a half our profession has become increasingly complex. Twenty years ago we had essentially two lamps, the F40 cool white and F40 warm white, and one ballast, the standard core and coil, for office lighting. Today there are over 50 combinations of lamps and ballasts, each with different characteristics, which might be employed. No one combination is right for all jobs. Additionally, offices were typically lighted to 100 fc to 200 fc, and only a few designers paid a great deal of attention to these designs. Today's offices are lighted to substantially lower levels, so the lighting designer must practice better lighting design. Our knowledge of design practices which improve the quality of illumination and enhance the seeing process has also increased, and continues to grow.

Equipment manufacturers have made tremendous advances in their equipment. Many new products with higher efficiencies and improved characteristics such as color, light control, electrical control, and aesthetics have been introduced, and the use of electronics in ballasts and controls has proliferated. Most of these new products work well, or reasonably so. Some do not.

There is a natural hesitancy to try new products when the old products work well. This is accentuated by the fact that some new products have performed poorly, or not at all. We live in a changing world, and those who refuse to accept new products and methods will soon be left behind. Caution is a good trait, however, since the designer is morally responsible and may be legally liable if a system does not work as represented.

The key to equipment selection lies in the evaluation of products to assure that they meet the job requirements and will function properly and reliably. Evaluate the manufacturer's technical data for the product, and ask if the claims are reasonable. Question claims which cannot be substantiated. For example, is it reasonable to reduce the power input to a fluorescent lamp by 33% with no reduction in light output simply by connecting a capacitor in the lamp circuit? Absolutely not. Can the installation of a specular reflector in a four-lamp fluorescent troffer permit the removal of two lamps with no reduction in lighting level? By itself, the reflector cannot, since most four-lamp troffers have efficiencies ranging from 65% to 75%, and in order to produce the same lighting levels with one half of the lamps would require fixture efficiencies in the range of 130% to 150%, which simply cannot happen. The operation of lighting equipment follows well defined laws of physics. *There is no magic.*

Consider the track record of the manufacturer when evaluating new products. If the company is well established and has a proven record of making sure that new products are reasonably well perfected before taking them to market, then the probability is good that the products will perform as represented. This does not mean that everything will always work properly, but the odds are greatly improved. This also does not mean that new products made by new companies will not perform well; however, they should be subjected to close scrutiny. Many products are introduced before they are ready, out of necessity, to improve the company's immediate cash flow. Production lines generate no revenue unless they are producing and the products are sold.

The bottom line is that there is no magic in the design of a lighting system. The job requirements must be thoroughly analyzed to understand the design objective for the lighting system, equipment must be evaluated based on its suitability for the application, and the system designed in accordance with sound engineering practice. Energy is an important consideration, but it is only one of the factors which dictate the success of the designer's efforts. Above all, the lighting professional must continually improve his or her knowledge of the field, and stay abreast of our rapidly changing industry.

INDEX